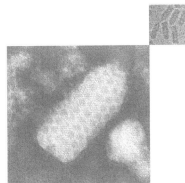

Negative Strand
RNA Virus

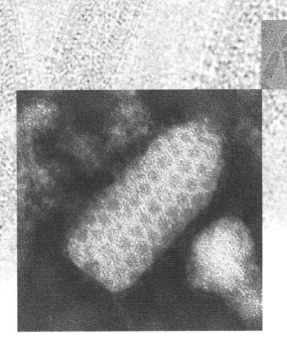

Negative Strand
RNA Virus

Editor

Ming Luo

The University of Alabama at Birmingham, USA

 **World Scientific**

NEW JERSEY · LONDON · SINGAPORE · BEIJING · SHANGHAI · HONG KONG · TAIPEI · CHENNAI

Published by

World Scientific Publishing Co. Pte. Ltd.

5 Toh Tuck Link, Singapore 596224

USA office: 27 Warren Street, Suite 401-402, Hackensack, NJ 07601

UK office: 57 Shelton Street, Covent Garden, London WC2H 9HE

British Library Cataloguing-in-Publication Data
A catalogue record for this book is available from the British Library.

ISBN-13 978-981-4317-22-1
ISBN-10 981-4317-22-5

Typeset by Stallion Press
Email: enquiries@stallionpress.com

Printed in Singapore.

Contents

Foreword

The Negative Strand RNA Viruses are the causative agents of a large number of human and animal diseases. Several of the Negative Strand RNA Viruses have been classified by the United States National Institutes of Allergy and Infectious Diseases as category A, B or C priority pathogens, resulting in increased attention to both research on these viruses as well as to the development of vaccines and antiviral drugs.

The Negative Strand RNA Viruses can be divided into those viruses with non-segment genomes (order *Mononegavirales*) and those viruses with segment RNA genomes. The Mononegavirales include bornavirus, vesicular stomatitis virus (VSV) and rabies virus, Marburg virus and Ebola virus, measles virus, mumps virus, Nipah virus, Hendra virus, canine distemper virus, Sendai virus, Newcastle disease virus, human parainfluenza viruses 1-4, parainfluenza virus 5 (formerly known as simian virus 5), human respiratory syncytial virus and human metapneumovirus. The segmented Negative Strand RNA Viruses include influenza A, B and C viruses, the bunyaviruses and the new and old world arenaviruses.

In addition to studies on the mechanism of replication of these viruses and also studies on viral pathogenesis and epidemiology, several of the Negative Strand RNA Viruses have been used as model systems in cell biology, biochemistry and structural biology. The VSV glycoprotein (G) has served as a model to study the cellular exocytic pathway. The VSV RNA polymerase has been studied intensively as a model RNA polymerase. The influenza virus glycoproteins, hemagglutinin (HA) and neuraminidase, were among the earliest glycoproteins to have their structure determined at atomic resolution. The influenza virus HA together with the VSV G protein and parainfluenza virus 5 fusion (F) protein have served as model metastable proteins that undergo on triggering dramatic protein refolding and in doing so mediate the fusion of the viral envelope with a host cell membrane, enabling the viral nucleocapsid to enter the cytoplasm.

In this book are timely chapters on important topics to elucidate the biology of Negative Strand RNA Viruses. In two chapters the process of entry into the host cell by rhabdoviruses and parainfluenza virus are described. As befitting an important model system, four chapters describe different aspects of transcription and replication of VSV. For parainfluenza viruses the proteins involved in transcription are natively unfolded and this is described together with functional implications for transcription and replication. The assembly and budding of VSV and paramyxoviruses are reviewed together with virus–host interactions of VSV and Ebola virus. The innate immune system and the mechanism by which paramyxoviruses counteract this cellular pathway are described. One chapter is devoted to discussing the molecular and cellular biology of emerging bunyaviruses and lastly a chapter describes the current status of potential therapeutics to Ebola virus. All of these chapters bring together a great deal of knowledge about the cellular, structural and molecular biology of Negative Strand RNA Viruses.

Despite our current knowledge of Negative Strand RNA Viruses we still have not fully understood the switch between RNA transcription to synthesize mRNAs and RNA replication to make anti-genome RNA strands and genomic RNA. The atomic structure of the RNA-dependent RNA polymerase has yet to be determined for any Negative Strand RNA virus. For those viruses that use their receptor binding protein to activate the fusion protein, a molecular description of how binding lowers the energy barrier for the fusion protein to cause virus entry into cells is only understood at a rudimentary level. Similarly, there is only fragmentary knowledge of how host-cell proteins interact with viral proteins to mediate the assembly and budding of most of the Negative Strand RNA Viruses. Although many scientists have made impressive advances in studying the innate immune response to Negative Strand RNA Viruses, there is still much to be learned about this complex cellular pathway. We still lack vaccines to important human pathogens such as respiratory syncytial virus, new and old world arenaviruses (particularly Lassa fever virus) and Ebola virus. Furthermore, anti-viral drugs are currently mostly a mirage on the horizon, the one major success story being inhibitors of influenza A and B virus neuraminidase, such as oseltamivir phosphate. Hopefully, a full understanding of the biology and the atomic structure of proteins for Negative Strand RNA Viruses will aid in devising new vaccines and provide new targets for the development of anti-viral drugs.

Robert A. Lamb, Ph.D., Sc.D.
Department of Molecular Biosciences
Northwestern University

Chapter 1

Overview of Negative-Strand RNA Viruses

Biao He*

1. Non-Segmented Negative-Strand RNA Viruses

There are four families of non-segmented negative-strand RNA viruses (NNSV): *Paramyxoviridae*, *Filoviridae*, *Rhabdoviridae*, and *Bornaviridae*. The single-stranded RNA genomes of the *Mononegavirales* range from approximately 11,000 to 19,000 nucleotides in length and contain a series of tandemly-linked genes separated by non-transcribed sequences.

For paramyxoviruses, the gene order is 3'-NP-P(V/W/C)-M-F-(SH)-HN-L-5' where genes in parenthesis are not found in all species (reviewed in Ref. 1). The viral RNA-dependent RNA polymerase (vRdRp), which transcribes the nucleocapsid protein (NP)-encapsidated RNA into 5' capped and 3' polyadenylated mRNAs, minimally consists of two proteins, phosphoprotein (P) and the large (L) polymerase protein.[2] The vRdRp is thought to bind the genomic RNA at a single 3' entry site and to transcribe the genome by a sequential and polar process. The vRdRp also replicates the viral RNA genome.[3–6] The functional template for transcription of non-segmented negative-sense RNA viruses is the helical nucleocapsid (NC) (reviewed in Ref. 1). Although the exact details of mRNA production are unknown, the process is currently believed to involve termination and reinitiation (stop and start) at each gene junction and these junctions consist of three nucleotide sequence elements. At the gene end (GE) sequence, polyadenylation occurs through

*Department of Infectious Diseases, College of Veterinary Medicine, University of Georgia, Athens, GA 30602, USA. E-mail: bhe@uga.edu

reiterative copying of a 4–7 uridyl (U) residue tract, and transcription terminates releasing a polyadenylated RNA. At this juncture, the vRdRp either leaves the template (attenuation) or passes over an intergenic sequence (IGS) region not found in mRNAs, and the vRdRp reinitiates mRNA synthesis at a downstream transcriptional gene start (GS) sequence. Viral mRNAs are capped, presumably by vRdRp.

The L proteins of paramyxoviruses have masses of 220 to 250 KD. They are believed to have the capacity to initiate, elongate, and terminate transcription. The P protein of paramyxoviruses is essential for synthesis of viral RNA, and it is phosphorylated (hence the name phosphoprotein).[1] Among all viruses in *Mononegavirales*, the P protein of VSV is the best studied. The P protein of VSV contains two major phosphorylated regions: the N-terminal region and the C-terminal region. The N-terminal phosphorylation is believed to be important for viral RNA transcription and the C-terminal phosphorylation is believed to be important for viral RNA genome replication. It is known that the P protein can interact with both N and L proteins. It is thought that the P–L interaction can stabilize L, and the N–P interaction enables P to serve as a chaperone allowing N to encapsidate nascent viral RNA genomes during viral RNA genome replication.

Regulation of the phosphorylation status of P is thought to be key in regulating the switch by vRdRp from a viral mRNA synthesis (transcription) configuration to a viral RNA replication configuration.[7,8] It is thought that phosphorylation of viral proteins is carried out by host kinases.[9] A host kinase that phosphorylates the N-terminal region of VSV P has been identified as casein kinase II (CKII),[10] while the host kinase for the C-terminal region has not yet been identified.[11] A host kinase, Akt, has been reported to play a critical and direct role in viral RNA synthesis in virus-infected cells.

1.1. *Paramyxoviridae*

Viruses in the *Paramyxoviridae* family of *Mononegavirales* include many important human and animal pathogens, such as human parainfluenza viruses (PIVs), Sendai virus (SeV), mumps virus (MuV), Newcastle disease virus (NDV), measles virus (MeV), rinderpest virus, and human respiratory syncytial virus (RSV), as well as the emerging viruses, Nipah and Hendra virus (HeV). *Paramyxoviridae* has two subfamilies: *Paramyxovirinae* and *Pneumovirinae*. *Paramyxovirinae* has five genuses: *Avulavirus*, *Henipavirus*, *Morbillivirus*, *Respirovirus*, and *Rubulavirus*. *Pneumovirinae* has two genuses: *Metapneumovirus* and *Pneumovirus*.

1.1.1. *Rubulavirus*

PIV5, formerly known as simian virus 5 (SV5),[12] is a prototypical member of the *Rubulavirus* genus of the family *Paramyxoviridae*.[1] Although PIV5 was originally isolated from cultured primary monkey cells, its natural host is the dog, in which it may cause kennel cough.[13] PIV5 can infect humans,[14] but no known symptoms or diseases in humans have been associated with PIV5.[15] PIV5 encodes eight viral proteins.[1] NP, phosphoprotein (P), and large RNA polymerase (L) protein are essential for viral RNA synthesis (mRNA transcription and genome RNA replication). The V protein plays important roles in viral pathogenesis. V inhibits interferon (IFN) signaling as well as production. In addition, overexpression of V inhibits viral gene expression.[16] The fusion (F) protein, a glycoprotein, mediates both cell-to-cell and virus-to-cell fusion in a pH-independent manner that is essential for virus entry into cells. The hemagglutinin-neuraminidase (HN), another viral glycoprotein, is also involved in virus entry and release from the host cells. The matrix (M) protein plays an important role in virus assembly and budding.[17,18] The small hydrophobic (SH) protein is a 44-residue hydrophobic integral membrane protein and plays a role in blocking the TNF-α-mediated, extrinsic apoptotic pathway.[19–22]

MuV, a rubulavirus with an identical genome structure to PIV5, causes acute inflammatory infections in humans involving most organ systems.[23] MuV is most notable as a highly neurotropic and neurovirulent agent causing a number of central nervous system (CNS) lesions ranging from mild meningitis to severe, and occasionally fatal, encephalitis. Although sequences of different vaccine strains and several clinical isolates are known, the molecular bases of MuV virulence are not well understood.[24] No single nucleotide change, amino acid change, or even any particular virus gene is wholly responsible for the overall virulence or attenuation of a specific virus strain.[25,26] A number of reports found no simple pattern of genomic mutations capable of discriminating virulent from attenuated MuV strains.[24,27,28] MuV infection was the most common cause of viral meningitis and encephalitis until the arrival of mass immunization with MuV vaccine.[23] While MuV infection has been dramatically reduced due to vaccination, safety concerns associated with the vaccination as well as effectiveness of the vaccines have surfaced.[29] The Jeryl Lynn (JL) vaccine has been proven to be highly efficacious and produces few adverse reactions. The rate of aseptic meningitis following vaccination with JL (one case per 1.8 million doses) is below background levels. However, other live attenuated MuV vaccines have had much higher incidents of vaccine-associated meningitis. The Urabe vaccine, which was widely distributed in Japan, Europe, and Canada, is estimated to cause one case of meningitis in every 1000 to 11,000 doses of the vaccine distributed in the UK and one case of meningitis in every 62,000 doses of the vaccine distributed in Canada. The Urabe vaccine was therefore withdrawn due to safety concerns. Interestingly, even with widespread vaccination programs in place, mumps outbreaks continue to occur. Causes cited range from failure to vaccinate, vaccine failure, and emergence of MuV strains capable of escaping vaccine-induced immunity.[30–33]

Human PIV2 generally causes a relatively mild infection of the upper respiratory tract of infants. Unlike PIV5 and MuV, HPIV2 does not have an SH protein.

1.1.2. *Henipavirus*

Viruses in the *Henipavirus* genus of the family *Paramyxoviridae*, HeV and Nipah virus (NiV), emerged during the past decade. HeV causes a febrile respiratory illness in humans and animals and was responsible for the deaths of two humans and 17 horses in three separate incidents in Australia between 1994 and 1999.[34–36] The first known human infections with NiV were detected during an outbreak of severe febrile encephalitis in peninsular Malaysia and Singapore in 1998–1999. A total of 276 patients (105 fatal) with viral encephalitis due to NiV disease were reported in Malaysia and Singapore, mostly among adult males who were involved in pig farming or pork production activities. The outbreak was controlled by culling more than 1 million pigs.[37–39] More recently, NiV has been established as the cause of fatal, febrile encephalitis that occurred in humans in Bangladesh and India during the winters of 2001, 2003, and 2004.[40–43] Fruit-eating bats are a natural reservoir for NiV and humans are infected via intermediate hosts such as pigs, by exposure to infected fruit bats or material contaminated by infected bats, or by direct human-to-human transmission. Seropositive bats have been detected in Malaysia, India, Bangladesh, Cambodia, and Thailand, and bat isolates have been obtained from Malaysia and Cambodia.[40,44–46]

1.1.3. *Avulavirus*

NDV is a prototypical virus in this genus. NDV is highly infectious and causes fatal infections in birds. It is a major threat to the commercial poultry industry.

1.1.4. *Morbillivirus*

MeV is one of the most contagious viruses. MeV infection has historically been a very common childhood infection. MeV infects through the respiratory tract and causes a maculopapular rash and conjunctivitis. MeV infection causes immunosuppression, leading to secondary infections. Successful vaccination programs have mostly controlled the infection. However, due to non-compliance with vaccination programs in some developed countries and lack of vaccination programs in developing countries, MeV infection remains a major health threat.[47] In addition, MeV causes rare, but fatal, subacute, sclerosing panencephalitis.

1.1.5. *Respirovirus*

Respirovirus includes human parainfluenza virus 1 (HPIV1) and 3 (HPIV3), and SeV. HPIV1 and 3 are common respiratory pathogens that infect young children, mainly causing mild upper respiratory tract infection. They can cause "croup" and occasionally pneumonia. SeV causes a highly infectious and fatal respiratory disease in mice.

1.1.6. *Pneumovirinae*

Pneumovirinae includes *Pneumovirus* genus and *Metapneumovirus* genus. RSV is the prototype virus of the *Pneumovirus* genus.[48,49] The RNA genome of RSV is 15,222 nt long and contains a linear array of 10 transcription units from which 11 proteins are translated. A number of these proteins are homologous to proteins from other paramyxoviruses, including those involved in viral RNA synthesis (N, nucleocapsid; P, phosphoprotein; L, polymerase), assembly (M, matrix), and attachment and fusion (G and F). In addition, RSV encodes two proteins in the M2 gene (M2-1 and M2-2) that regulate RNA synthesis by the RSV polymerase complex.[50–53] The non-structural NS1 and NS2 proteins are important for IFN antagonism. Although they are dispensable for viral replication *in vitro* (i.e., viable RSV lacking NS1 and/or NS2 can be obtained through reverse genetics), deletion of either NS gene attenuates RSV significantly *in vitro* and markedly *in vivo*, even in cells lacking IFN expression,[54,55] indicating that NS1 and NS2 play additional roles in regulating viral replication.

RSV is the most important cause of pediatric viral respiratory infection and is a major cause of morbidity and mortality among infants as well as immunocompromised subjects and the elderly.[48] In addition, severe RSV infection can result in wheezing and asthma later in life. Unlike infection by other respiratory viruses, RSV does not induce long-lasting protective immunity against subsequent infection. Thus, most individuals are infected multiple times throughout the course of their lives. In the early 1960s, vaccination of infants with a formalin-inactivated RSV vaccine not only failed to protect against RSV disease during the following RSV season, but some vaccines developed enhanced disease upon infection with RSV, resulting in increased rates of severe pneumonia and 2 deaths.[56] A number of different approaches have been evaluated, including subunit vaccines, vectored vaccines, and live attenuated vaccines; however, there remains no currently licensed RSV vaccine. Challenges to RSV vaccine development, in addition to avoiding enhanced disease, include induction of balanced immune responses in the presence of maternal antibodies and avoidance of reactogenicity in infants. Aerosolized ribavirin and prophylactic immunoglobulin therapy are used in the clinical setting. However, the high cost of palivizumab prophylaxis raises the question of cost-effectiveness relative to health benefits due to the need for monthly injections during RSV season. Even though vaccine development has not been successful for RSV, the fact that antibody against F is effective in preventing RSV in infant indicates that developing an RSV vaccine is feasible.

Human metapneumovirus (hMPV) was discovered in young children with respiratory illness in 2001. It causes upper and lower respiratory infections in young children, leading to mild infection, as well as severe bronchiolitis and pneumonia. HMPV causes a disease very similar to RSV infection. It is thought that cases of HMPV infection may have been misdiagnosed as RSV infection due to the similarity of symptoms.

1.1.7. *Emerging paramyxoviruses*

In addition to the emerging *Henipavirus*, there are many newly isolated paramyxoviruses that do not fit into the five genuses. J paramyxovirus (JPV) was first isolated from cultures of rodent kidneys in Australia in the 1970s by Jun,[57,58] after whom the virus was named. Antibodies against JPV were detected in wild mice (27 out of 96), wild rats (17 out of 106), pigs (13 out of 107), and even in humans (2 out of 91).[57,58] Although infections of laboratory animal mice and rats, intranasally or subcutaneously, are not associated with clinical signs of disease, the infections do cause varying degrees of hemorrhagic interstitial pneumonia, and virus is detected in blood, lung, liver, kidney, and spleen at 3-weeks post-infection. Morphological studies using electron microscopy and analysis of nucleocapsid structure indicate that JPV is a paramyxovirus.[57,59] Recently, the genome structure and sequence of the virus have been determined. JPV is indeed a paramyxovirus and has the longest genome (18,954 nucleotides) in the family.[60] JPV also encodes a novel transmembrane protein of 258 amino acid residues, which has no homology to any known proteins. Interestingly, the matrix (M) protein has high homology to a putative human protein, Angrem 52 (81% identity).[60] In contrast, the closest M protein to the M protein of JPV in the paramyxovirus family is NiV M, which has 50% identity. Angrem 52 has been identified as a gene that is upregulated by angiotensin II in cultured human mesangial cells.[61] Because of high homologies of Angrem 52 to a paramyxovirus M protein, it has been predicted that this putative human protein is actually from a novel paramyxovirus that infects human cells.[62,63] This putative virus was identified and named Beilong virus.[64] Analysis of Beilong virus indicates that its genome structure is identical to JPV. Due to their unique genome structure, JPV and Beilong virus have been proposed to form new genuses in the subfamily of *Paramyxovirinae*. In addition to JPV and Beilong virus, other emerging paramyxoviruses include: Fer-de-lance virus Menangle virus, Mossman virus, Salem virus, and Tupaia paramyxovirus.

1.2. *Rhabdoviridae*

Rhabdo- is a prefix meaning "rod"-like in Greek and the Rhabdovirus name reflects the "rod"-shaped, lipid membrane in which the viral genome is packed. This family of viruses has the most diverse hosts within the negative-stranded RNA viruses, including plants and mammals. There are six genuses: *Cytorhabdovirus* includes many plant viruses, such as lettuce necrotic yellow virus; *Ephemerovirus* includes bovine ephemeral fever virus; *Lyssavirus* includes rabies virus; *Novirhabdovirus* includes many fish viruses, such as viral hemorrhagic septicemia virus; *Nucleorhabdovirus* includes potato yellow dwarf virus; and *Vesiculovirus* includes vesicular stomatitis virus (VSV).

1.2.1. *Vesiculovirus*

VSV is a prototypical rhabdovirus. Gene order and process of RNA synthesis of rhabdoviruses are similar to viruses in the *Paramyxoviridae*.[65] Many important discoveries regarding replication of NNSVs have been made using VSV as a model system. VSV infects many mammals. It causes acute diseases in cattle, horses, and pigs, and causes major economic losses. Humans who have close contacts with infected animals can get infected through contact with vesicular fluid and tissues. The symptoms of VSV infection in humans are similar to influenza virus infection: fever, chills, and muscle pain.

1.2.2. *Lyssavirus*

Rabies virus causes fatal infections in humans, with 100% fatality if not treated. Rabies virus is spread through bites from infected animal, mostly infected dogs in endemic areas. It is thought that the initial infected tissue is muscle. The virus spreads within the infected host centripetally through neurons and eventually reaches the brain, where symptoms of rabies virus infection manifest. It may take several weeks to as long as years for the virus to reach the brain, thus there is a latent period of infection, which provides a window of opportunity to treat rabies virus infection in humans. Treating human immediately after exposure with anti-rabies virus antibody injection and immunization with rabies virus vaccine has been effective in preventing rabies.

1.3. *Filoviridae*

Filoviridae includes two genuses: *Ebolavirus* and *Marburgvirus*. *Ebolavirus* contains Zaire, Sudan, Reston, and Cote d'Ivoire strains based on the location of original outbreaks. Filoviruses are among the most virulent and lethal pathogens known to exist, with the more pathogenic species having fatality rates higher than 80% in humans. Natural outbreaks of hemorrhagic fever are likely to continue and perhaps even accelerate in tropical Africa.

1.4. *Bornaviridae*

Bornaviridae contains one genus, *Borna disease virus* (BDV), which has been suspected of contributing to human mental illness. The virus is unique among all NNSVs in that its genome replication occurs in the nucleus. BDV is neurotropic. Infection of vertebrate animals with bornavirus results in damage to the CNS and causes behavioral abnormalities. Serologic and epidemiologic studies indicate that humans have been exposed to BDV, or a BDV-like virus.

2.   Segmented Negative-Stranded RNA Viruses

There are three families of segmented negative-stranded RNA viruses: *Arenaviridae*, *Bunyaviridae*, and *Orthomyxoviridae*. These viruses have two to eight segments of negative-stranded RNA genomes.

   Orthomyxoviridae contains five genera: *Influenzavirus A, B*, and *C, Isavirus*, and *Thogotovirus*. The genomes of these viruses range from six segments to eight segments. Influenza virus is often shown as a spherical form of about 100 nm in diameter; however, pleomorphic and filamentous forms exist as well. Influenza viruses are classified into three genuses: A, B, and C. Influenza A and B viruses have eight segments of RNA genomes, while Influenza C virus has only seven segments. They all contain segments encoding for polymerase proteins PA, PB1, and PB2 that are vRdRp and are responsible for replicating and transcribing viral RNA. The NP is encoded by a different segment and encapsidates the viral genome RNA. Viral genome RNA that is not encapsidated by NP is non-functional as a template for replication or transcription. HA and NA are expressed from two different segments and are important for virus entry and egress. HA binds to the target cell surface via sialic acid residues on host proteins in the plasma membrane triggering endocytosis. HA undergoes a conformation change in the late endosome due to lowered pH, promoting fusion between the viral and endosomal membranes and exposing the encapsidated viral RNA genome (RNP) to the cytosol. The viral RNP is then transported to the nucleus, where viral RNA synthesis (replication and mRNA transcription) occurs. NA is essential for virus release from infected cells because it cleaves the sialic acid residue off of adjacent molecules to prevent binding of HA to the already infected cells. There are two segments encoding two different viral proteins each through splicing. One segment encodes M1 and M2, M1 being a product from non-spliced mRNA and M2 being a product of spliced mRNA. M1 is a matrix protein underlining the viral membrane in the virion and plays a critical role in virus egress from the plasma membrane. M2 is an ion channel that facilitates release of the RNP-fused endosomal membrane, and recently it has been found that M2 plays a critical role in virus egress as well. NS1 and NS2 are also made from the same segment; they are the products of unspliced and spliced mRNA, respectively. NS1 binds to RNA and plays a critical role in regulating virus–host interactions. NS2 plays a critical role in virus egress by facilitating export of the RNP from the nucleus to the cytosol. In some instances, PB1-F2, a protein of 79 to 101 amino acid residues, is encoded by the segment that encodes the PB1 protein through an alternative open reading frame. PB1-F2 is located in the mitochondrial membrane and is thought to play a role in virus-induced apoptosis. Influenza A and B have very similar genome structure and encode similar proteins with similar functions. Influenza C virus has only seven segments. It does not have two segments encoding HA and NA; instead, one segment encodes a HEF protein that binds to the receptor (9-*O*-acetylneuraminic acid), promotes viral membrane and cell membrane fusion, and has a neuraminate-*O*-acetyl esterase activity to remove the receptor on already infected cells to facilitate virus release.

Influenza A and C viruses infect a variety of species including humans and birds, while type B infects only humans. Symptoms of influenza virus infection include fever, sore throat, dry cough, muscle pain, headache, and malaise. Symptoms often last for seven days. Influenza A virus causes significant morbidity and mortality each year, mostly due to secondary infection and contributing to the deaths of immuno-compromised populations such as elderly and HIV patients. Only influenza A virus is associated with pandemics. Influenza A virus is classified by its two major surface glycoproteins, HA and NA. There are 16 HA[66] and nine NA subtypes, differing by $\geq 30\%$ in protein homology, which are used to categorize influenza A virus into subtypes (e.g., H1N1, H3N2, H5N1, etc.). Point mutations during virus replication result in changes in protein sequences. Changes in the antibody-binding sites of surface glycoproteins allow viruses to evade antibody-mediated immunity and to re-infect humans and animals (antigenic drift). As influenza virus has a segmented genome, if different influenza A virus subtypes infect the same host, exchange of gene segments can occur producing a new virus with a unique combination of viral proteins (antigenic shift). It is antigenic shifts that result in new virus subtypes and may give rise to pandemics.[67]

Drugs targeting M2 and NA have been approved for use to treat influenza virus infection. NA inhibitors, when applied early in infection, can shorten the days of symptomatic infection. Both kinds of drugs generate drug-resistant strains of influenza A virus. Vaccination is the best means to prevent influenza virus infection.

Infectious Salmon Anemia Virus (ISAV) is economically important because it causes high mortality in salmon farms on the Atlantic coast. It has eight segments. Interestingly, the segment encodes a protein with receptor binding, as well as esterase, activity, HE. ISAV is the only known orthomyxovirus that infects fish. Thogoto virus was first isolated from ticks removed from cattle in the Thogoto forest in Kenya in 1960. Thogoto virus has a six-segmented RNA genome. In addition to segments encoding vRdRp, PA, PB1, PB2, and NP, one segment encodes a glycoprotein (G) responsible for virus entry, and another segment encodes two proteins M and ML by splicing and non-splicing mRNA. Thogoto virus is not known to cause disease in humans. Thogoto virus is an arbovirus that infects both mammalian hosts and ticks.

References

1. Lamb, R.A. and D. Kolakofsky. *Paramyxoviridae*: The viruses and their replication, in Fields Virology (4th edition), D.M. Knipe and P.M. Howley, Editors. Lippincott Williams and Wilkins, Philadelphia, 2001.
2. Emerson, S.U. and Y.-H. Yu. Both NS and L proteins are required for *in vitro* RNA synthesis by vesicular stomatitis virus. *J Virol*, 1975. **15**: 1348–56.
3. Abraham, G. and A.K. Banerjee. Sequential transcription of the genes of vesicular stomatitis virus. *Proc Natl Acad Sci U S A*, 1976. **73**: 1504–8.
4. Ball, L.A. and C.N. White. Order of transcription of genes of vesicular stomatitis virus. *Proc Natl Acad Sci U S A*, 1976. **73**: 442–6.
5. Emerson, S.U. Reconstitution studies detect a single polymerase entry site on the vesicular stomatitis virus genome. *Cell*, 1982. **31**: 635–42.
6. Iverson, L.E. and J.K. Rose. Sequential synthesis of 5′-proximal vesicular stomatitis virus mRNA sequences. *J Virol*, 1982. **44**: 356–65.
7. Chen, J.L., T. Das and A.K. Banerjee. Phosphorylated states of vesicular stomatitis virus P protein *in vitro* and *in vivo*. *Virology*, 1997. **228**: 200–12.
8. Spadafora, D., D.M. Canter, R.L. Jackson and J. Perrault. Constitutive phosphorylation of the vesicular stomatitis virus P protein modulates polymerase complex formation but is not essential for transcription or replication. *J Virol*, 1996. **70**: 4538–48.

9. Lenard, J. Host cell protein kinases in nonsegmented negative-strand virus (mononegavirales) infection. *Pharmacol Ther*, 1999. **83**: 39–48.

10. Barik, S. and A.K. Banerjee. Phosphorylation by cellular casein kinase II is essential for transcriptional activity of vesicular stomatitis virus phosphoprotein P. *Proc Natl Acad Sci U S A*, 1992. **89**: 6570–4.

11. Das, S.C. and A.K. Pattnaik. Role of the hypervariable hinge region of phosphoprotein P of vesicular stomatitis virus in viral RNA synthesis and assembly of infectious virus particles. *J Virol*, 2005. **79**: 8101–12.

12. Chatziandreou, N., N. Stock, D. Young, J. Andrejeva, K. Hagmaier, D.J. McGeoch and R.E. Randall. Relationships and host range of human, canine, simian and porcine isolates of simian virus 5 (parainfluenza virus 5). *J Gen Virol*, 2004. **85**: 3007–16.

13. McCandlish, I.A., H. Thompson, H.J. Cornwell and N.G. Wright. A study of dogs with kennel cough. *Vet Rec*, 1978. **102**: 293–301.

14. Cohn, M.L., E.D. Robinson, D. Thomas, M. Faerber, S. Carey, R. Sawyer, K.K. Goswami, A.H. Johnson and J.R. Richert. T cell responses to the paramyxovirus simian virus 5: Studies in multiple sclerosis and normal populations. *Pathobiology*, 1996. **64**: 131–5.

15. Hsiung, G.D., P.W. Chang, R.R. Cuadrado and P. Isacson. Studies of parainfluenza viruses. III. Antibody responses of different animal species after immunization. *J Immunol*, 1965. **94**: 67–73.

16. Lin, Y., F. Horvath, J.A. Aligo, R. Wilson and B. He. The role of simian virus 5 V protein on viral RNA synthesis. *Virology*, 2005. **338**: 270–80.

17. Schmitt, A.P., B. He and R.A. Lamb. Involvement of the cytoplasmic domain of the hemagglutinin-neuraminidase protein in assembly of the paramyxovirus simian virus 5. *J Virol*, 1999. **73**: 8703–12.

18. Schmitt, A.P., G.P. Leser, D.L. Waning and R.A. Lamb. Requirements for budding of paramyxovirus simian virus 5 virus-like particles. *J Virol*, 2002. **76**: 3952–64.

19. He, B., G.P. Leser, R.G. Paterson and R.A. Lamb. The paramyxovirus SV5 small hydrophobic (SH) protein is not essential for virus growth in tissue culture cells. *Virology*, 1998. **250**: 30–40.

20. He, B., G.Y. Lin, J.E. Durbin, R.K. Durbin and R.A. Lamb. The sh integral membrane protein of the paramyxovirus simian virus 5 is required to block apoptosis in mdbk cells. *J Virol*, 2001. **75**: 4068–79.

21. Hiebert, S.W., C.D. Richardson and R.A. Lamb. Cell surface expression and orientation in membranes of the 44 amino acid SH protein of simian virus 5. *J Virol*, 1988. **62**: 2347–57.

22. Lin, Y., A.C. Bright, T.A. Rothermel and B. He. Induction of apoptosis by paramyxovirus simian virus 5 lacking a small hydrophobic gene. *J Virol*, 2003. **77**: 3371–83.

23. Carbone, K.M. and J.S. Wolinsky. Mumps Virus, in Field's Virology (4th edition, vol. 1), D.M. Knipe and P.M. Howley, editors. Lippincott Williams and Wilkins: Philadelphia, 2001, pp. 1381–1400.

24. Amexis, G., S. Rubin, V. Chizhikov, F. Pelloquin, K. Carbone and K. Chumakov. Sequence diversity of Jeryl Lynn strain of mumps virus: Quantitative mutant analysis for vaccine quality control. *Virology*, 2002. **300**: 171–9.

25. Rubin, S.A., G. Amexis, M. Pletnikov, Z. Li, J. Vanderzanden, J. Mauldin, C. Sauder, T. Malik, K. Chumakov and K.M. Carbone. Changes in mumps virus gene sequence associated with variability in neurovirulent phenotype. *J Virol*, 2003. **77**: 11616–24.

26. Sauder, C.J., K.M. Vandenburgh, R.C. Iskow, T. Malik, K.M. Carbone and S.A. Rubin. Changes in mumps virus neurovirulence phenotype associated with quasispecies heterogeneity. *Virology*, 2006. **350**: 48–57.

27. Cusi, M.G., L. Santini, S. Bianchi, M. Valassina and P.E. Valensin. Nucleotide sequence at position 1081 of the hemagglutinin-neuraminidase gene in wild-type strains of mumps virus is the most relevant marker of virulence. *J Clin Microbiol*, 1998. **36**: 3743–4.

28. Ivancic, J., T.K. Gulija, D. Forcic, M. Baricevic, R. Jug, M. Mesko-Prejac and R. Mazuran. Genetic characterization of L-Zagreb mumps vaccine strain. *Virus Res*, 2005. **109**: 95–105.

29. Furesz, J. Safety of live mumps virus vaccines. *J Med Virol*, 2002. **67**: 299–300.

30. Crowley, B. and M.A. Afzal. Mumps virus reinfection — clinical findings and serological vagaries. *Commun Dis Public Health*, 2002. **5**: 311–3.

31. Lim, C.S., K.P. Chan, K.T. Goh and V.T. Chow. Hemagglutinin-neuraminidase sequence and phylogenetic analyses of mumps virus isolates from a vaccinated population in Singapore. *J Med Virol*, 2003. **70**: 287–92.

32. Strohle, A., C. Bernasconi and D. Germann. A new mumps virus lineage found in the 1995 mumps outbreak in western Switzerland identified by nucleotide sequence analysis of the SH gene. *Arch Virol*, 1996. **141**: 733–41.

33. Utz, S., J.L. Richard, S. Capaul, H.C. Matter, M.G. Hrisoho and K. Muhlemann. Phylogenetic analysis of clinical mumps virus isolates from vaccinated and non-vaccinated patients with mumps during an outbreak, Switzerland 1998–2000. *J Med Virol*, 2004. **73**: 91–6.

34. Murray, K., R. Rogers, L. Selvey, P. Selleck, A. Hyatt, A. Gould, L. Gleeson, P. Hooper and H. Westbury. A novel morbillivirus pneumonia of horses and its transmission to humans. *Emerg Infect Dis*, 1995. **1**: 31–3.

35. Murray, K., P. Selleck, P. Hooper, A. Hyatt, A. Gould, L. Gleeson, H. Westbury and L.E.A. Hiley. A morbillivirus that caused fatal disease in horses and humans. *Science*, 1995. **268**: 94–7.

36. Selvey, L.A., R.M. Wells, J.G. McCormack, A.J. Ansford, K. Murray, R.J. Rogers, P.S. Lavercombe, P. Selleck and J.W. Sheridan. Infection of humans and horses by a newly described morbillivirus. *Med J Aust*, 1995. **162**: 642–5.

37. CDC. Outbreak of Hendra-like virus — Malaysia and Singapore, 1998–1999. *Morbid Mortal Wkly Rep*, 1999. **48**: 265–9.

38. CDC. Update: Outbreak of Nipah virus — Malaysia and Singapore, 1999. *Morbid Mortal Wkly Rep*, 1999. **48**: 335–7.

39. Chua, K.B., W.J. Bellini, P.A. Rota, B.H. Harcourt, A. Tamin, S.K. Lam, T.G. Ksiazek, P.E. Rollin, S.R. Zaki, W. Shieh, C.S. Goldsmith, D.J. Gubler, J.T. Roehrig, B. Eaton, A.R. Gould, J. Olson, H. Field, P. Daniels, A.E. Ling, C.J. Peters, L.J. Anderson and B.W. Mahy. Nipah virus: A recently emergent deadly paramyxovirus. *Science*, 2000. **288**: 1432–5.

40. ICDDRB. Nipah encephalitis outbreak over a wide area of Bangladesh, 2004. *Health Sci Bull*, 2004. **2**: 7–11.

41. ICDDRB. Person-to-person transmission of Nipah virus during outbreak in Faridpur District. *Health Sci Bull*, 2004. **2**: 5–9.

42. WHO. Nipah virus outbreak(s) in Bangladesh, January-April 2004. *Weekly Epidemiology Record*, 2004. **79**: 168–71.

43. Wong, K.T., I. Grosjean, C. Brisson, B. Blanquier, M. Fevre-Montange, A. Bernard, P. Loth, M.C. Georges-Courbot, M. Chevallier, H. Akaoka, P. Marianneau, S.K. Lam, T.F. Wild and V. Deubel. A golden hamster model for human acute Nipah virus infection. *Am J Pathol*, 2003. **163**: 2127–37.

44. Butler, D. Fatal fruit bat virus sparks epidemics in southern Asia. *Nature*, 2004. **429**: 7.

45. Chan, Y.P., K.B. Chua, C.L. Koh, M.E. Lim and S.K. Lam. Complete nucleotide sequences of Nipah virus isolates from Malaysia. *J Gen Virol*, 2001. **82**: 2151–5.

46. Reynes, J.M., D. Counor, S. Ong, C. Faure, V. Seng, S. Molia, J. Walston, M.C. Georges-Courbot, V. Deubel and J.L. Sarthou. Nipah virus in Lyle's flying foxes, Cambodia. *Emerg Infect Dis*, 2005. **11**: 1042–7.

47. Yanagi, Y., M. Takeda and S. Ohno. Measles virus: Cellular receptors, tropism and pathogenesis. *J Gen Virol*, 2006. **87**: 2767–79.

48. Collins, P.L., R.M. Chanock and B.R. Murphy. Respiratory syncytial virus, in *Fields Virology* (4th edition), D.M. Knipe and P.M. Howley, Editors. Lippincott Williams and Wilkins: Philadelphia, 2001, pp. 1443–85.

49. DeFeo-Jones, D., S.F. Barnett, S. Fu, P.J. Hancock, K.M. Haskell, K.R. Leander, E. McAvoy, R.G. Robinson, M.E. Duggan, C.W. Lindsley, Z. Zhao, H.E. Huber and R.E. Jones. Tumor cell sensitization to apoptotic stimuli by selective inhibition of specific Akt/PKB family members. *Mol Cancer Ther*, 2005. **4**: 271–9.

50. Bermingham, A. and P.L. Collins. The M2-2 protein of human respiratory syncytial virus is a regulatory factor involved in the balance between RNA replication and transcription. *Proc Natl Acad Sci U S A*, 1999. **96**: 11259–64.

51. Fearns, R. and P.L. Collins. Model for polymerase access to the overlapped L gene of respiratory syncytial virus. *J Virol*, 1999. **73**: 388–97.

52. Lindsley, C.W., Z. Zhao, W.H. Leister, R.G. Robinson, S.F. Barnett, D. Defeo-Jones, R.E. Jones, G.D. Hartman, J.R. Huff, H.E. Huber and M.E. Duggan. Allosteric Akt (PKB) inhibitors: Discovery and SAR of isozyme selective inhibitors. *Bioorg Med Chem Lett*, 2005. **15**: 761–4.

53. Zhao, Z., W.H. Leister, R.G. Robinson, S.F. Barnett, D. Defeo-Jones, R.E. Jones, G.D. Hartman, J.R. Huff, H.E. Huber, M.E. Duggan and C.W. Lindsley. Discovery of 2,3,5-trisubstituted pyridine derivatives as potent Akt1 and Akt2 dual inhibitors. *Bioorg Med Chem Lett*, 2005. **15**: 905–9.

54. Spann, K.M., K.C. Tran, B. Chi, R.L. Rabin and P.L. Collins. Suppression of the induction of alpha, beta, and lambda interferons by the NS1 and NS2 proteins of human respiratory syncytial virus in human epithelial cells and macrophages [corrected]. *J Virol*, 2004. **78**: 4363–9.

55. Spann, K.M., K.C. Tran and P.L. Collins. Effects of nonstructural proteins NS1 and NS2 of human respiratory syncytial virus on interferon regulatory factor 3, NF-kappaB, and proinflammatory cytokines. *J Virol*, 2005. **79**: 5353–62.

56. Kim, H.W., J.G. Canchola, C.D. Brandt, G. Pyles, R.M. Chanock, K. Jensen and R.H. Parrott. Respiratory syncytial virus disease in infants despite prior administration of antigenic inactivated vaccine. *Am J Epidemiol*, 1969. **89**: 422–34.

57. Jun, M.H., N. Karabatsos and R.H. Johnson. A new mouse paramyxovirus (J virus). *Aust J Exp Biol Med Sci*, 1977. **55**: 645–7.

58. Mesina, J.E., R.S. Campbell, J.S. Glazebrook, D.B. Copeman and R.H. Johnson. The pathology of feral rodents in North Queensland. *Tropenmed Parasitol*, 1974. **25**: 116–27.

59. Jun, M.H. Studies on a virus isolated from wild mice (Mus musculus). M.S. James Cook University, Townsville, Queensland, Australia, 1976.

60. Jack, P.J., D.B. Boyle, B.T. Eaton and L.F. Wang. The complete genome sequence of J virus reveals a unique genome structure in the family Paramyxoviridae. *J Virol*, 2005. **79**: 10690–700.

61. Liang, X., H. Zhang, A. Zhou, P. Hou and H. Wang. Screening and identification of the up-regulated genes in human mesangial cells exposed to angiotensin II. *Hypertens Res*, 2003. **26**: 225–35.

62. Basler, C.F., A. Garcia-Sastre and P. Palese. A novel paramyxovirus? *Emerg Infect Dis*, 2005. **11**: 108–12.

63. Schomacker, H., P.L. Collins and A.C. Schmidt. In silico identification of a putative new paramyxovirus related to the Henipavirus genus. *Virology*, 2004. **330**: 178–85.

64. Li, Z., M. Yu, H. Zhang, D.E. Magoffin, P.J. Jack, A. Hyatt, H.Y. Wang and L.F. Wang. Beilong virus, a novel paramyxovirus with the largest genome of non-segmented negative-stranded RNA viruses. *Virology*, 2006. **346**: 219–28.

65. Barr, J.N., S.P. Whelan and G.W. Wertz. Transcriptional control of the RNA-dependent RNA polymerase of vesicular stomatitis virus. *Biochim Biophys Acta*, 2002. **1577**: 337–53.

66. Fouchier, R.A., V. Munster, A. Wallensten, T.M. Bestebroer, S. Herfst, D. Smith, G.F. Rimmelzwaan, B. Olsen and A.D. Osterhaus. Characterization of a novel influenza A virus hemagglutinin subtype (H16) obtained from black-headed gulls. *J Virol*, 2005. **79**: 2814–22.

67. Wright, P.F.W. Orthomyxoviruses, in *Fields Virology*, D.M. Knipe, P.M. Griffin, D.E. Martin, M.A. Lamb, R.A. Roizman and S.E. Straus, Editors. Lippincott Williams and Wilkins: Philadelphia, 2001, pp. 1533–79.

Chapter 2

Rhabdovirus Entry into the Host Cell

Aurélie Albertini* and Yves Gaudin*

1. Introduction

Rhabdoviruses are widespread among a great diversity of organisms (such as plants, insects, fishes, mammals, reptiles, and crustaceans).[1] The prototypes and best studied viruses of this family are vesicular stomatitis virus (VSV), a member of the *Vesiculovirus* genus, and rabies virus (RV), a member of the *Lyssavirus* genus. Other genera of the family include the *Novirhabdovirus* genus (containing many fish viruses such as infectious hematopoietic necrosis virus, IHNV, and viral hemorrhagic septicemia virus, VHSV), the *Ephemerovirus* genus (prototype: bovine ephemeral fever virus, BEFV), the *Cytorhabdovirus* genus (prototype: lettuce necrotic yellows virus), the *Nucleorhabdovirus* genus (prototype: potato yellow dwarf virus), and the newly discovered[2] genus *Dichorhabdovirus* (prototype: orchid fleck virus).

Rhabdoviruses are enveloped viruses and have in common a rigid bullet shape. Their genome is a single RNA molecule of negative polarity. It associates with the nucleoprotein N, the viral polymerase L, and the phosphoprotein P to form the nucleocapsid. The nucleocapsid is condensed by the matrix protein M into a tightly coiled helical structure, which is surrounded by a lipid bilayer containing the viral glycoprotein G. G plays a critical role during the initial steps of virus infection. It mediates both virus attachment to specific receptors and, after virion endocytosis, the fusion between viral and endosomal membranes.

Glycoprotein G is a type I membrane glycoprotein. After cleavage of the amino-terminal signal peptide, the complete mature glycoprotein is about 500 amino acids long (495 for VSV and 505 for RV). The bulk of the mass of G is located outside the

*Centre de Recherche de Gif, Laboratoire de Virologie Moléculaire et Structurale, CNRS, UPR 3296, Avenue de la Terrasse, 91198 Gif-sur-Yvette Cedex, France.

viral membrane and constitutes the amino-terminal ectodomain. This ectodomain is the target of neutralizing antibodies.[3–7] The protein is anchored in the membrane by a single transmembrane (TM) hydrophobic segment, which probably adopts an α-helical conformation. The rest of the protein is intraviral and probably interacts with internal proteins.

For both VSV and RV, it has been shown that G forms trimers.[8,9] This oligomeric organization seems less stable than that of other viral glycoproteins (such as influenza hemagglutinin, HA) and is sensitive to detergent solubilization.[8–10] In the case of VSV, there exists a dynamic equilibrium between monomers and trimers of G, both *in vitro* after detergent solubilization[10] and *in vivo*.[11,12]

2.   Rhabdoviruses Receptors

Glycoprotein G is responsible for adsorption of virus onto the host cell and therefore determines the tissue tropism of the virus. The nature of the receptor remains a matter of debate for both VSV and RV.

VSV has a large host spectrum: it infects many cell types and organisms (such as insect cells). This suggests that the receptor is a rather ubiquitous molecule. Phosphatidylserine has been considered to be the viral receptor for a long time[13] despite the fact that it is only present at the surface of apoptotic cells. Indeed, more recent results indicate that phosphatidylserine is not a receptor for VSV.[14] Other results have quite convincingly indicated the participation of gangliosides in the receptor structure of VSV in CER (chicken embryo related) cells.[15]

In the case of RV, apart from the very beginning and the end of the infectious process, non-adapted isolates (street rabies) exclusively multiply and propagate in neurons. *In vitro*, such isolates can only infect established cell lines of neuronal origin. Several passages are required to select a fixed strain fully adapted to the multiplication in established non-neuronal cell lines (such as BHK21 and Vero cells).[16–18] Most of the fixed laboratory strains are the results of such an adaptation process. Evelyn Rokitnicki Abelseth, Pasteur Virus, or Challenge Virus Standard are fixed RV strains that have been selected according to this procedure. All have kept their neurotropism and propagate in the nervous system like street viruses. It is probable that adaptation is partly due to the ability of RV-fixed strains to use ubiquitous receptors present at the surface of non-neuronal cell type.[19] Thus, although many molecules have been proposed to be RV receptors,[20] it is not clear whether they are really used by natural isolates during animal infection.

Host cell treatment with different phospholipases has been shown to reduce fixed RV strains binding suggesting that some phospholipids can play the role of viral receptors.[21] Similarly, cells pretreated with neuraminidase were shown to be non-susceptible to viral infection.[22] After incorporation of exogenous gangliosides, cells recovered their susceptibility to RV infection. When purified gangliosides were incorporated, GT1b and GQ1b were the most effective, whereas GM1 and GM3 were poorly active and GD3 was inactive.[23] These results indicate that highly sialylated gangliosides are a part of the cellular membrane receptor structure for the attachment of fixed RV strains.

In addition to phospholipids and gangliosides, three proteins have been proposed to play the role of viral receptors. The first protein that has been proposed to serve as a receptor for RV is the nicotinic acetylcholine receptor (nAChR).[24] This idea was reinforced when sequence similarity was found between a segment of RV glycoprotein and the entire sequence of snake venom curaremimetic neurotoxins that are potent ligands of the acetylcholine receptor.[25] Later, an interaction between RV and purified Torpedo acetylcholine receptor was demonstrated.[26] Finally, purified RV was shown to bind the α subunit of nAChR in an overlay assay[27] but a direct evidence that this molecule is a receptor in animal is still lacking. Furthermore, RV can infect neurons that do not express nAChR.[28] Finally, nAChR is located mainly on muscle cells: although it could account for the ability of street RV to multiply locally in myotubes at the site of inoculation,[29] which would facilitate subsequent penetration into neurons, other molecules are needed to mediate viral entry into neurons.

The second protein that has been proposed to play the role of RV receptor is the neural cell adhesion molecule (NCAM).[30] The evidence for this is that a preincubation of RV with soluble NCAM inhibited its ability to infect susceptible cells. Moreover, transfection of resistant L fibroblasts with the NCAM-encoding gene induces RV susceptibility. Finally, the infection of NCAM-deficient mice by RV resulted in a slightly delayed mortality and a restricted brain invasion. This suggests the *in vivo* relevance for the use of NCAM as a receptor.

Quite simultaneously, the murine low-affinity nerve growth factor receptor, p75NTR was identified as a ligand of a soluble form of RV glycoprotein.[31] The ability of the RV glycoprotein to bind p75NTR was dependent on the presence of a lysine and arginine in positions 330 and 333, which were known to control virus penetration into motor and sensory neurons of adult mice. Furthermore, p75NTR-expressing BSR cells were permissive for a non-adapted fox RV isolate (street virus). Finally, the glycoprotein from another genotype of lyssavirus (GT 6, European bat lyssavirus type 2) was also shown to bind p75NTR.[32] Nevertheless, mice lacking all extracellular receptor domains were still susceptible to infection (of which the rate and specificity were unchanged) indicating that RVG-p75(NTR) interaction is not necessary for RV infection of primary neurons.[33]

Very few is known concerning the receptor of other rhabdoviruses, the only exception is VHSV, a salmonid rhabdovirus, for which it has been shown that monoclonal antibodies (MAbs) directed against a fibronectin containing complex protect cells from the infection. As the purified rainbow trout fibronectin was able to bind specifically to VHSV, fibronectin was proposed to be a receptor for VHSV and some other fish rhabdoviruses.[34]

3. Fusion Properties of Rhabdoviruses

After binding, the virion enters the cell by the endocytic pathway, as described in detail, and subsequently the viral envelope fuses with the membrane of the endosome after its acidification. Fusion is triggered by the low pH of the endosomal compartment and is mediated by the viral glycoprotein. The pH dependence is very similar from one rhabdovirus to another.[35–37] Fusion is maximal around pH 6. For VSV, the threshold for fusion activity (pH 6.6) is slightly higher than for RV (pH 6.3). For VHSV, it has been demonstrated that the lower is the pH for fusion, the more attenuated is the virus.[36] This suggests that the pH threshold for fusion could be a determinant for virulence. This is also consistent with data obtained on VSV indicating that a shift toward lower values of the pH threshold of fusion leads to a loss of infectivity for cell culture.[38,39]

As for other viruses fusing at low pH, preincubation of the virus at low pH in the absence of a target membrane leads to inhibition of the viral fusion activity. However, this inhibition is reversible: readjusting the pH above 7 leads to the recovery of the initial fusion activity.[40,41] This is the main difference between rhabdoviruses and other viruses fusing at low pH, for which low pH-induced fusion inactivation is irreversible.[42]

It has been demonstrated that G can assume at least three different conformational states having different biochemical and biophysical properties.[35,41] The native prefusion state is detected at the viral surface above pH 7. Under this conformation, G is supposed to interact with its receptor. Immediately after acidification, the virions appear to be more hydrophobic.[41] In the absence of a target membrane, this increased hydrophobicity results in viral aggregation.[40] This hydrophobic interaction is mediated by G in an activated state that interacts with the target membrane as a first step of the fusion reaction.[43] After longer incubation at low pH, a third conformation of G is detected, the postfusion conformation that is antigenically distinct from both the prefusion and activated conformations. In electron microscopy images, the ectodomain of the postfusion conformation appears to be more elongated than that of the prefusion (11 nm versus 8 nm).[41]

There is a pH-dependent thermodynamic equilibrium between different states of G that is shifted toward the postfusion state at low pH.[44] Characterization of this equilibrium using MAbs recognizing specifically the native state of RV G reveals the cooperativity of the transition upon proton binding. Analysis of the data using a Hill plot indicated that about 2.8 protons bind simultaneously to G to induce the transition toward the postfusion state.[44] As G is a trimer, this suggested that a single specific residue per monomer, having a pKa much lower in the prefusion than in the postfusion state, is implicated in the structural transition.

The equilibrium between different conformations of G explained why the low pH-induced fusion inactivation is reversible: differently from fusion glycoproteins of viruses from other families, the native conformation is not metastable and thus the structural transition is not irreversible. In fact, the reversibility of the fusogenic low pH-induced conformational change is essential to allow G to be transported through the acidic compartment of the Golgi apparatus and to recover its native prefusion state at the viral surface.[45]

The fusion pathway of rhabdoviruses with artificial liposomes has been studied in detail. Neither RV nor VSV has a specific lipid requirement for fusion.[40,46,47] An investigation of the effects of lipids having various dynamic molecular shapes on RV fusion process has suggested that as other enveloped viruses,[48,49] RV-induced fusion proceeds via the formation of intermediate stalks that are local lipidic connections between outer leaflets of the fusion membrane. Stalk formation is followed by the formation of a transient hemifusion diaphragm (i.e., a local bilayer made by the two initial inner leaflets of the viral and target membranes) in which the formation of a pore and its further enlargement lead to complete fusion.[50]

For viral fusion proteins for which the structural transition is irreversible, it has been proposed that the energy released during the conformational change is used to achieve the energetically expensive membrane-fusion reaction[51] (the activation energy of the fusion process has been estimated to be in the range of 40 kcal/mol[35,52,53]). Nevertheless, even in these cases, experimental data suggest that a large number of spikes are involved in the fusion process. In the case of rhabdoviruses, the existence of a pH-dependent equilibrium between pre- and postfusion conformations of G implies that the energy released during the structural transition of one trimer is small compared with the energetic barrier of the fusion reaction. This indicated that a concerted action of several glycoprotein trimers is needed. Indeed, for RV, the minimal number of spikes involved in the formation of a fusion complex has been estimated to be about 15.[44]

4. Structure of G

The crystalline structures of both pre- and postfusion forms of VSV-G ectodomain have been solved[54,55] (Fig. 1). The soluble ectodomain (Gth, aa residues 1–422) had been generated by limited proteolysis with themolysin directly on purified virus. Both conformations revealed a structural organization that is very different from that of the other viral fusion proteins previously described.

Remarkably, the fold of Gth in its postfusion state was the same as that of HSV 1 glycoprotein gB of which the structure was published at the same time.[56] This defined a new class (III) of fusion proteins. Epstein Barr Virus gB and baculovirus gp64, of which the structures were further determined, have also been demonstrated to belong to this new class.[57,58] Alignment of the amino acid sequences of G proteins from animal rhabdoviruses belonging to different genera reveals that all of them have the same fold except possibly the C-terminal part of ephemeroviruses glycoprotein.[54]

The polypeptide chain of Gth folds into four distinct domains: a β-sheet rich lateral domain (domain I), a central domain that is involved in the trimerization of the molecule (domain II), a pleckstrin homology domain (domain III), and the fusion domain (domain IV). After the end of the trimerization domain (after aa 405), there remain 40 amino acids for the polypeptide chain to reach G TM domain but the structure of the prefusion conformation is unknown after amino acid residue 413 and that of the postfusion conformation is unknown after residue 410 (Fig. 1).

Major antigenic sites are located in both domains I and III. For RV G, p75NTR binding domain is located in lateral domain I, whereas the putative binding domain of nAChR is located in domain III.

The structures revealed that the conformational change from pre- to postfusion state involves a dramatic reorganization of the glycoprotein (Fig. 2). During the structural transition, domains I, III, and IV retain their tertiary structure but they undergo large rearrangements in their relative orientation. This is due to both secondary structure modifications occurring in the hinge regions between the pleckstrin homology domain (domain III) and the fusion domain (domain IV) and complete refolding of the trimerization domain (Fig. 2, panel 2).

Global refolding of G exhibits striking similarities to that of class I proteins such as paramyxovirus fusion protein F[59] and influenza virus HA.[60] The pre- and postfusion states of a protomer are related by flipping both the fusion domain and a C-terminal segment relative to a rigid block made by domain I and part of domain II (Fig. 2, panel 1). During this movement, both the fusion loops and the TM domain move approximately 160 Å from one end of the molecule to the other (Fig. 2, panel 3). The fusion domain is projected toward the membrane through the combination of two movements: a rotation around the hinge between the pleckstrin homology domain and the fusion domain and the lengthening of the central helix (that is involved in the formation of the trimeric central core of the postfusion conformation). The movement of the TM domain is due to the refolding of the

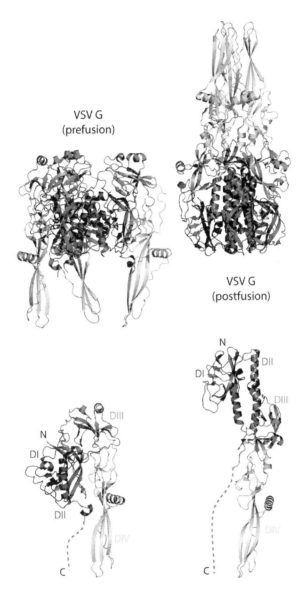

VSV G
(prefusion)

VSV G
(postfusion)

Fig. 1. Overall structures of the pre- and postfusion forms of VSV glycoprotein. Ribbon diagrams of the prefusion and postfusion structures of G trimers (top) (pdb codes: 2J6J and 2CMZ, respectively). Ribbon diagrams of the prefusion and postfusion structures of G protomers (bottom). G is colored by domains in different levels of grey (DI is the lateral domain, DII the trimerization domain, DIII is the pleckstrin homology domain, and DIV is the fusion domain). The C-terminus segment reaching the trans-membrane domain (not observed in both x-ray structures) is represented as a dashed line.

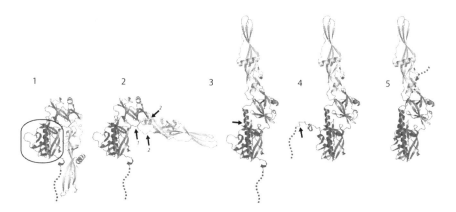

Fig. 2. A possible pathway for the transition from the prefusion protomer to the postfusion protomer by keeping invariant the rigid block orientation (indicated in the first drawing). 1 is the initial state (prefusion conformation, pdb code: 2J6J) and 5 is the final state (postfusion conformation, pdb code 2CMZ). 2, 3, and 4 are putative intermediates. For more clarity pre- and postfusion states are represented by one G protomer, the missing fragment reaching the transmembrane domain is depicted by a dashed line. In intermediate 2, the fusion domain swings around the DIV–DII hinge (arrow 1) while refolding of the DIII–DII connection (arrows 2). In state 3 the lengthening of the central helix (arrow) projects the fusion domain toward the target membrane. Then in 4 the refolding of the C-terminal fragment leads to the final postfusion state (5).

C-terminal segment into an α-helix (Fig. 2, panel 4). In the trimeric postfusion state, the three C-terminal helices position themselves into the groove of the trimeric central core in an antiparallel manner to form a six-helix bundle.

The resulting overall organization of the postfusion state is reminiscent of that of class I protein. As in class I proteins, the fusion domains are N-terminal to the central helices and the TM domains are located at the C-terminus of the antiparallel outer helices (Fig. 3).

Although both the pre- and postfusion states are trimers, there are large differences between the trimeric interfaces of both conformations. In the postfusion conformation, the buried interface between two subunits is about $3860 \, \text{Å}^2$ (the main part being located in domain II at the top of the molecule)[54] whereas it is only $1600 \, \text{Å}^2$ in the prefusion state.[55] This explains the increased stability of the oligomeric structure of G at low pH.[9] Considering the change of trimeric interface during the structural transition and the topological difficulty to go from one structure to another at the viral surface without breaking the threefold symmetry, it has been suggested that a possible scheme for the transition is that it goes through monomeric intermediates.[61]

Finally, the prefusion form of Gth is organized in a p6 lattice inside the crystal (Fig. 4). Remarkably, this organization is similar to the local hexagonal lattice of

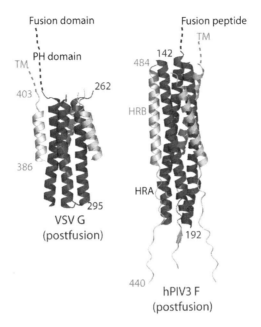

Fig. 3. Comparison of the trimerization motif of VSV G in postfusion conformation with the six-helix bundle of PIV3 F (parainfluenza virus type 3 fusion protein). Heptad-repeats HRA (aa 142–192) and HRB (aa 440–484) regions are indicated on the PIV3 F protein, and similar regions on the VSV G protein are colored in the same level of grey. The position of fusion domain and pleckstrin homology (PH) domain (on VSV G), fusion peptide (on PIV3 F), and transmembrane domain (TM) are indicated.

some RV mutants affected in the kinetics of their structural transition when they are briefly incubated at intermediate pH on ice.[62] Whether this organization reflects the structure of a fusion relevant complex at the viral surface is still an open question.

5. The Fusion Domain

The fusion domain is made of two loops located at the tip of an elongated three-stranded beta sheet (Fig. 5). This structural organization resembles that found in the fusion domain of class II fusion proteins encoded by flaviviruses and alphaviruses.[63–65] In VSV G, four hydrophobic residues are present W72, Y73, Y116, and A117 (Fig. 5A). Sequence alignments allowed the identification of the fusion loops of other rhabdoviral glycoproteins and have revealed that they all contain many aromatic residues. The fact that the residues in the loops are an essential part of the membrane interacting motif is consistent with data obtained from directed mutagenesis experiments on both VSV[66,67] and VHSV.[68]

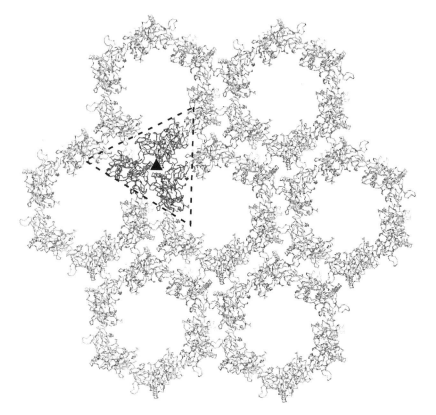

Fig. 4. Hexagonal lattice of glycoproteins inside the prefusion crystals. One VSV G trimer is colored in dark gray. The 3-fold axis is indicated by the black triangle (perpendicular to the plane of the drawing).

In the postfusion structure, the tip of the glycoprotein has a bowl-like shape made of the association of three fusion domains (Fig. 5C). In this conformation, the fusion loops and the TM domains located at the same end of the molecule and, the fusion loops are inserted inside the viral membrane.

In the prefusion structure, in striking contrast to class I and class II viral fusion proteins, the fusion loops are not buried at an oligomeric interface but point toward the viral membrane. It is therefore possible that in the prefusion conformation, the fusion loops also interact with the membrane.

Nevertheless, hydrophobic photolabeling experiments have demonstrated the ability of G fusion domain to insert into the target membrane as a first step of the fusion process. This strongly suggests that during the structural transition, an intermediate state on the fusion pathway exposes the fusion domain at the top of the molecule.

(A)

```
          68            77    111           121
  VSV   CDFRWYGPKY          PQSCGYATVTD
          88            97    131           141
  BEFV  CSETWYFSTS          PAGCFWNTEMN
          114          123    147           157
 SIGMA  CDMPWYFSPT          PEDCSWNSVNT
          89            98    136           146
RABIES  TYTNFVGYVT          PDYRWLRTVKT
          110          119    150           160
  VHSV  CSTSFFGGQT          PSCIWMKNNVH
```

(C)

(B)

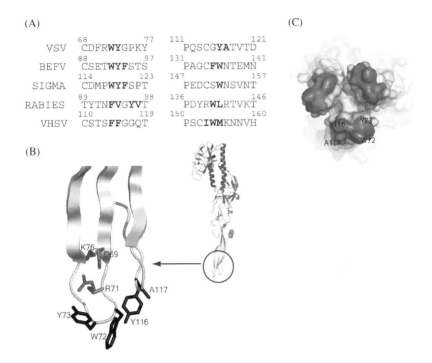

Fig. 5. A: Sequence alignment of the fusion loops of several rhabdovirus glycoproteins; hydrophobic conserved residues are in bold (BEFV = Bovine Ephemeral Fever Virus, VHSV = Viral Hemorrhagic Septicemia Virus). B: close-up view of the tip of the VSV G fusion loops. Hydrophobic residues are represented as black sticks; charged residues that impede deep penetration of the fusion domain in membranes are represented as grey sticks. C: Full atom representation of the fusion loops of VSV G postfusion trimer, hydrophobic residues are indicated.

It has to be noted that in any of these structures (pre- and postfusion interacting with the viral membrane of intermediate state interacting with the target membrane), any deep penetration of the fusion domain inside the membrane is precluded. Clearly, the presence of charged residues and polar groups in the vicinity of the fusion loops rather indicates that the polar aromatic residues (tryptophans and tyrosines) position themselves at the interface between the fatty acid chains and the polar head group layers of lipids.

6.   Residues Playing a Key Role during the Structural Transition

Both the pre- and postfusion structures allow the identification of residues that play the role of pH sensitive molecular switches. In the prefusion state, three conserved histidines (H60 and H162, both located in the fusion domain, and H407, located in the C-terminal segment of the protein) cluster together. Protonation of these residues at low pH is likely to destabilize the interaction between the C-terminal segment of Gth and the fusion domain in the prefusion conformation. This might trigger the movement of the fusion domain toward the target membrane. Indeed, modification of G histidines with diethylpyrocarbonate inhibits its fusion properties.[69]

Conversely, in the postfusion form, a large number of acidic amino acids are brought close together in the six-helix bundle. In the postfusion state, these residues are protonated and form hydrogen bonds. The deprotonation of these residues at higher pH induces strong repulsive forces that destabilize the trimer and triggers the conformational change back to the prefusion state. These residues are buried in the postfusion state but solvent exposed in the prefusion state. As a consequence, they have a pKa that is much higher in the postfusion than in the prefusion conformation (particularly D268 that is fully buried at the trimer interface in the postfusion conformation). This stronger affinity of the postfusion state for the protons explains the cooperativity of rhabdoviral structural transition as a function of pH.[44]

A large number of mutations have been described in VSV, RV, and VHSV that affect either the fusion properties or the structural transition. As mentioned above, replacement of a hydrophobic residue by a polar one or replacement of an aromatic residue by an aliphatic one that is less prone to destabilize the interface between fatty acid chains and the polar head group layers of lipids abolishes fusion activity.

For both VSV and rabies, natural mutants for which the prefusion conformation of G is either thermodynamically or kinetically stabilized have been selected.[62,70] These mutants are located in the hinge region between the fusion domain and the pleckstrin homology domain (mutations M44I and M44V in RV G), in the long central helix (mutations Q285R for VSV and E282K for RV), in the neighborhood of H397, the equivalent of H407 in VSV (e.g., V393G and M396T), and in the amino-terminal part of VSV G (mutation F2L). The phenotype of these mutants can be easily explained at a molecular level by looking at the structures.

The phenotype of some other mutants is more difficult to explain. For example, for both VSV and VHSV, many mutations in residues that are located in the polyproline helix (residues 108–111 of VSV G) upstream the second fusion loop decrease the fusion efficiency and/or lower the pH threshold for fusion.[66,68,71] Although these mutations affect a region that keeps together three fusion domains in the trimeric postfusion state and underline the importance of this segment during the structural transition, it is not clear how they affect the fusion phenotype.

7. Other Domains of G Involved in Membrane Fusion

Two other domains have been postulated to play a role in the fusion process. The first one is the membrane proximal domain. Indeed, deletion of the 13 membrane proximal amino acids (residues 433–445) dramatically reduced cell–cell fusion activity and reduced virus infectivity approximately 100-fold.[72] However, mutations of conserved aromatic residues (W441, F442, and W445) indicated that these residues are not important for fusion.[72] The structure of this segment, absent in the crystal structure, is not known but in the case of RV G, it has been proposed to adopt an α-helical conformation having a strong amphipatic signature.[73]

The second domain is the TM domain. It has been shown that VSV G anchored in the membrane by a glycerophosphatidylinositol instead of its TM domain has no fusion activity. Furthermore, mutations of TM glycine residues block fusion at a hemifusion stage.[74]

Although the membrane proximal domain[75,76] and the TM domain[77–79] of other viral fusion proteins have also been shown to modulate membrane fusion, the role played by both domains is still far from being understood.

8. Cellular Aspects of VSV Entry

After binding at the cell surface, rhabdoviruses enter the host cell by the endocytic pathway. The exact cellular mechanism, however, was largely unknown but recent significant reports on VSV have enlightened our understanding of this crucial step.

These studies have confirmed that the majority of VSV particles are endocytosed in a clathrin-based, dynamin-2-dependent manner[80] consistent with previous electron microscopy studies showing VSV particles within coated pits and coated vesicles.[81] Indeed, entry of VSV into the host cell is sensitive to siRNA targeting clathrin heavy chain,[82] to chlorpromazin (a drug blocking clathrin-mediated endocytosis),[82] and to dynasore (an inhibitor of the GTPase activity of dynamin).[80] Endocytosis of VSV is a fast process with a half time of 2.5–3 min,[80] and it appears that the virions are able to induce the nucleation of clathrin for their uptake.[83]

Remarkably, VSV is significantly larger than the dimensions of a typical clathrin-coated vesicle. On average, virus-containing vesicles contain more clathrin and clathrin adaptor molecules than classical vesicles, but this increase is insufficient to permit full coating of the vesicle.[83] In fact, the vesicles used by the virus for entry are only partially clathrin-coated. These clathrin structures that internalize VSV require actin polymerization for efficient uptake into cells and treatment of cells with either cytochalasin D or latrunculin B inhibited entry by blocking the transition of virus-containing pits to completed vesicles.[83]

After endocytosis, fusion occurred rapidly within 1–2 min after internalization in early endosomes.[80] This is consistent with experiments using dominant negative forms of RAB GTPases that have shown that an RAB 5 mutant that blocks the maturation of endocytic vesicle to early endosomes inhibits VSV infection, whereas an RAB 7 mutant that blocks the traffic from early to late endosome was without effect on VSV infection.[84]

In contradiction with the classical view, it has been proposed that the acid-activated fusion of VSV and the release of the nucleocapsid into the cytosol occur in two distinct steps[85]; the viral envelope first would fuse with an internal vesicle within multivesicular bodies. The viral nucleocapsid would be released into the cytosol by back-fusion of the internal vesicle with the limiting membrane through the action of cellular fusion machinery. Although it is difficult to fully exclude that VSV uses different entry pathways depending on the cells and their physiological state, most of the evidences obtained so far do not support this latter view.

9. RV Entry into Neurons

RV infection occurs at the periphery and the virus is rapidly transported to the central nervous system. Two different pathways have been proposed for the transport of RV through the axon to the cell body where transcription and replication take place: either fusion occurs early and the nucleocapsid is transported alone or the whole virion is transported inside a vesicle.

It has been demonstrated that lyssavirus P proteins interact with a component of the dynein motor complex, dynein light chain LC8[86–88] suggesting that the ribonucleocapsid might be transported by the cytoplasmic dynein motor complex. Nevertheless, as deletion of the LC8-binding site in the phosphoprotein does not affect viral transport from a peripheral site to the CNS, the contribution of dynein to transport of viral RNP complexes has been questioned.[89,90] Furthermore, retroviruses pseudotyped with RV G are transported in a retrograde way and reach the central nervous system after peripheral delivery in a similar manner to RV.[91,92] This suggests that long distance transport along axons depends on RV G and that the whole virion is transported inside a vesicle. This hypothesis has been challenged by infecting *in vitro*-differentiated NS20Y neuroblastoma cells with double-labeled virus particles comprising a red fluorescent envelope and a green fluorescent phosphoprotein. These experiments indicated that enveloped RV particles are transported as a cargo inside vesicles by typical retrograde axonal transport.[93]

10. Early Post-entry Events

Very little is known about the events, immediately following the release of the nucleocapsid into the cytoplasm, which precede (and might even trigger) primary transcription. Recently, it has been proposed that low induced pH conformational changes in G protein promote acidification of the virus interior, which facilitates the release of M from ribonucleoprotein particles during uncoating[94] in agreement with previous observations suggesting that G was able to induce pore formation in the viral membrane at low pH.[95]

Following the release of nucleocapsid in the cytoplasm, RV hijacks cellular machineries to induce the formation of inclusion bodies (similar to the Negri bodies found in the cytoplasm of infected neurons).[96,97] These structures that are detected very early after infection have been recently demonstrated to be the site of viral transcription and replication.[97] This observation seems to be a specificity of RV: in the case of VSV, replication and transcription do not seem to be confined to any particular region of the cytoplasm.[98]

11. Final Remarks

The structure determination of VSV G has brought important insight into the mechanism of rhabdovirus-induced membrane fusion. Nevertheless, many questions remain unanswered: the structures of intermediate conformations of G and the way they cooperate to interact with and deform both the viral and target membranes are still very elusive (as for other enveloped viruses). Determining the structures of these intermediates in association with membranes is a major challenge in the field. Together with the progress in super resolution microscopy and the design of recombinant fluorescent viruses that allow a direct visualization of viral particles in live cells, this should give an integrated view of the entry pathway of rhabdoviruses into their host cell.

References

1. Rose, J.K. and M.A. Whitt, *Rhabdoviridae: The Viruses and Their Replication.* Fields (4th edition), 2001, pp. 1221–4.
2. Kondo, H., *et al.*, Orchid fleck virus is a rhabdovirus with an unusual bipartite genome. *J Gen Virol*, 2006. **87**(Pt 8): 2413–21.
3. Lefrancois, L. and D.S. Lyles, Antigenic determinants of vesicular stomatitis virus: Analysis with antigenic variants. *J Immunol*, 1983. **130**(1): 394–8.

4. Seif, I., *et al.*, Rabies virulence: Effect on pathogenicity and sequence characterization of rabies virus mutations affecting antigenic site III of the glycoprotein. *J Virol*, 1985. **53**(3): 926–34.

5. Vandepol, S.B., L. Lefrancois and J.J. Holland, Sequences of the major antibody binding epitopes of the Indiana serotype of vesicular stomatitis virus. *Virology*, 1986. **148**(2): 312–25.

6. Prehaud, C., *et al.*, Antigenic site II of the rabies virus glycoprotein: Structure and role in viral virulence. *J Virol*, 1988. **62**(1): 1–7.

7. Flamand, A., *et al.*, Mechanisms of rabies virus neutralization. *Virology*, 1993. **194**(1): 302–13.

8. Gaudin, Y., *et al.*, Rabies virus glycoprotein is a trimer. *Virology*, 1992. **187**(2): 627–32.

9. Doms, R.W., *et al.*, Role for adenosine triphosphate in regulating the assembly and transport of vesicular stomatitis virus G protein trimers. *J Cell Biol*, 1987. **105**(5): 1957–69.

10. Lyles, D.S., V.A. Varela and J.W. Parce, Dynamic nature of the quaternary structure of the vesicular stomatitis virus envelope glycoprotein. *Biochemistry*, 1990. **29**(10): 2442–9.

11. Zagouras, P. and J.K. Rose, Dynamic equilibrium between vesicular stomatitis virus glycoprotein monomers and trimers in the Golgi and at the cell surface. *J Virol*, 1993. **67**(12): 7533–8.

12. Zagouras, P., A. Ruusala and J.K. Rose, Dissociation and reassociation of oligomeric viral glycoprotein subunits in the endoplasmic reticulum. *J Virol*, 1991. **65**(4): 1976–84.

13. Schlegel, R., *et al.*, Inhibition of VSV binding and infectivity by phosphatidylserine: Is phosphatidylserine a VSV-binding site? *Cell*, 1983. **32**(2): 639–46.

14. Coil, D.A. and A.D. Miller, Phosphatidylserine is not the cell surface receptor for vesicular stomatitis virus. *J Virol*, 2004. **78**(20): 10920–6.

15. Sinibaldi, L., *et al.*, Gangliosides in early interactions between vesicular stomatitis virus and CER cells. *Microbiologica*, 1985. **8**(4): 355–65.

16. Kissling, R.E., Growth of rabies virus in non-nervous tissue culture. *Proc Soc Exp Biol Med*, 1958. **98**(2): 223–5.

17. Wiktor, T.J., M.V. Fernandes and H. Koprowski, Cultivation of Rabies Virus in Human Diploid Cell Strain Wi-38. *J Immunol*, 1964. **93**: 353–66.

18. Schneider, L.G., M. Horzinek and H.D. Matheka, Purification of rabies virus from tissue culture. *Arch Gesamte Virusforsch*, 1971. **34**(4): 351–9.

19. Seganti, L., *et al.*, Susceptibility of mammalian, avian, fish, and mosquito cell lines to rabies virus infection. *Acta Virol*, 1990. **34**(2): 155–63.

20. Lafon, M., Rabies virus receptors. *J Neurovirol*, 2005. **11**(1): 82–7.

21. Superti, F., *et al.*, Role of phospholipids in rhabdovirus attachment to CER cells. Brief report. *Arch Virol*, 1984. **81**(3–4): 321–8.

22. Conti, C., F. Superti and H. Tsiang, Membrane carbohydrate requirement for rabies virus binding to chicken embryo related cells. *Intervirology*, 1986. **26**(3): 164–8.

23. Superti, F., *et al.*, Involvement of gangliosides in rabies virus infection. *J Gen Virol*, 1986. **67**(Pt 1): 47–56.

24. Lentz, T.L., *et al.*, Is the acetylcholine receptor a rabies virus receptor? *Science*, 1982. **215**(4529): 182–4.

25. Lentz, T.L., *et al.*, Amino acid sequence similarity between rabies virus glycoprotein and snake venom curaremimetic neurotoxins. *Science*, 1984. **226**(4676): 847–8.

26. Lentz, T.L., *et al.*, Binding of rabies virus to purified Torpedo acetylcholine receptor. *Brain Res*, 1986. **387**(3): 211–9.

27. Gastka, M., J. Horvath and T.L. Lentz, Rabies virus binding to the nicotinic acetylcholine receptor alpha subunit demonstrated by virus overlay protein binding assay. *J Gen Virol*, 1996. **77**(Pt 10): 2437–40.

28. McGehee, D.S. and L.W. Role, Physiological diversity of nicotinic acetylcholine receptors expressed by vertebrate neurons. *Annu Rev Physiol*, 1995. **57**: 521–46.

29. Burrage, T.G., G.H. Tignor and A.L. Smith, Rabies virus binding at neuromuscular junctions. *Virus Res*, 1985. **2**(3): 273–89.

30. Thoulouze, M.I., *et al.*, The neural cell adhesion molecule is a receptor for rabies virus. *J Virol*, 1998. **72**(9): 7181–90.

31. Tuffereau, C., *et al.*, Low-affinity nerve-growth factor receptor (P75NTR) can serve as a receptor for rabies virus. *Embo J*, 1998. **17**(24): 7250–9.

32. Tuffereau, C., *et al.*, Interaction of lyssaviruses with the low-affinity nerve-growth factor receptor p75NTR. *J Gen Virol*, 2001. **82**(Pt 12): 2861–7.

33. Tuffereau, C., *et al.*, The rabies virus glycoprotein receptor p75NTR is not essential for rabies virus infection. *J Virol*, 2007. **81**(24): 13622–30.

34. Bearzotti, M., *et al.*, Fish rhabdovirus cell entry is mediated by fibronectin. *J Virol*, 1999. **73**(9): 7703–9.

35. Clague, M.J., *et al.*, Gating kinetics of pH-activated membrane fusion of vesicular stomatitis virus with cells: Stopped-flow measurements by dequenching of octadecyl-rhodamine fluorescence. *Biochemistry*, 1990. **29**(5): 1303–8.

36. Gaudin, Y., P. de Kinkelin and A. Benmansour, Mutations in the glycoprotein of viral haemorrhagic septicaemia virus that affect virulence for fish and the pH threshold for membrane fusion. *J Gen Virol*, 1999. **80**(Pt 5): 1221–9.

37. Roche, S. and Y. Gaudin, Evidence that rabies virus forms different kinds of fusion machines with different pH thresholds for fusion. *J Virol*, 2004. **78**(16): 8746–52.

38. Fredericksen, B.L. and M.A. Whitt, Mutations at two conserved acidic amino acids in the glycoprotein of vesicular stomatitis virus affect pH-dependent conformational changes and reduce the pH threshold for membrane fusion. *Virology*, 1996. **217**(1): 49–57.

39. Fredericksen, B.L. and M.A. Whitt, Attenuation of recombinant vesicular stomatitis viruses encoding mutant glycoproteins demonstrate a critical role for maintaining a high pH threshold for membrane fusion in viral fitness. *Virology*, 1998. **240**(2): 349–58.

40. Gaudin, Y., *et al.*, Reversible conformational changes and fusion activity of rabies virus glycoprotein. *J Virol*, 1991. **65**(9): 4853–9.

41. Gaudin, Y., *et al.*, Low-pH conformational changes of rabies virus glycoprotein and their role in membrane fusion. *J Virol*, 1993. **67**(3): 1365–72.

42. Gaudin, Y., Reversibility in fusion protein conformational changes. The intriguing case of rhabdovirus-induced membrane fusion. *Subcell Biochem*, 2000. **34**: 379–408.

43. Durrer, P., *et al.*, Photolabeling identifies a putative fusion domain in the envelope glycoprotein of rabies and vesicular stomatitis viruses. *J Biol Chem*, 1995. **270**(29): 17575–81.

44. Roche, S. and Y. Gaudin, Characterization of the equilibrium between the native and fusion-inactive conformation of rabies virus glycoprotein indicates that the fusion complex is made of several trimers. *Virology*, 2002. **297**(1): 128–35.

45. Gaudin, Y., *et al.*, Biological function of the low-pH, fusion-inactive conformation of rabies virus glycoprotein (G): G is transported in a fusion-inactive state-like conformation. *J Virol*, 1995. **69**(9): 5528–34.

46. Herrmann, A., *et al.*, Effect of erythrocyte transbilayer phospholipid distribution on fusion with vesicular stomatitis virus. *Biochemistry*, 1990. **29**(17): 4054–8.

47. Yamada, S. and S. Ohnishi, Vesicular stomatitis virus binds and fuses with phospholipid domain in target cell membranes. *Biochemistry*, 1986. **25**(12): 3703–8.

48. Chernomordik, L.V., *et al.*, The pathway of membrane fusion catalyzed by influenza hemagglutinin: Restriction of lipids, hemifusion, and lipidic fusion pore formation. *J Cell Biol*, 1998. **140**(6): 1369–82.

49. Zaitseva, E., *et al.*, Class II fusion protein of alphaviruses drives membrane fusion through the same pathway as class I proteins. *J Cell Biol*, 2005. **169**(1): 167–77.

50. Gaudin, Y., Rabies virus-induced membrane fusion pathway. *J Cell Biol*, 2000. **150**(3): 601–12.

51. Carr, C.M., C. Chaudhry and P.S. Kim, Influenza hemagglutinin is spring-loaded by a metastable native conformation. *Proc Natl Acad Sci U S A*, 1997. **94**(26): 14306–13.

52. Lee, J. and B.R. Lentz, Secretory and viral fusion may share mechanistic events with fusion between curved lipid bilayers. *Proc Natl Acad Sci U S A*, 1998. **95**(16): 9274–9.

53. Kuzmin, P.I., *et al.*, A quantitative model for membrane fusion based on low-energy intermediates. *Proc Natl Acad Sci U S A*, 2001. **98**(13): 7235–40.

54. Roche, S., *et al.*, Crystal structure of the low-pH form of the vesicular stomatitis virus glycoprotein G. *Science*, 2006. **313**(5784): 187–91.

55. Roche, S., *et al.*, Structure of the prefusion form of the vesicular stomatitis virus glycoprotein g. *Science*, 2007. **315**(5813): 843–8.

56. Heldwein, E.E., *et al.*, Crystal structure of glycoprotein B from herpes simplex virus 1. *Science*, 2006. **313**(5784): 217–20.

57. Kadlec, J., *et al.*, The postfusion structure of baculovirus gp64 supports a unified view of viral fusion machines. *Nat Struct Mol Biol*, 2008. **15**(10): 1024–30.

58. Backovic, M., R. Longnecker and T.S. Jardetzky, Structure of a trimeric variant of the Epstein-Barr virus glycoprotein B. *Proc Natl Acad Sci U S A*, 2009. **106**(8): 2880–5.

59. Yin, H.S., *et al.*, Structure of the parainfluenza virus 5 F protein in its metastable, prefusion conformation. *Nature*, 2006. **439**(7072): 38–44.

60. Bullough, P.A., *et al.*, Structure of influenza haemagglutinin at the pH of membrane fusion. *Nature*, 1994. **371**(6492): 37–43.

61. Roche, S., *et al.*, Structures of vesicular stomatitis virus glycoprotein: Membrane fusion revisited. *Cell Mol Life Sci*, 2008. **65**(11): 1716–28.

62. Gaudin, Y., *et al.*, Identification of amino acids controlling the low-pH-induced conformational change of rabies virus glycoprotein. *J Virol*, 1996. **70**(11): 7371–8.

63. Bressanelli, S., *et al.*, Structure of a flavivirus envelope glycoprotein in its low-pH-induced membrane fusion conformation. *Embo J*, 2004. **23**(4): 728–38.

64. Gibbons, D.L., *et al.*, Conformational change and protein–protein interactions of the fusion protein of Semliki Forest virus. *Nature*, 2004. **427**(6972): 320–5.

65. Modis, Y., *et al.*, Structure of the dengue virus envelope protein after membrane fusion. *Nature*, 2004. **427**(6972): 313–9.

66. Fredericksen, B.L. and M.A. Whitt, Vesicular stomatitis virus glycoprotein mutations that affect membrane fusion activity and abolish virus infectivity. *J Virol*, 1995. **69**(3): 1435–43.

67. Sun, X., S. Belouzard and G.R. Whittaker, Molecular architecture of the bipartite fusion loops of vesicular stomatitis virus glycoprotein G, a class III viral fusion protein. *J Biol Chem*, 2008. **283**: 6418–27.

68. Rocha, A., *et al.*, Conformation- and fusion-defective mutations in the hypothetical phospholipid-binding and fusion peptides of viral hemorrhagic septicemia salmonid rhabdovirus protein G. *J Virol*, 2004. **78**(17): 9115–22.

69. Carneiro, F.A., *et al.*, Membrane fusion induced by vesicular stomatitis virus depends on histidine protonation. *J Biol Chem*, 2003. **278**(16): 13789–94.

70. Martinez, I. and G.W. Wertz, Biological differences between vesicular stomatitis virus Indiana and New Jersey serotype glycoproteins: Identification of amino acid residues modulating pH-dependent infectivity. *J Virol*, 2005. **79**(6): 3578–85.

71. Zhang, L. and H.P. Ghosh, Characterization of the putative fusogenic domain in vesicular stomatitis virus glycoprotein G. *J Virol*, 1994. **68**(4): 2186–93.

72. Jeetendra, E., *et al.*, The membrane-proximal region of vesicular stomatitis virus glycoprotein G ectodomain is critical for fusion and virus infectivity. *J Virol*, 2003. **77**(23): 12807–18.

73. Maillard, A., *et al.*, Spectroscopic characterization of two peptides derived from the stem of rabies virus glycoprotein. *Virus Res*, 2003. **93**(2): 151–8.

74. Cleverley, D.Z. and J. Lenard, The transmembrane domain in viral fusion: Essential role for a conserved glycine residue in vesicular stomatitis virus G protein. *Proc Natl Acad Sci U S A*, 1998. **95**(7): 3425–30.

75. Howard, M.W., *et al.*, Aromatic amino acids in the juxtamembrane domain of severe acute respiratory syndrome coronavirus spike glycoprotein are important for receptor-dependent virus entry and cell–cell fusion. *J Virol*, 2008. **82**(6): 2883–94.

76. Vishwanathan, S.A. and E. Hunter, Importance of the membrane-perturbing properties of the membrane-proximal external region of human immunodeficiency virus type 1 gp41 to viral fusion. *J Virol*, 2008. **82**(11): 5118–26.

77. Helseth, E., *et al.*, Changes in the transmembrane region of the human immunodeficiency virus type 1 gp41 envelope glycoprotein affect membrane fusion. *J Virol*, 1990. **64**(12): 6314–8.

78. Kemble, G.W., T. Danieli and J.M. White, Lipid-anchored influenza hemagglutinin promotes hemifusion, not complete fusion. *Cell*, 1994. **76**(2): 383–91.

79. Melikyan, G.B., *et al.*, A point mutation in the transmembrane domain of the hemagglutinin of influenza virus stabilizes a hemifusion intermediate that can transit to fusion. *Mol Biol Cell*, 2000. **11**(11): 3765–75.

80. Johannsdottir, H.K., *et al.*, Host cell factors and functions involved in vesicular stomatitis virus entry. *J Virol*, 2009. **83**(1): 440–53.

81. Superti, F., *et al.*, Entry pathway of vesicular stomatitis virus into different host cells. *J Gen Virol*, 1987. **68** (Pt 2): 387–99.

82. Sun, X., *et al.*, Role of clathrin-mediated endocytosis during vesicular stomatitis virus entry into host cells. *Virology*, 2005. **338**(1): 53–60.

83. Cureton, D.K., *et al.*, Vesicular stomatitis virus enters cells through vesicles incompletely coated with clathrin that depend upon actin for internalization. *PLoS Pathog*, 2009. **5**(4): e1000394.

84. Sieczkarski, S.B. and G.R. Whittaker, Differential requirements of Rab5 and Rab7 for endocytosis of influenza and other enveloped viruses. *Traffic*, 2003. **4**(5): 333–43.

85. Le Blanc, I., *et al.*, Endosome-to-cytosol transport of viral nucleocapsids. *Nat Cell Biol*, 2005. **7**(7): 653–64.

86. Raux, H., A. Flamand and D. Blondel, Interaction of the rabies virus P protein with the LC8 dynein light chain. *J Virol*, 2000. **74**(21): 10212–6.

87. Jacob, Y., *et al.*, Cytoplasmic dynein LC8 interacts with lyssavirus phosphoprotein. *J Virol*, 2000. **74**(21): 10217–22.

88. Poisson, N., *et al.*, Molecular basis for the interaction between rabies virus phosphoprotein P and the dynein light chain LC8: Dissociation of dynein-binding properties and transcriptional functionality of P. *J Gen Virol*, 2001. **82**(Pt 11): 2691–6.

89. Mebatsion, T., Extensive attenuation of rabies virus by simultaneously modifying the dynein light chain binding site in the P protein and replacing Arg333 in the G protein. *J Virol*, 2001. **75**(23): 11496–502.

90. Tan, G.S., *et al.*, The dynein light chain 8 binding motif of rabies virus phosphoprotein promotes efficient viral transcription. *Proc Natl Acad Sci U S A*, 2007. **104**(17): 7229–34.

91. Mazarakis, N.D., *et al.*, Rabies virus glycoprotein pseudotyping of lentiviral vectors enables retrograde axonal transport and access to the nervous system after peripheral delivery. *Hum Mol Genet*, 2001. **10**(19): 2109–21.

92. Mentis, G.Z., *et al.*, Transduction of motor neurons and muscle fibers by intramuscular injection of HIV-1-based vectors pseudotyped with select rabies virus glycoproteins. *J Neurosci Methods*, 2006. **157**(2): 208–17.

93. Klingen, Y., K.K. Conzelmann and S. Finke, Double-labeled rabies virus: Live tracking of enveloped virus transport. *J Virol*, 2008. **82**(1): 237–45.

94. Mire, C.E., *et al.*, Glycoprotein-dependent acidification of vesicular stomatitis virus enhances release of matrix protein. *J Virol*, 2009. **83**(23): 12139–50.

95. Kasermann, F. and C. Kempf, Low pH-induced pore formation by spike proteins of enveloped viruses. *J Gen Virol*, 1996. **77**(Pt 12): 3025–32.

96. Menager, P., *et al.*, Toll-like receptor 3 (TLR3) plays a major role in the formation of rabies virus Negri Bodies. *PLoS Pathog*, 2009. **5**(2): e1000315.

97. Lahaye, X., *et al.*, Functional characterization of Negri bodies (NBs) in rabies virus-infected cells: Evidence that NBs are sites of viral transcription and replication. *J Virol*, 2009. **83**(16): 7948–58.

98. Das, S.C., *et al.*, Visualization of intracellular transport of vesicular stomatitis virus nucleocapsids in living cells. *J Virol*, 2006. **80**(13): 6368–77.

Chapter 3

Virus Entry: Parainfluenza Viruses

Masato Tsurudome*

1. Introduction

The constituents of family *Paramyxoviridae* are basically spherical enveloped viruses, which harbor nonsegmented single-strand RNA genomes of negative polarity. This family consists of two subfamilies: *Paramyxovirinae* and *Pneumovirinae*. The subfamily *Paramyxovirinae* includes five genera: *Respirovirus*, *Rubulavirus*, *Avulavirus*, *Morbillivirus*, and *Henipavirus*, whereas the subfamily *Pneumovirinae* is composed of two genera: *Pneumovirus* and *Metapneumovirus*.

Parainfluenza viruses are classified into three genera: *Respirovirus*, *Rubulavirus*, and *Avulavirus* in the subfamily *Paramyxovirinae*. The parainfluenza viruses that belong to the genus Respirovirus are human parainfluenza virus 1 (HPIV1), murine PIV1 (Sendai virus, SeV), HPIV3, and bovine PIV3, while those belonging to the genus *Rubulavirus* are HPIV2, simian PIV2 (simian virus 41, SV41), HPIV4A, HPIV4B, and canine PIV2 (parainfluenza virus 5, PIV5, SV5). The parainfluenza viruses that belong to the genus *Avulavirus* are nine serotypes of avian paramyxoviruses that include Newcastle disease virus (NDV) as the prototype. Infection of humans by HPIVs (1, 2, 3, 4A, and 4B) and that of animals by PIV5, SeV, and bovine PIV3 are mostly limited to the epithelial cells of the respiratory tract. Similarly, infection of birds by avirulent strains of NDV is mostly confined to the respiratory tract. However, virulent NDV strains can cause systemic infection in birds.

---

*Department of Microbiology and Molecular Genetics, Mie University Graduate School of Medicine, 2-174 Edobashi, Tsu, Mie 514-8507, Japan.

In the viral envelope, there are two kinds of glycoprotein spikes: a hemagglutinin-neuraminidase (HN) tetramer and a fusion protein (F) trimer, which are also present on the surface of infected cells. HN is involved in binding to the cellular receptor and destruction of the receptor, while F is responsible for induction of membrane fusion. The entry of parainfluenza viruses into cells takes place at the plasma membrane independently of pH, where membrane fusion between the viral envelope and the plasma membrane enables the invasion of viral genome into cytoplasm. In cultured cells, some parainfluenza viruses mediate cell–cell fusion, which sometimes results in extensive syncytium formation that leads to cell death. It is apparent that there are mechanistic differences between virus–cell fusion and cell–cell fusion, but a majority of the studies on the parainfluenza virus-mediated fusion have been done by analyzing cell–cell fusion and have brought lots of findings that would be essentially applicable to virus–cell fusion. It is considered that attachment of HN to the receptors on the cell surface somehow triggers a series of conformational changes of F through an HN–F interaction, which eventually results in membrane fusion. Currently, however, the molecular basis underlying the HN–F interaction is not fully understood and the mechanism by which HN triggers F activation remains to be clarified. This chapter reviews the general concept of parainfluenza virus-mediated fusion that requires two viral glycoproteins: HN and F.

2.   Cellular Receptor

Parainfluenza viruses use sialoconjugate moieties on the cell surface as their receptors.[1,2] It is thought that either proteins with N-linked carbohydrate side chains or gangliosides serve as the receptors.[3–5] The viral attachment protein, HN, is responsible for the binding to the sialoconjugate receptors. HN recognizes the terminal *N*-acetylneuraminic acid (NeuAc) that is attached to the penultimate galactose (Gal) in the sugar chain through an α2-6 or α2-3 linkage.[6,7] Importantly, HN has a neuraminidase activity, thereby cleaving the NeuAc-Gal linkage in the sialoconjugates.[1,2] As HN recognizes the sialoconjugates either of the cellular components or of the viral glycoproteins, its neuraminidase activity is considered beneficial for viral release and for preventing self-aggregation of progeny viruses after being released from the infected cells.

Distribution of the two species of sialoconjugates that are present in the human respiratory tract has been investigated with the aid of lectins specific for each species.[8] The epithelial cells in the nasal mucosa, paranasal sinuses, pharynx, trachea, and bronchi mainly express NeuAcα2-6Gal-containing sialoconjugates, with NeuAcα2-3Gal-containing sialoconjugates being occasionally expressed in the nasal mucosa. The epithelial cells in the terminal and respiratory bronchioles also express NeuAcα2-6Gal-containing sialoconjugates. It is worth noting that a substantial number of cells, most likely alveolar type II cells, express NeuAcα2-3Gal-containing sialoconjugates.

Virus overlay assays, combined with thin-layer chromatography, have revealed that HPIV1 and HPIV3 preferentially bind to the oligosaccharides of the neolacto-series gangliosides containing *N*-acetyllactosaminoglycans with terminal NeuAcα2-3Gal; HPIV3 can also bind to NeuAcα2-6Gal-containing oligosaccharides of the neolacto-series gangliosides.[9] Consistent with these results, HPIV1 exhibits three orders of magnitude higher neuraminidase activity against NeuAcα2-3Gal-containing sialoconjugates than NeuAcα2-6Gal-containing sialoconjugates, while HPIV3 shows only eight-fold more neuraminidase activity against the former than the latter.[10] As for HPIV2, its neuraminidase activity against NeuAcα2-3Gal-containing sialoconjugates is about 40 fold higher compared with that against NeuAcα2-6Gal-containing sialoconjugates, showing a clear preference for the former sialoconjugates.[10] Thus, though the epithelial cells in the human respiratory tract mostly express NeuAcα2-6Gal-containing sialoconjugates as mentioned above, human parainfluenza viruses analyzed so far show more or less a prominent preference for NeuAcα2-3Gal-containing sialoconjugates. It should be pointed out, in this context, that NeuAcα2-3Gal-containing sialoconjugates are known to be predominantly expressed in the intestine of ducks.[11] Ducks are natural hosts of avian influenza viruses, which preferentially bind to NeuAcα2-3Gal-containing sialoconjugates, while human influenza A viruses preferentially bind to NeuAcα2-6Gal-containing sialoconjugates.[12,13] The murine PIV1 (SeV) exhibits a receptor-binding activity which is similar to HPIV3, except that it also recognizes ganglio-series gangliosides with terminal NeuAcα2-3Gal.[9] On the other hand, the avian parainfluenza virus, NDV, specifically binds to NeuAcα2-3Gal-containing sialoconjugates.[14,15] The above-mentioned findings thus indicate that parainfluenza viruses prefer NeuAcα2-3Gal-containing sialoconjugates irrespective of their natural host.

In SeV, intriguingly, the asialoglycoprotein receptor (ASGP-R) can serve as the receptor for F and mediates entry of the virus in an HN-independent manner.[16] Although the physiological significance of the ASGP-R-mediated viral entry is not clear, the infection through ASGP-R is nearly as efficient as that through conventional sialoconjugate receptor. The fusion activity of F can thus be triggered even by a somewhat "nonspecific" interaction between the carbohydrate side chains of F and the lectin-like proteins on the cell surface.

3.   Viral Attachment Protein, HN

HN is a type II integral membrane glycoprotein.[17] The HN monomers of SeV and PIV5 form disulfide-linked dimers and then assemble to form disulfide-linked or noncovalently associated tetramers, respectively,[18,19] in which Cys111 at the distal end of the PIV5 HN stalk region is involved in the formation of disulfide-linked HN dimers.[20] In some strains of NDV, Cys123 at the membrane distal end of the HN stalk region mediates the formation of disulfide-linked HN dimers.[21,22] In contrast, the HN proteins of the other NDV strains form noncovalently associated dimers,[21] and HPIV3 HN forms noncovalently associated oligomers.[23]

The ectodomain of the HN tetramer consists of two regions: a box-shaped head and a, presumably long and slender, stalk that connects the head region to the membrane (Fig. 1).[24,25] Two functions of HN, receptor binding and neuraminidase,

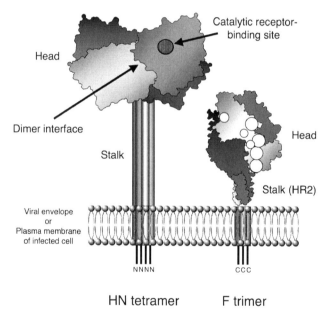

Fig. 1.   Schematic diagram of parainfluenza virus HN and F. The HN tetramer and the F timer in viral envelope (or plasma membrane of the infected cell) are shown. The head region of the HN tetramer and the ectodomain of F trimer in its prefusion form are depicted on the basis of the atomic structures of PIV5 glycoproteins.[20,44] The stalk region of the F trimer consists of three HR2 domains, which together with three HR1 domains become the components of the six-helix bundle structure when the F trimer has adopted the postfusion form.[101] Circles in the head region of F indicate the positions of amino acid domains in the "middle region" that are exposed on the trimer surface (those only in one monomer are shown for simplicity). The middle region of HPIV2 F contains the sites that are involved in the functional interaction with HPIV2 HN.[122]

are carried by the head region,[25,26] while the stalk region is inferred to be involved in the third function, the fusion promotion, as described in Sec. 5. The fusion-promoting function of HN appears to be dependent on being bound to its cellular receptor,[27–30] and it seems that HN promotes F-mediated fusion dependently on the balance between its inherent F-triggering efficiency and receptor-attachment regulatory functions (binding and destruction).[31]

Crystallographic studies of the HN proteins of NDV, PIV5, and HPIV3 have revealed that these HN proteins crystallize as dimers with similar symmetry, where the two receptor-binding sites are located at nearly 90° angles relative to each other.[20,32,33] PIV5 HN and NDV HN also form a tetramer in the crystal through the association of two dimers.[20,34] The atomic structures of the head regions of PIV5 HN and HPIV3 HN with or without a sialoconjugate ligand indicate that a single site per HN monomer is involved in both the receptor-binding and neuraminidase activities and that the ligand does not induce any major conformational change in the head region.[20,33] However, NDV HN has been proved to harbor a second non-catalytic receptor-binding site at the dimer interface in addition to the catalytic site[34]; mutation at this second site reduces the fusion-promoting activity but has no effects on the neuraminidase activity of HN.[35] Furthermore, crystallographic studies of NDV HN have indicated that ligand binding to the catalytic site induces a conformational change in HN, leading to a more stable dimer association.[32] It is thus hypothesized that the change in the dimer association creates the new binding site, which allows the virus to remain proximal to the cell surface as fusion proceeds and alters the conformation of the stalk region that eventually triggers the conformational changes of F.[34] The importance of the NDV HN dimer interface in fusion has also been demonstrated by mutational studies.[36,37] However, this notion has been refuted by the recent finding that the fusion-promoting function of NDV HN is not affected by the introduction of intermonomeric disulfide bond(s) across the dimer interface.[38]

Recently, the presence of a second receptor-binding site at the position similar to that of NDV HN has been suggested for the HN proteins of HPIV1 and HPIV3.[39,40] However, this putative second binding site is masked by an N-linked glycan and it awaits further investigation whether such a cryptic receptor-binding site plays some role in attachment and/or fusion functions of the viruses. Interestingly, HPIV3 mutants with an unmasked second site in the HN are circulating in nature, but the clinical significance of these viruses is not clear.[40] It is worth noting, on the other hand, that the presence of another receptor-binding site at the dimer interface, but at a position different from the second receptor-binding site of NDV HN, has been suggested for HPIV3 HN.[41] This putative third binding site is inferred to be also involved in the interaction with or activation of F molecule.[41]

As for the stalk region, the atomic structure has not been determined, but it is suggested for PIV5 HN that the stalk region forms a tetramer with a predominantly helical and flexible rod-like structure.[24]

4. Viral Fusion Protein, F

F is a type I integral membrane glycoprotein, which is synthesized as a precursor, F_0, and then noncovalently assembles to form homotrimers with rotational symmetry (Fig. 1).[42–44] The F_0 molecule is cleaved by cellular proteases and forms a disulfide-linked subunit structure consisting of the transmembrane subunit, F_1, and the peripheral subunit, F_2 (Fig. 2). This cleavage is followed by a conformational change of F,[45–49] together with the generation of a highly conserved fusion peptide at the F_1 amino terminus,[46,47] which is considered to play a direct role in the fusion event.[50,51] Besides the fusion peptide, the F_1 ectodomain has heptad repeat domains, HR1 and HR2[52]; the HR1 domain is immediately next to the carboxyl terminus of the fusion peptide, while the HR2 domain precedes the transmembrane domain.

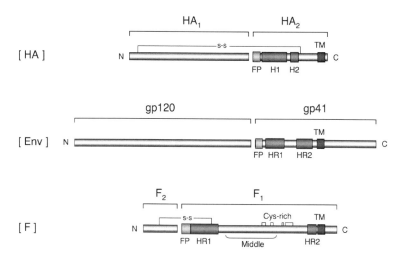

Fig. 2. Schematic diagram of representative class I fusion proteins. The positions of two heptad repeat domains, HR1 and HR2, that form the six-helix bundle structure in the postfusion form of the parainfluenza virus F and the human immunodeficiency virus Env are indicated on the basis of the secondary structures in the postfusion forms of these proteins.[101,161,162] Two α-helical regions in the postfusion form of human influenza A virus HA$_2$,[163] which correspond to the HR1 and HR2 domains of F1, are designated here as H1 and H2, respectively. The peripheral subunits HA$_1$ and gp120 harbor receptor-binding site(s).[164–166] FP, fusion peptide; TM, transmembrane domain; Cys-rich, cysteine-rich domain; Middle, the position of HPIV2 F middle region that determines the HN specificity in fusion induction.

Between the two heptad repeat domains, there is a long intervening region that harbors a highly conserved stretch of eight cysteine residues.[52] This cysteine-rich region includes four loops, each of which is formed with two cysteines connected by a disulfide bond.[53,54]

The cleavage of F_0 by cellular proteases is considered a prerequisite for the fusion activity of F.[55,56] In the case of the virulent strains of NDV and the prototype strains of HPIV2, HPIV3, SV41, and PIV5, the F_0 molecules are known to be cleaved by subtilisin-like serine proteases, including furin,[57–59] which are localized in the *trans*-Golgi network.[60,61] Accordingly, the cleavage site sequences of these F_0 molecules meet the consensus motif for furin, Arg-X-Arg/Lys-Arg[62]; the cleavage occurs at the carboxyl terminus of the multibasic cleavage site of the F_0 molecules. In contrast, the F_0 molecules of HPIV1, HPIV4A, HPIV4B, and SeV have a monobasic cleavage site and are not cleaved intracellularly.[58] Instead, they can be cleaved by the addition of trypsin under experimental conditions,[58] the same as the hemagglutinin (HA) precursor, HA_0, of the human influenza A viruses, whose cleavage site sequences are monobasic.[63,64] Intriguingly, the F_0 molecules of the avirulent strains of NDV require exogenous trypsin for cleavage,[65,66] even when the cleavage site sequences meet the minimal motif (Arg-X-X-Arg) cleavable by furin.[58,67–69]

In the rat respiratory tract, trypsin-like secretory proteins such as mini-plasmin or tryptase Clara are known to cleave the monobasic cleavage sites of SeV F_0 and human influenza virus HA_0.[70–72] As for the human respiratory tract, trypsin-like transmembrane proteins, TMPRSS2 and HAT, which are able to cleave the human influenza virus HA_0, are proved to be expressed in the epithelial cells.[73–75] TMPRSS2 is also able to cleave the F_0 of human metapneumovirus, which is not cleaved intracellularly.[76]

It is worth noting that the cleavage efficiencies of the F_0 molecules of some parainfluenza viruses are determined not only by the cleavage site sequences but also by other structural features in F (such as amino acid sequences or presence/absence of a carbohydrate side chain in the vicinity of the cleavage site) and the available host proteases.[77–80]

The parainfluenza virus F belongs to the class I viral fusion proteins, such as HA of the human influenza virus or Env of the human immunodeficiency virus type 1 (Fig. 2), which are considered to mediate fusion by undergoing a series of conformational changes.[44,81,82] However, unlike most of the class I fusion proteins reported thus far, F does not have receptor-binding function and requires the attachment protein HN for inducing fusion. It is thought that binding of HN to the receptor induces a conformational change of HN that, in turn, triggers conformational changes of F through a physical HN–F interaction.[37] As it was proposed previously,[83] the conformational changes of F appear to involve the following unique structures: (i) a fusion-inactive uncleaved F_0 structure, (ii) a metastable F_1 plus F_2 structure, (iii) an activated intermediate triggered by HN binding to its receptor, (iv) a prehairpin intermediate in which the fusion peptide interacts with target membranes, and (v) a fusogenic "postfusion" form that has a six-helix bundle structure as its core, where three HR1 domains form a three-stranded α-helical coiled coil to which three HR2 domains pack in antiparallel orientation.[84] Formation of the six-helix bundle structure is considered directly coupled to membrane fusion.[30,84] It is worth noting that, if these conformational changes take place in the absence of the target membrane, the fusion peptide may interact with its own membrane and F would be inactivated as it has been demonstrated for the human influenza virus HA with the aid of photocrosslinking technique using photoactivatable radioactive phospholipid analogs.[85] More importantly, the photocrosslinking analyses have also revealed that the fusion peptide of HA in the influenza virion inserts into the target cell membrane before the onset of acid-induced membrane fusion.[86–89] As for parainfluenza viruses, such membrane insertion of the fusion peptide prior to fusion has been suggested for SeV F.[51]

Formerly, it was regarded that the parainfluenza virus F can mediate membrane fusion without help of other viral proteins, because the F of PIV5 strain W3A could induce cell–cell fusion when expressed alone in cultured cells.[90,91] It turned out, however, that the F of PIV5 strain WR and those of most parainfluenza viruses require their homologous HN proteins for the induction of cell–cell fusion as described in Sec. 5. It is worth noting that the L22P mutant, in which the leucine at position 22 of PIV5 WR F has been replaced with the W3A F counterpart, mediates syncytium formation in the absence of HN.[92] Thus, a single amino acid substitution can bestow the HN-independent fusion activity on the otherwise fusion-inactive F molecule. Interestingly, WR F can mediate cell–cell fusion by itself at 53°C, while the S443P mutant of W3A F, which is functionally equivalent to the WR F mutant L22P, can do so even at 22°C, suggesting that the presence of proline residues at positions 22 and 443 destabilizes PIV5 F and thereby decreases the energy required for triggering the conformational changes in the absence of HN.[93] Interestingly, the G3A and G7A mutations in the fusion peptide also destabilize the PIV5 WR F and render it HN-independent in terms of fusion activity.[83] However, it has not been clarified how the presence of these amino acids mediates such destabilization.

Intriguingly, the F of PIV5 strain T1 cannot induce cell–cell fusion in the absence of HN despite that it harbors proline residues at positions 22 and 443.[94] Thus, chimeric analysis between L22P and PIV5 T1 F has been carried out, revealing that Glu132 in the HR1 domain is also required for the HN-independent fusion activity of L22P.[94] Furthermore, the L539A and L548A mutations in the cytoplasmic domain of the F of PIV5 strain SER, which has a long cytoplasmic tail similar to PIV5 T1 F,[95] are found to eliminate the HN requirement for cell–cell fusion.[96] In the case of NDV, on the other hand, it has been proved that the L289A mutation in F eliminates the HN requirement for cell–cell fusion.[97] These findings suggest that multiple amino acids can bestow the HN-independent fusion activity on F but a given amino acid appears to play this role only on certain F backgrounds.[98,99]

Another important feature of the parainfluenza virus F is that the transmembrane domain and/or cytoplasmic tail are required for the formation or stability of its prefusion form.[44] Correspondingly, in order to generate soluble form of PIV5 W3A F_0 in the prefusion conformation, the transmembrane domain and cytoplasmic tail should be replaced by a trimerization domain, GCNt.[44,100] Without GCNt appended to the carboxyl terminus, the anchorless soluble form of F_0 spontaneously folds as an F in the postfusion conformation, as reported for NDV F and HPIV3 F.[53,101] Clearly, this aberrant folding does not require the cleavage of F and takes place even for an F which does not exhibit the HN-independent fusion activity. Interestingly, furthermore, the soluble form of the hyperfusogenic S443P mutant of the W3A F_0 folds to the postfusion conformation even though it is appended with GCNt,[100] being consistent with its highly destabilized nature.[93] Also interestingly, when the soluble W3A F_0 appended with GCNt is cleaved by trypsin and then heated at 50°C, it converts to the postfusion form, organizes into fusion peptide-mediated rosettes as detected by electron microscopy, and becomes detectable by immunoprecipitation with a conformation-dependent monoclonal antibody.[100] It is worth noting, in this context, that this monoclonal antibody can efficiently immunoprecipitate PIV5 WR F after cleavage, while it poorly reacts to the same F in flow cytometry.[102] Thus, in the absence of attached GCNt, solubilization with detergent results in conversion of the prefusion form of WR F most likely into the postfusion form. This artificial conformational change is observed even when the solubilization of WR F is carried out with "mild" nonionic detergent such as 1% Triton X-100.[48]

Electron cryomicroscopy of the PIV5 W3A virions has recently revealed that the cleaved F not in complex with HN adopts postfusion form and thus it is inactivated, suggesting that the prefusion state of the cleaved W3A F may be stabilized by its association with HN.[103] Occurrence of such spontaneous inactivation has also been proposed for the hyperfusogenic mutants of W3A F that are expressed on the transfected cell surface in the absence of HN.[83]

5. HN–F Interaction in Membrane Fusion

It is generally accepted that the membrane fusion by parainfluenza viruses is induced through a series of conformational changes of F after it has been triggered by a physical HN–F interaction.[104,105] The existence of physical HN–F interactions has been certified by the finding that HPIV2 HN and F either in the transfected cells or in the infected cells can be coimmunoprecipitated.[106,107] Similarly, NDV HN and F either in the transfected cells or in the infected cells can be coprecipitated.[108,109] In the case of NDV, interestingly, it has been shown that mutant HNs that are defective in receptor binding fail to physically interact with F on the cell surface, suggesting that receptor binding by HN triggers the HN–F interaction.[110] Furthermore, the extent of fusion induced by coexpression of NDV HN and F is directly proportional to the strength of the HN–F interaction.[111] Although coimmunoprecipitation of PIV5 HN and F has so far been unsuccessful, bimolecular fluorescence complementation experiment has revealed that the complementary fragments of the yellow fluorescent protein, which have been appended to PIV5 HN and F, increase the HN–F interaction and promote cell–cell fusion.[112] Thus, the extent of fusion by the PIV5 glycoproteins is directly proportional to the strength of the HN–F interaction as it is the case with the NDV glycoproteins. Being consistent with these findings, endoplasmic reticulum (ER) coretention experiments suggest that the PIV5 HN–F complex is not formed intracellularly.[113] These findings in NDV and PIV5 indicate that the association of HN and F on the cell surface regulates the induction of fusion. It is assumed that this HN–F association is triggered by receptor binding and is quite transient (Fig. 3, Model II).[112,114]

On the other hand, another model has been proposed for the fusion induced by the glycoproteins of measles virus (MV), a member of genus Morbillivirus, in which the dissociation of H and F regulates the induction of fusion (Fig. 3, Model I)[114]: the extent of fusion is inversely related with the strength of the H–F interaction[115,116] and ER coretention experiments suggest that the MV H–F complex is formed in the ER.[117] It is not fully understood why MV uses the mechanism of fusion that is apparently distinct from that used by NDV and PIV5. Since MV H recognizes SLAM as a natural receptor,[118] difference in receptor moiety (protein or sialoconjugate) may account for the difference in mechanism of fusion.

In general, the HN–F interaction is virus type specific: for instance, the HN proteins of HPIV1, HPIV2, HPIV3, and PIV5 mediate cell–cell fusion only when expressed in combination with their homologous F molecules.[119,120] However, HPIV2 HN is able to substitute for SV41 HN or HPIV4A HN,[121] while the HN protein of mumps virus, a member of genus Rubulavirus, can substitute for HPIV2 HN or PIV5 HN (Fig. 4).[92,122] It is worth noting, in this context, that the converse

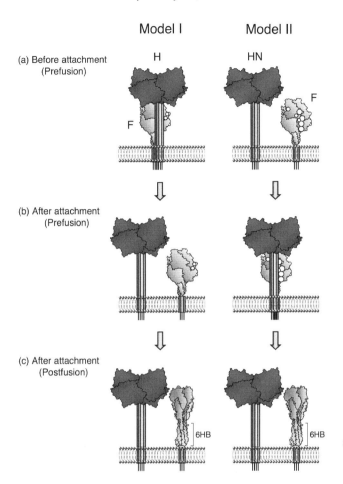

Fig. 3. Different models of membrane fusion by paramyxoviruses. Model I represents the dissociation model or clamp model proposed for fusion induced by morbilliviruses.[112,114] Two circles in the head region of F indicate the positions of amino acids that affect its physical and functional interaction with H.[128] Model II represents the association model or provocateur model proposed for rubulaviruses and avulaviruses.[112,114] Circles in the head region of F indicate the positions of amino acid domains that are candidates for the interacting sites with HN as described in the legend of Fig. 1. (a) Interaction between the attachment protein and fusion protein before receptor binding. The H–F complex is formed in the ER and transported to the cell surface. In the complex, H stabilizes F and prevents it from undergoing the conformational changes that lead to fusion. In contrast, HN and F are transported independently. (b) Interaction between the attachment protein and fusion protein on receptor binding. Attachment of H to the receptor results in the dissociation of F from the H–F complex, while attachment of HN promotes the formation of the HN–F complex. (c) Induction of fusion. After dissociation from H, F undergoes the conformational changes. On the other hand, the HN in the transiently formed HN–F complex transmits a signal to F, thereby it destabilizes F. As the result, the conformational changes of F and its dissociation from HN take place. The postfusion form of the F trimer was drawn on the basis of the atomic structure of HPIV3 F.[101] 6HB: six-helix bundle.

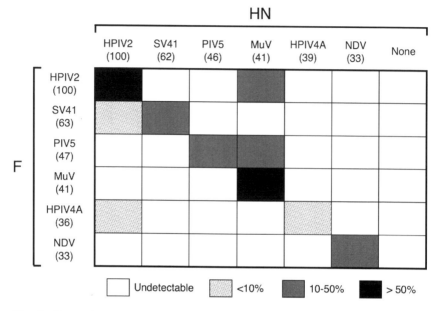

Fig. 4. Extent of cell–cell fusion induced by various combinations of HN and F. Cell–cell fusion was induced by coexpression of the viral glycoproteins in HeLa cells or in BHK cells and the fusion indices (%) were morphometrically quantified at 24 h post-transfection.[92,121,122] As it has been found that the F protein of PIV5 strain W3A has HN-independent fusion activity,[90,91] the F protein of PIV5 strain WR that requires HN for fusion was adopted.[92] Numbers in parentheses indicate the amino acid sequence identities (%) of the proteins compared with HPIV2 HN or F. MuV, mumps virus, a member of the genus *Rubulavirus*.

heterotypic combinations of HN and F usually result in no fusion (Fig. 4). Therefore, the "cross-reactivity" of some HN proteins cannot simply be explained by their proximity in primary structure to the substitutable HN proteins. Rather, similarity in the tertiary and/or quaternary structures between F-interacting sites on the HN proteins appears to be crucial for their compatibility and a subtle difference in the structure might result in the "one-way" cross-reactivity. It is important to note, in this context, that a given HN has a potential to substitute for another HN (and to interact with its partner F) when the overall amino acid sequence identity between the HN proteins is 39% or greater (Fig. 4).

Chimeric analyses of the HN proteins of HPIV2, SV41, HPIV3, NDV, and SeV have indicated that the HN stalk region contains the site that determines the specificity for the homologous F in promoting cell–cell fusion and thus is responsible for the functional HN–F interaction.[121,123,124] These studies do not exclude the possibility, however, that several amino acids in the transmembrane domain, which precede the stalk region, may play some role in determining the F specificity. The importance of the stalk region in the fusion-promoting function of HN is supported by the finding that an anti-HN monoclonal antibody that recognizes the HPIV2 HN stalk region can completely inhibit cell–cell fusion and virus–cell fusion (or virus entry) without affecting hemagglutinating and neuraminidase activities.[125,126] Finally, in addition to the functional HN–F interaction in the fusion induction, involvement of the HN stalk region in the physical HN–F interaction has been certified for the NDV glycoproteins by coimmunoprecipitation.[127]

A chimeric study between the F proteins of HPIV2 and SV41 has indicated, on the other hand, that the middle region (aa 227–360) of HPIV2 F determines the specificity for HPIV2 HN, suggesting that this region is responsible for the physical interaction with homologous HN.[122] This F middle region is located in the intervening region between the two heptad repeats (Fig. 2), a great part of which is exposed on the lateral surface of the F trimer (Fig. 1). As described in Sec. 3, the HN stalk region may exist as a straightforward stalk with the globular head region at its membrane distal end. If this configuration is true, then the HN stalk region could get access to the F head region (Figs. 1 and 3), yet whether the F middle region is directly involved in the physical interaction with HN awaits further investigation.

It should be noted, in this context, that the F head region of MV has recently proved to be involved in both the functional and physical interactions with the attachment protein H (Fig. 3).[128] More recently, furthermore, involvement of the MV H stalk region in both the functional and physical interactions with MV F has been demonstrated.[129]

The above findings strongly suggest that physical interaction between HN(H) and F is mediated between the HN(H) stalk region and the F head region, but this notion has not been verified as yet and would require a methodology which is more appropriate than coimmunoprecipitation. There is a possibility, on the other hand, that other regions on the glycoproteins are also involved in the physical interaction. In HPIV2, chimeric analyses suggested that the HN head region and the F stalk region (or the HR2 domain) play subsidiary roles only in the cross-reactive HN–F interaction.[121,122] Interestingly, peptide-based studies on NDV and SeV glycoproteins have indicated that the synthetic peptides corresponding to the F HR2 domain bind to the HN head region, raising the possibility that a physical HN–F interaction takes place between the F stalk region and the HN head region.[130,131] However, this assumption is contradictory to the notion that the HN stalk region forms a rod-like structure as mentioned in Sec. 3 (Fig. 1) and the proposed involvement of the NDV HN head region in the HN–F interaction has been refuted by a mutational study.[127]

6.   Cellular Factors that Influence Membrane Fusion

Cellular factors such as sialoconjugate receptors or proteolytic enzymes are essential for parainfluenza viruses to enter cells via virus–cell fusion as described so far. Accumulating evidence suggests, furthermore, that a variety of other cellular factors play important roles in membrane fusion caused by parainfluenza viruses.

It is generally thought that virus–cell fusion and cell–cell fusion are not entirely identical in terms of mechanism. In the case of parainfluenza viruses, it was previously suggested for HPIV3 that cell–cell fusion requires more receptors than are required for virus–cell fusion and for spread of infection.[29,132] It is of interest to note, in this context, that the PIV5 strain T1 induces minimal cell–cell fusion despite that it can spread from cell to cell,[94] though viral and cellular factor(s) that would regulate the T1 virus-mediated cell–cell fusion remains to be determined. Similarly, the PIV5 strain SER induces minimal cell–cell fusion,[95] whereas virus–cell fusion occurs at nearly wild-type levels.[133] Intriguingly, treating the SER-infected cells with low pH (4.8–6.0) results in induction of prominent cell–cell fusion and the SER infection is completely inhibited in the presence of bafilomycin A1 or ammonium chloride, suggesting that the entry of SER virus is mediated predominantly by acid-induced fusion between viral and endosomal membranes, which could explain the inability of SER virus to mediate cell–cell fusion at neutral pH.[134] However, since the requirement for low-pH triggering of SER virus-mediated fusion could not be corroborated,[135] the mechanism underlying the "low-fusion" phenotype of SER virus is currently obscure. In the case of a lentogenic NDV designated clone 30, on the other hand, cell–cell fusion is induced in the infected cells at neutral pH.[136] Nonetheless, virus–cell fusion is inhibited by 30% in the presence of ammonium chloride and preincubation of virus at pH 5 yields a mild inhibition of fusion activity, indicating that the clone 30 may use the endocytic pathway as a complementary way in addition to the direct fusion with the plasma membrane.[136]

A monoclonal antibody neutralization escape mutant of HPIV2, with two mutations, N83Y and M186I, in the HN and no mutation in the F, induces minimal cell–cell fusion, whereas it can spread from cell to cell.[126] Recently, it has been shown that cell–cell fusion induced by H83/186 virus, a recombinant HPIV2 that harbors the N83Y and M186I mutations in the HN and induces minimal cell–cell fusion, is greatly promoted in the presence of lysophosphatidic acid (LPA).[106] LPA is a bioactive phospholipid present in serum, which accounts for much of the cellular proliferative effect of serum,[137–139] and is known to activate multiple signal transduction pathways including those initiated by Rho family GTPases such as RhoA, Rac1, or Cdc42, thereby regulating cell proliferation, migration, and morphogenesis.[140,141] As cell–cell fusion mediated by wild-type HPIV2 or

LPA-promoted cell–cell fusion mediated by H83/186 is markedly blocked by Y-27632, a highly selective inhibitor of Rho kinases that are activated by the Rho family GTPases,[142,143] it is proposed that HPIV2 HN has a novel function that is involved in Rho-mediated signal transduction, thereby regulating cell–cell fusion and cytopathicity without significantly affecting virus–cell fusion.[106] This notion may not be applicable to the HN proteins of other parainfluenza viruses and the mechanism by which HPIV2 HN transduces the signal is not clear. However, the putative signal-transducing function of HN would be an effective strategy of HPIV2 to cope with the host cells, leading to immediate spread of infection by promoting cell–cell fusion or otherwise leading to prolonged virus replication by suppressing cell–cell fusion. It is worth noting, in this context, that RhoA and its downstream signaling cascades are activated in cells infected with respiratory syncytial virus (RSV),[144] the prototype of genus Pneumovirus. Correspondingly, either Y-27632 or *Clostridium botulinum* C3 protein, which specifically ADP-ribosylates Rho GTPases,[145] effectively reduces cell–cell fusion induced by RSV.[146] The activation of RhoA results in formation of vinculin-mediated focal adhesions and assembly of actin stress fibers,[137] and it has been suggested that RhoA-mediated actin polymerization is associated with the syncytium-inducing phenotype of RSV.[146] Intriguingly, RhoA binds to RSV F and an RhoA-derived peptide is able to inhibit cell–cell fusion induced by RSV or HPIV3.[147,148] The RhoA-binding site on RSV F resides in the carboxyl terminus of the fusion peptide as it has been mapped with the aid of yeast two-hybrid system,[147] but it seems unlikely that the cytoplasmic protein RhoA is accessible to the fusion peptide in the F ectodomain. However, association of RhoA with the fusion peptide would become reality provided that RSV F exists in more than one topological forms with respect to membranes, similar to NDV F.[149,150] It should be pointed out, on the other hand, that the effects of Rho family GTPases on cell–cell fusion are dependent on both the viral glycoprotein and the cell type.[151] Furthermore, in the case of Env of human immunodeficiency virus type 1, induction of cell–cell fusion and virus–cell fusion relies on Rac1 activation but not on Rho or Cdc42, in which the coreceptor CCR5 transduces a signal from Env.[152]

The current model of fusion induced by the class I fusion proteins indicates that the F trimer must undergo bending in spite of their rotational symmetry after it has formed a prehairpin intermediate and before it adopts postfusion conformation.[44,81,82] In the case of the F proteins of morbilliviruses, it has been demonstrated that the ectodomain of approximately one subunit of the F trimer, either in the viral envelope or in the plasma membrane of the infected cell, is cleaved proximal to the membrane by unknown cellular protease(s).[153] Prevention of the cleavage by mutation of F results in marked reduction in fusion, suggesting that this restrictive cleavage enables the F trimer to undergo bending that facilitates close apposition of two opposing membranes and thus fusion. It is worth noting, on the other hand, that free thiols are generated in NDV F through reduction of the disulfide bonds by thiol/disulfide isomerases on the cell surface.[154] Generation of free thiols is required for entry of NDV into cells as well as for cell–cell fusion by the viral glycoproteins,[154,155] and the attachment of HN to its receptor facilitates the interaction between F and host cell isomerases.[156] It is suggested that free thiols on NDV F generated by the reduction allow F to alter its conformation and the subsequent reoxidation of the thiols completes the conformational changes of F, eventually resulting in fusion.[155]

The primary culture of human airway tracheobronchial epithelial (HAE) cells grown at an air–liquid interface generates a well-differentiated pseudostratified mucociliary epithelium that closely models human airway epithelium *in vivo*.[157] The entry and release of HPIV3 occur exclusively at the apical surface of the ciliated epithelial cells of HAE.[158] HPIV3 infection of the ciliated epithelial cells takes place via NeuAcα2-6Gal-containing sialoconjugates, not via α2-3- or α2-8-linked sialoconjugates,[158]whereas binding assays *in vitro* indicate, as noted in Sec. 2, that HPIV3 can bind either to NeuAcα2-3Gal-containing sialoconjugates or to NeuAcα2-6Gal-containing sialoconjugates, with slight preference for the former sialoconjugates. Interestingly, the progeny virus does not spread beyond the ciliated cell types and there is no evidence of cell–cell fusion for the infected cells.[158] The absence of cell–cell fusion is presumably due to polarized trafficking of F to the apical surface that would restrict its interaction with neighboring cells. These findings also indicate that the polarity as well as the integrity of tight junction of the epithelial cells is preserved in spite of HPIV3 infection, which allows spread of infection in the epithelial cell monolayer without inducing cell–cell fusion.

7. Perspectives

Binding to receptors and following induction of virus–cell fusion by the viral fusion protein are general functions of the enveloped viruses, which enable the viral genome to enter host cells and to initiate replication. In the case of the viruses belonging to subfamily Paramyxovirinae, however, the viral fusion protein (F) requires homologous attachment protein (HN, H, or G) for the induction of fusion.

The molecular basis underlying the interaction between F and the attachment protein has been extensively investigated but there remain several questions to be solved. Elucidation of the three-dimensional structure of the HN ectodomain including the stalk region and identification of the site on the F ectodomain that is involved in the interaction with the HN stalk region will be major tasks in the future. Importantly, in this context, investigation of the physical interactions between HN and F should be performed with a method that can detect the interaction at defined time points under physiological conditions. One of the promising candidates for such method would be the photocrosslinking technique,[159] which has recently become available for detection of protein–protein interactions in living cells in a residue-specific manner.[160] On the other hand, the virus-mediated fusion is regulated not only by the viral proteins but also by multiple cellular factors, which requires further investigation.

Improvement of knowledge concerning the mechanism of fusion by which parainfluenza viruses accomplish entry and exhibit cytopathicity would be extremely useful not only for understanding the parainfluenza virus pathology but also for prevention and control of parainfluenza virus infection.

References

1. Scheid A, Caliguiri LA, Compans RW, Choppin PW. (1972) Isolation of paramyxovirus glycoproteins. Association of both hemagglutinating and neuraminidase activities with the larger SV5 glycoprotein. *Virology* **50**: 640–652.
2. Tozawa H, Watanabe M, Ishida N. (1973) Structural components of Sendai virus. Serological and physicochemical characterization of hemagglutinin subunit associated with neuraminidase activity. *Virology* **55**: 242–253.
3. Suzuki T, Harada M, Suzuki Y, Matsumoto M. (1984) Incorporation of sialoglyco-protein containing lacto-series oligosaccharides into chicken asialoerythrocyte membranes and restoration of receptor activity toward hemagglutinating virus of Japan (Sendai virus). *J Biochem* **95**: 1193–1200.
4. Wybenga LE, Epand RF, Nir S, *et al.* (1996) Glycophorin as a receptor for Sendai virus. *Biochemistry* **35**: 9513–9518.
5. Ferreira L, Villar E, Muñoz-Barroso I. (2004) Gangliosides and N-glycoproteins function as Newcastle disease virus receptors. *Int J Biochem Cell Biol* **36**: 2344–2356.
6. Holmgren J, Svennerholm L, Elwing H, Fredman P, Strannegard O. (1980) Sendai virus receptor: Proposed recognition structure based on binding to plastic-adsorbed gangliosides. *Proc Natl Acad Sci U S A* **77**: 1947–1950.
7. Markwell MA, Paulson JC. (1980) Sendai virus utilizes specific sialyloligosaccharides as host cell receptor determinants. *Proc Natl Acad Sci U S A* **77**: 5693–5697.
8. Shinya K, Ebina M, Yamada S, Ono M, Kasai N, Kawaoka Y. (2006) Avian flu: Influenza virus receptors in the human airway. *Nature* **440**: 435–436.
9. Suzuki T, Portner A, Scroggs RA, *et al.* (2001) Receptor specificities of human respiroviruses. *J Virol* **75**: 4604–4613.
10. Ah-Tye C, Schwartz S, Huberman K, Carlin E, Moscona A. (1999) Virus–receptor interactions of human parainfluenza viruses types 1, 2 and 3. *Microb Pathog* **27**: 329–336.

11. Ito T, Couceiro JN, Kelm S, *et al.* (1998) Molecular basis for the generation in pigs of influenza A viruses with pandemic potential. *J Virol* **72**: 7367–7373.

12. Connor RJ, Kawaoka Y, Webster RG, Paulson JC. (1994) Receptor specificity in human, avian, and equine H2 and H3 influenza virus isolates. *Virology* **205**: 17–23.

13. Rogers GN, Paulson JC. (1983) Receptor determinants of human and animal influenza virus isolates: Differences in receptor specificity of the H3 hemagglutinin based on species of origin. *Virology* **127**: 361–373.

14. Paulson JC, Weinstein J, Dorland L, van Halbeek H, Vliegenthart JF. (1982) Newcastle disease virus contains a linkage-specific glycoprotein sialidase. Application to the localization of sialic acid residues in N-linked oligosaccharides of α1-acid glycoprotein. *J Biol Chem* **257**: 12734–12738.

15. Suzuki Y, Suzuki T, Matsunaga M, Matsumoto M. (1985) Gangliosides as paramyxovirus receptor. Structural requirement of sialo-oligosaccharides in receptors for hemagglutinating virus of Japan (Sendai virus) and Newcastle disease virus. *J Biochem* **97**: 1189–1199.

16. Bitzer M, Lauer U, Baumann C, Spiegel M, Gregor M, Neubert WJ. (1997) Sendai virus efficiently infects cells via the asialoglycoprotein receptor and requires the presence of cleaved F_0 precursor proteins for this alternative route of cell entry. *J Virol* **71**: 5481–5486.

17. Hiebert SW, Paterson RG, Lamb RA. (1985) Hemagglutinin-neuraminidase protein of the paramyxovirus simian virus 5: Nucleotide sequence of the mRNA predicts an N-terminal membrane anchor. *J Virol* **54**: 1–6.

18. Markwell MA, Fox CF. (1980) Protein–protein interactions within paramyxoviruses identified by native disulfide bonding or reversible chemical cross-linking. *J Virol* **33**: 152–166.

19. Ng DT, Randall RE, Lamb RA. (1989) Intracellular maturation and transport of the SV5 type II glycoprotein hemagglutinin-neuraminidase: Specific and transient association with GRP78-BiP in the endoplasmic reticulum and extensive internalization from the cell surface. *J Cell Biol* **109**: 3273–3289.

20. Yuan P, Thompson TB, Wurzburg BA, Paterson RG, Lamb RA, Jardetzky TS. (2005) Structural studies of the parainfluenza virus 5 hemagglutinin-neuraminidase tetramer in complex with its receptor, sialyllactose. *Structure* **13**: 803–815.

21. Sheehan JP, Iorio RM, Syddall RJ, Glickman RL, Bratt MA. (1987) Reducing agent-sensitive dimerization of the hemagglutinin-neuraminidase glycoprotein of Newcastle disease virus correlates with the presence of cysteine at residue 123. *Virology* **161**: 603–606.

22. McGinnes LW, Morrison TG. (1994) The role of the individual cysteine residues in the formation of the mature, antigenic HN protein of Newcastle disease virus. *Virology* **200**: 470–483.

23. Collins PL, Mottet G. (1991) Homooligomerization of the hemagglutinin-neuraminidase glycoprotein of human parainfluenza virus type 3 occurs before the acquisition of correct intramolecular disulfide bonds and mature immunoreactivity. *J Virol* **65**: 2362–2371.

24. Yuan P, Leser GP, Demeler B, Lamb RA, Jardetzky TS. (2008) Domain architecture and oligomerization properties of the paramyxovirus PIV 5 hemagglutinin-neuraminidase (HN) protein. *Virology* **378**: 282–291.

25. Thompson SD, Laver WG, Murti KG, Portner A. (1988) Isolation of a biologically active soluble form of the hemagglutinin-neuraminidase protein of Sendai virus. *J Virol* **62**: 4653–4660.
26. Mirza AM, Sheehan JP, Hardy LW, Glickman RL, Iorio RM. (1993) Structure and function of a membrane anchor-less form of the hemagglutinin-neuraminidase glyco-protein of Newcastle disease virus. *J Biol Chem* **268**: 21425–21431.
27. McGinnes LW, Morrison TG. (2006) Inhibition of receptor binding stabilizes Newcastle disease virus HN and F protein-containing complexes. *J Virol* **80**: 2894–2903.
28. Moscona A, Peluso RW. (1991) Fusion properties of cells persistently infected with human parainfluenza virus type 3: Participation of hemagglutinin-neuraminidase in membrane fusion. *J Virol* **65**: 2773–2777.
29. Moscona A, Peluso RW. (1992) Fusion properties of cells infected with human parain-fluenza virus type 3: Receptor requirements for viral spread and virus-mediated mem-brane fusion. *J Virol* **66**: 6280–6827.
30. Russell CJ, Jardetzky TS, Lamb RA. (2001) Membrane fusion machines of paramyx-oviruses: Capture of intermediates of fusion. *EMBO J* **20**: 4024–4034.
31. Porotto M, Murrell M, Greengard O, Doctor L, Moscona A. (2005) Influence of the human parainfluenza virus 3 attachment protein's neuraminidase activity on its capacity to activate the fusion protein. *J Virol* **79**: 2383–2392.
32. Crennell S, Takimoto T, Portner A, Taylor G. (2000) Crystal structure of the multifunc-tional paramyxovirus hemagglutinin-neuraminidase. *Nat Struct Biol* **7**: 1068–1074.
33. Lawrence MC, Borg NA, Streltsov VA, *et al.* (2004) Structure of the haemagglutinin-neuraminidase from human parainfluenza virus type III. *J Mol Biol* **335**: 1343–1357.
34. Zaitsev V, von Itzstein M, Groves D, *et al.* (2004) Second sialic acid binding site in Newcastle disease virus hemagglutinin-neuraminidase: Implications for fusion. *J Virol* **78**: 3733–3741.
35. Bousse TL, Taylor G, Krishnamurthy S, Portner A, Samal SK, Takimoto T. (2004) Biological significance of the second receptor binding site of Newcastle disease virus hemagglutinin-neuraminidase protein. *J Virol* **78**: 13351–13355.
36. Corey EA, Mirza AM, Levandowsky E, Iorio RM. (2003) Fusion deficiency induced by mutations at the dimer interface in the Newcastle disease virus hemagglutinin-neuraminidase is due to a temperature-dependent defect in receptor binding. *J Virol* **77**: 6913–6922.
37. Takimoto T, Taylor GL, Connaris HC, Crennell SJ, Portner A. (2002) Role of the Hemagglutinin-neuraminidase protein in the mechanism of paramyxovirus–cell mem-brane fusion. *J Virol* **76**: 13028–13033.
38. Mahon PJ, Mirza AM, Musich TA, Iorio RM. (2008) Engineered intermonomeric disulfide bonds in the globular domain of Newcastle disease virus hemagglutinin-neuraminidase protein: Implications for the mechanism of fusion promotion. *J Virol* **82**: 10386–10396.
39. Alymova IV, Taylor G, Mishin VP, *et al.* (2008) Loss of the N-linked glycan at residue 173 of human parainfluenza virus type 1 hemagglutinin-neuraminidase exposes a sec-ond receptor-binding site. *J Virol* **82**: 8400–8410.
40. Mishin VP, Watanabe M, Taylor G, *et al.* (2010) N-linked glycan at residue 523 of human parainfluenza virus type 3 hemagglutinin-neuraminidase masks a second receptor-binding site. *J Virol* **84**: 3094–3100.

41. Porotto M, Fornabaio M, Kellogg GE, Moscona A. (2007) A second receptor binding site on human parainfluenza virus type 3 hemagglutinin-neuraminidase contributes to activation of the fusion mechanism. *J Virol* **81**: 3216–3228.

42. Paterson RG, Harris TJ, Lamb RA. (1984) Fusion protein of the paramyxovirus simian virus 5: Nucleotide sequence of mRNA predicts a highly hydrophobic glycoprotein. *Proc Natl Acad Sci U S A* **81**: 6706–6710.

43. Russell R, Paterson RG, Lamb RA. (1994) Studies with cross-linking reagents on the oligomeric form of the paramyxovirus fusion protein. *Virology* **199**: 160–168.

44. Yin HS, Wen X, Paterson RG, Lamb RA, Jardetzky TS. (2006) Structure of the parainfluenza virus 5 F protein in its metastable, prefusion conformation. *Nature* **439**: 38–44.

45. Dutch RE, Hagglund RN, Nagel MA, Paterson RG, Lamb RA. (2001) Paramyxovirus fusion protein: A conformational change on cleavage activation. *Virology* **281**: 138–150.

46. Hsu M, Scheid A, Choppin PW. (1981) Activation of the Sendai virus fusion protein (F) involves a conformational change with exposure of a new hydrophobic region. *J Biol Chem* **256**: 3557–3563.

47. Kohama T, Garten W, Klenk HD. (1981) Changes in conformation and charge paralleling proteolytic activation of Newcastle disease virus glycoproteins. *Virology* **111**: 364–376.

48. Tsurudome M, Ito M, Nishio M, Kawano M, Komada H, Ito Y. (2006) A mutant fusion (F) protein of simian virus 5 induces hemagglutinin-neuraminidase-independent syncytium formation despite the internalization of the F protein. *Virology* **347**: 11–27.

49. Umino Y, Kohama T, Sato TA, Sugiura A, Klenk H-D, Rott R. (1990) Monoclonal antibodies to three structural proteins of Newcastle disease virus: Biological characterization with particular reference to the conformational change of envelope glycoproteins associated with proteolytic cleavage. *J Gen Virol* **71**: 1189–1197.

50. Gething MJ, White JM, Waterfield MD. (1978) Purification of the fusion protein of Sendai virus: Analysis of the NH_2-terminal sequence generated during precursor activation. *Proc Natl Acad Sci U S A* **75**: 2737–2740.

51. Novick SL, Hoekstra D. (1988) Membrane penetration of Sendai virus glycoproteins during the early stages of fusion with liposomes as determined by hydrophobic photoaffinity labeling. *Proc Natl Acad Sci U S A* **85**: 7433–7437.

52. Chambers P, Pringle CR, Easton AJ. (1990) Heptad repeat sequences are located adjacent to hydrophobic regions in several types of virus fusion glycoproteins. *J Gen Virol* **71**: 3075–3080.

53. Chen L, Gorman JJ, McKimm-Breschkin J, *et al.* (2001) The structure of the fusion glycoprotein of Newcastle disease virus suggests a novel paradigm for the molecular mechanism of membrane fusion. *Structure* **9**: 255–266.

54. Iwata S, Schmidt AC, Titani K, *et al.* (1994) Assignment of disulfide bridges in the fusion glycoprotein of Sendai virus. *J Virol* **68**: 3200–3206.

55. Homma M, Ohuchi M. (1973) Trypsin action on the growth of Sendai virus in tissue culture cells. 3. Structural difference of Sendai viruses grown in eggs and tissue culture cells. *J Virol* **12**: 1457–1465.

56. Scheid A, Choppin PW. (1974) Identification of biological activities of paramyxovirus glycoproteins. Activation of cell fusion, hemolysis, and infectivity of proteolytic cleavage of an inactive precursor protein of Sendai virus. *Virology* **57**: 475–490.

57. Gotoh B, Ohnishi Y, Inocencio NM, *et al.* (1992) Mammalian subtilisin-related proteinases in cleavage activation of the paramyxovirus fusion glycoprotein: Superiority of furin/PACE to PC2 or PC1/PC3. *J Virol* **66**: 6391–6397.

58. Karron RA, Collins PL. (2007) Parainfluenza viruses. In: Knipe DM, Howley PM, Griffin DE, Martin MA, Lamb RA, Roizman B, Straus SE (eds), *Fields Virology*, pp. 1497–1526. Lippincott Williams & Wilkins, Philadelphia.

59. Ortmann D, Ohuchi M, Angliker H, Shaw E, Garten W, Klenk HD. (1994) Proteolytic cleavage of wild type and mutants of the F protein of human parainfluenza virus type 3 by two subtilisin-like endoproteases, furin and Kex2. *J Virol* **68**: 2772–2776.

60. Molloy SS, Thomas L, VanSlyke JK, Stenberg PE, Thomas G. (1994) Intracellular trafficking and activation of the furin proprotein convertase: Localization to the TGN and recycling from the cell surface. *EMBO J* **13**: 18–33.

61. Schäfer W, Stroh A, Berghöfer S, *et al.* (1995) Two independent targeting signals in the cytoplasmic domain determine *trans*-Golgi network localization and endosomal trafficking of the proprotein convertase furin. *EMBO J* **14**: 2424–2435.

62. Hosaka M, Nagahama M, Kim WS, *et al.* (1991) Arg-X-Lys/Arg-Arg motif as a signal for precursor cleavage catalyzed by furin within the constitutive secretory pathway. *J Biol Chem* **266**: 12127–12130.

63. Klenk HD, Rott R, Orlich M, Blödorn J. (1975) Activation of influenza A viruses by trypsin treatment. *Virology* **68**: 426–439.

64. Lazarowitz SG, Choppin PW. (1975) Enhancement of the infectivity of influenza A and B viruses by proteolytic cleavage of the hemagglutinin polypeptide. *Virology* **68**: 440–454.

65. Nagai Y, Klenk HD. (1977) Activation of precursors to both glycoproteins of Newcastle disease virus by proteolytic cleavage. *Virology* **77**: 125–134.

66. Nagai Y, Klenk HD, Rott R. (1976) Proteolytic cleavage of the viral glycoproteins and its significance for the virulence of Newcastle disease virus. *Virology* **72**: 494–508.

67. Nagai Y. (1995) Virus activation by host proteinases. A pivotal role in the spread of infection, tissue tropism and pathogenicity. *Microbiol Immunol* **39**: 1–9.

68. Molloy SS, Bresnahan PA, Leppla SH, Klimpel KR, Thomas G. (1992) Human furin is a calcium-dependent serine endoprotease that recognizes the sequence Arg-X-X-Arg and efficiently cleaves anthrax toxin protective antigen. *J Biol Chem* **267**: 16396–16402.

69. Toyoda T, Sakaguchi T, Imai K, *et al.* (1987) Structural comparison of the cleavage-activation site of the fusion glycoprotein between virulent and avirulent strains of Newcastle disease virus. *Virology* **158**: 242–247.

70. Kido H, Yokogoshi Y, Sakai K, *et al.* (1992) Isolation and characterization of a novel trypsin-like protease found in rat bronchiolar epithelial Clara cells. A possible activator of the viral fusion glycoprotein. *J Biol Chem* **267**: 13573–13579.

71. Murakami M, Towatari T, Ohuchi M, *et al.* (2001) Mini-plasmin found in the epithelial cells of bronchioles triggers infection by broad-spectrum influenza A viruses and Sendai virus. *Eur J Biochem* **268**: 2847–2855.

72. Tashiro M, Yokogoshi Y, Tobita K, Seto JT, Rott R, Kido H. (1992) Tryptase Clara, an activating protease for Sendai virus in rat lungs, is involved in pneumopathogenicity. *J Virol* **66**: 7211–7216.

73. Böttcher E, Matrosovich T, Beyerle M, Klenk HD, Garten W, Matrosovich M. (2006) Proteolytic activation of influenza viruses by serine proteases TMPRSS2 and HAT from human airway epithelium. *J Virol* **80**: 9896–9898.

74. Donaldson SH, Hirsh A, Li DC, *et al.* (2002) Regulation of the epithelial sodium channel by serine proteases in human airways. *J Biol Chem* **277**: 8338–8345.

75. Takahashi M, Sano T, Yamaoka K, *et al.* (2001) Localization of human airway trypsin-like protease in the airway: An immunohistochemical study. *Histochem Cell Biol* **115**: 181–187.

76. Shirogane Y, Takeda M, Iwasaki M, *et al.* (2008). Efficient multiplication of human metapneumovirus in Vero cells expressing the transmembrane serine protease TMPRSS2. *J Virol* **82**: 8942–8946.

77. Bando H, Kawano M, Kondo K, *et al.* (1991) Growth properties and F protein cleavage site sequences of naturally occurring human parainfluenza type 2 viruses. *Virology* **184**: 87–92.

78. Coelingh KV, Winter CC. (1990) Naturally occurring human parainfluenza type 3 viruses exhibit divergence in amino acid sequence of their fusion protein neutralization epitopes and cleavage sites. *J Virol* **64**: 1329–1334.

79. Okada H, Seto JT, McQueen NL, Klenk H, Rott R, Tashiro M. (1998) Determinants of pantropism of the F1-R mutant of Sendai virus: Specific mutations involved are in the F and M genes. *Arch Virol* **143**: 2343–2352.

80. Tashiro M, Seto JT, Choosakul S, Hegemann H, Klenk HD, Rott R. (1992). Changes in specific cleavability of the Sendai virus fusion protein: Implications for pathogenicity in mice. *J Gen Virol* **73**: 1575–1579.

81. Jardetzky TS, Lamb RA. 2004. Virology: A class act. *Nature* **427**: 307–308.

82. Russell CJ, Luque LE. (2006) The structural basis of paramyxovirus invasion. *Trends Microbiol* **14**: 243–246.

83. Russell CJ, Jardetzky TS, Lamb RA. (2004) Conserved glycine residues in the fusion peptide of the paramyxovirus fusion protein regulate activation of the native state. *J Virol* **78**: 13727–13742.

84. Baker KA, Dutch RE, Lamb RA, Jardetzky TS. (1999) Structural basis for paramyxovirus-mediated membrane fusion. *Mol Cell* **3**: 309–319.

85. Weber T, Paesold G, Galli C, Mischler R, Semenza G, Brunner J. (1994) Evidence for H^+-induced insertion of influenza hemagglutinin HA_2 N-terminal segment into viral membrane. *J Biol Chem* **269**: 18353–18358.

86. Durrer P, Galli C, Hoenke S, *et al.* (1996) H^+-induced membrane insertion of influenza virus hemagglutinin involves the HA_2 amino-terminal fusion peptide but not the coiled coil region. *J Biol Chem* **271**: 13417–13421.

87. Harter C, James P, Bächi T, Semenza G, Brunner J. (1989) Hydrophobic binding of the ectodomain of influenza hemagglutinin to membranes occurs through the "fusion peptide". *J Biol Chem* **264**: 6459–6464.

88. Stegmann T, Delfino JM, Richards FM, Helenius A. (1991) The HA_2 subunit of influenza hemagglutinin inserts into the target membrane prior to fusion. *J Biol Chem* **266**: 18404–18410.

89. Tsurudome M, Glück R, Graf R, Falchetto R, Schaller U, Brunner J. (1992) Lipid interactions of the hemagglutinin HA_2 NH_2-terminal segment during influenza virus-induced membrane fusion. *J Biol Chem* **267**: 20225–20232.

90. Horvath CM, Paterson RG, Shaughnessy MA, Wood R, Lamb RA. (1992) Biological activity of paramyxovirus fusion proteins: Factors influencing formation of syncytia. *J Virol* **66**: 4564–4569.

91. Paterson RG, Hiebert SW, Lamb RA. (1985) Expression at the cell surface of biologically active fusion and hemagglutinin/neuraminidase proteins of the paramyxovirus simian virus 5 from cloned cDNA. *Proc Natl Acad Sci U S A* **82**: 7520–7524.

92. Ito M, Nishio M, Kawano M, *et al.* (1997) Role of a single amino acid at the amino terminus of the simian virus 5 F2 subunit in syncytium formation. *J Virol* **71**: 9855–9858.

93. Paterson RG, Russell CJ, Lamb RA. (2000). Fusion protein of the paramyxovirus SV5: Destabilizing and stabilizing mutants of fusion activation. *Virology* **270**: 17–30.

94. Ito M, Nishio M, Komada H, Ito Y, Tsurudome M. (2000) An amino acid in the heptad repeat 1 domain is important for the haemagglutinin-neuraminidase-independent fusing activity of simian virus 5 fusion protein. *J Gen Virol* **81**: 719–727.

95. Tong S, Li M, Vincent A, *et al.* (2002) Regulation of fusion activity by the cytoplasmic domain of a paramyxovirus F protein. *Virology* **301**: 322–333.

96. Seth S, Vincent A, Compans RW. (2003) Mutations in the cytoplasmic domain of a paramyxovirus fusion glycoprotein rescue syncytium formation and eliminate the hemagglutinin-neuraminidase protein requirement for membrane fusion. *J Virol* **77**: 167–178.

97. Sergel TA, McGinnes LW, Morrison TG. (2000) A single amino acid change in the Newcastle disease virus fusion protein alters the requirement for HN protein in fusion. *J Virol* **74**: 5101–5107.

98. Ito M, Nishio M, Kawano M, Komada M, Ito Y, Tsurudome M. (2009) Effects of multiple amino acids of the parainfluenza virus 5 fusion protein on the haemagglutinin-neuraminidase-independent fusion activity. *J Gen Virol* **90**: 405–413.

99. Terrier O, Durupt F, Cartet G, Thomas L, Lina B, Rosa-Calatrava M. (2009) Engineering of a parainfluenza virus type 5 fusion protein (PIV-5 F): Development of an autonomous and hyperfusogenic protein by a combinational mutagenesis approach. *Virus Res* **146**: 115–124.

100. Connolly SA, Leser GP, Yin HS, Jardetzky TS, Lamb RA. (2006) Refolding of a paramyxovirus F protein from prefusion to postfusion conformations observed by liposome binding and electron microscopy. *Proc Natl Acad Sci U S A* **103**: 17903–17908.

101. Yin HS, Paterson RG, Wen X, Lamb RA, Jardetzky T. (2005) Structure of the uncleaved ectodomain of the paramyxovirus (hPIV3) fusion protein. *Proc Natl Acad Sci U S A* **102**: 9288–9293.

102. Tsurudome M, Ito M, Nishio M, Kawano M, Komada H, Ito Y. (2001) Hemagglutinin-neuraminidase-independent fusion activity of simian virus 5 fusion (F) protein: Difference in conformation between fusogenic and nonfusogenic F proteins on the cell surface. *J Virol* **75**: 8999–9009.

103. Ludwig K, Schade B, Böttcher C, *et al.* (2008) Electron cryomicroscopy reveals different F1+F2 protein States in intact parainfluenza virions. *J Virol* **82**: 3775–3781.

104. Lamb RA, Jardetzky TS. (2007) Structural basis of viral invasion: Lessons from paramyxovirus F. *Curr Opin Struct Biol* **17**: 427–436.

105. Morrison TG. (2003) Structure and function of a paramyxovirus fusion protein. *Biochim Biophys Acta* **1614**: 73–84.

106. Tsurudome M, Nishio M, Ito M, *et al.* (2008) Effects of hemagglutinin-neuraminidase protein mutations on cell–cell fusion mediated by human parainfluenza type 2 virus. *J Virol* **82**: 8283–8295.

107. Yao Q, Hu X, Compans RW. (1997) Association of the parainfluenza virus fusion and hemagglutinin-neuraminidase glycoproteins on cell surfaces. *J Virol* **71**: 650–656.

108. Stone-Hulslander J, Morrison TG. (1997) Detection of an interaction between the HN and F proteins in Newcastle disease virus-infected cells. *J Virol* **71**: 6287–6295.

109. Deng R, Wang Z, Mahon PJ, Marinello M, Mirza A, Iorio RM. (1999) Mutations in the Newcastle disease virus hemagglutinin-neuraminidase protein that interfere with its ability to interact with the homologous F protein. *Virology* **253**: 43–54.

110. Iorio RM, Field GM, Sauvron JM, Mirza AM, Deng R, Mahon PJ, Langedijk JP. (2001) Structural and functional relationship between the receptor recognition and neuraminidase activities of the Newcastle disease virus hemagglutinin-neuraminidase protein: Receptor recognition is dependent on neuraminidase activity. *J Virol* **75**: 1918–1927.

111. Melanson VR, Iorio RM. (2004) Amino acid substitutions in the F-specific domain in the stalk of the Newcastle disease virus HN protein modulate fusion and interfere with its interaction with the F protein. *J Virol* **78**: 13053–13061.

112. Connolly SA, Leser GP, Jardetzky TS, Lamb RA. (2009) Bimolecular complementation of paramyxovirus fusion and hemagglutinin-neuraminidase proteins enhances fusion: Implications for the mechanism of fusion triggering. *J Virol* **83**: 10857–10868.

113. Paterson RG, Johnson ML, Lamb RA. (1997) Paramyxovirus fusion (F) protein and hemagglutinin-neuraminidase (HN) protein interactions: Intracellular retention of F and HN does not affect transport of the homotypic HN or F protein. *Virology* **237**: 1–9.

114. Iorio RM, Melanson VR, Mahon PJ. (2009) Glycoprotein interactions in paramyxovirus fusion. *Future Virol* **4**: 335–351.

115. Plemper RK, Hammond AL, Gerlier D, Fielding AK, Cattaneo R. (2002) Strength of envelope protein interaction modulates cytopathicity of measles virus. *J Virol* **76**: 5051–5061.

116. Corey EA, Iorio RM. (2007) Mutations in the stalk of the measles virus hemagglutinin protein decrease fusion but do not interfere with virus-specific interaction with the homologous fusion protein. *J Virol* **81**: 9900–9910.

117. Plemper RK, Hammond AL, Cattaneo R. (2001) Measles virus envelope glycoproteins hetero-oligomerize in the endoplasmic reticulum. *J Biol Chem* **276**: 44239–44246.

118. Tatsuo H, Ono N, Tanaka K, Yanagi Y. (2000) SLAM (CDw150) is a cellular receptor for measles virus. *Nature* **406**: 893–897.

119. Heminway BR, Yu Y, Galinski MS. (1994) Paramyxovirus mediated cell fusion requires co-expression of both the fusion and hemagglutinin-neuraminidase glycoproteins. *Virus Res* **31**: 1–16.

120. Hu X, Ray R, Compans RW. (1992) Functional interactions between the fusion protein and hemagglutinin-neuraminidase of human parainfluenza viruses. *J Virol* **66**: 1528–1534.

121. Tsurudome M, Kawano M, Yuasa T, *et al.* (1995) Identification of regions on the hemagglutinin-neuraminidase protein of human parainfluenza virus type 2 important for promoting cell fusion. *Virology* **213**: 190–203.

122. Tsurudome M, Ito M, Nishio M, *et al.* (1998) Identification of regions on the fusion protein of human parainfluenza type 2 virus which are required for haemagglutinin-neuraminidase proteins to promote cell fusion. *J Gen Virol* **79**: 279–289.

123. Deng R, Mirza AM, Mahon PJ, Iorio RM. (1997) Functional chimeric HN glycoproteins derived from Newcastle disease virus and human parainfluenza virus-3. *Arch Virol* **13** (Suppl): 115–130.

124. Tanabayashi K, Compans RW. (1996) Functional interaction of paramyxovirus glycoproteins: Identification of a domain in Sendai virus HN which promotes cell fusion. *J Virol* **70**: 6112–6118.

125. Tsurudome M, Nishio M, Komada H, Bando H, Ito Y. (1989) Extensive antigenic diversity among human parainfluenza type 2 virus isolates and immunological relationships among paramyxoviruses revealed by monoclonal antibodies. *Virology* **171**: 38–48.

126. Yuasa T, Kawano M, Tabata N, *et al.* (1995) A cell fusion-inhibiting monoclonal antibody binds to the presumed stalk domain of the human parainfluenza type 2 virus hemagglutinin-neuraminidase protein. *Virology* **206**: 1117–1125.

127. Melanson VR, Iorio RM. (2006) Addition of N-glycans in the stalk of the Newcastle disease virus HN protein blocks its interaction with the F protein and prevents fusion. *J Virol* **80**: 623–633.

128. Lee JK, Prussia A, Paal T, White LK, Snyder JP, Plemper RK. (2008) Functional interaction between paramyxovirus fusion and attachment proteins. *J Biol Chem* **283**: 16561–16572.

129. Paal T, Brindley MA, St Clair C, *et al.* (2009) Probing the spatial organization of measles virus fusion complexes. *J Virol* **83**: 10480–10493.

130. Gravel KA, Morrison TG. (2003) Interacting domains of the HN and F proteins of Newcastle disease virus. *J Virol* **77**: 11040–11049.

131. Tomasi M, Pasti C, Manfrinato C, Dallocchio F, Bellini T. (2003) Peptides derived from the heptad repeat region near the C-terminal of Sendai virus F protein bind the hemagglutinin-neuraminidase ectodomain. *FEBS Lett* **536**: 56–60.

132. Moscona A, Peluso RW. (1993) Relative affinity of the human parainfluenza virus type 3 hemagglutinin-neuraminidase for sialic acid correlates with virus-induced fusion activity. *J Virol* **67**: 6463–6468.

133. Connolly SA, Lamb RA. (2006) Paramyxovirus fusion: Real-time measurement of parainfluenza virus 5 virus–cell fusion. *Virology* **355**: 203–212.

134. Seth S, Vincent A, Compans RW. (2003) Activation of fusion by the SER virus F protein: A low-pH-dependent paramyxovirus entry process. *J Virol* **77**: 6520–6527.

135. Bissonnette ML, Connolly SA, Young DF, Randall RE, Paterson RG, Lamb RA. (2006) Analysis of the pH requirement for membrane fusion of different isolates of the paramyxovirus parainfluenza virus 5. *J Virol* **80**: 3071–3077.

136. San Román K, Villar E, Muñoz-Barroso I. (1999) Acidic pH enhancement of the fusion of Newcastle disease virus with cultured cells. *Virology* **260**: 329–341.

137. Ridley AJ, Hall A. (1992) The small GTP-binding protein rho regulates the assembly of focal adhesions and actin stress fibers in response to growth factors. *Cell* **70**: 389–399.

138. Tigyi G, Miledi R. (1992) Lysophosphatidates bound to serum albumin activate membrane currents in Xenopus oocytes and neurite retraction in PC12 pheochromocytoma cells. *J Biol Chem* **267**: 21360–21367.

139. Eichholtz T, Jalink K, Fahrenfort I, Moolenaar WH. (1993) The bioactive phospholipid lysophosphatidic acid is released from activated platelets. *Biochem J* **291**: 677–680.
140. Moolenaar WH, van Meeteren LA, Giepmans BN. (2004) The ins and outs of lysophosphatidic acid signaling. *Bioessays* **26**: 870–881.
141. Panetti TS. (2002) Differential effects of sphingosine 1-phosphate and lysophosphatidic acid on endothelial cells. *Biochim Biophys Acta* **1582**: 190–196.
142. Narumiya S, Ishizaki T, Uehata M. (2000) Use and properties of ROCK-specific inhibitor Y-27632. *Meth Enzymol* **325**: 273–284.
143. Uehata M, Ishizaki T, Satoh H, *et al.* (1997) Calcium sensitization of smooth muscle mediated by a Rho-associated protein kinase in hypertension. *Nature* **389**: 990–994.
144. Gower TL, Peeples ME, Collins PL, Graham BS. (2001) RhoA is activated during respiratory syncytial virus infection. *Virology* **283**: 188–196.
145. Sekine A, Fujiwara M, Narumiya S. (1989) Asparagine residue in the rho gene product is the modification site for botulinum ADP-ribosyltransferase. *J Biol Chem* **264**: 8602–8605.
146. Gower TL, Pastey MK, Peeples ME, *et al.* (2005) RhoA signaling is required for respiratory syncytial virus-induced syncytium formation and filamentous virion morphology. *J Virol* **79**: 5326–5336.
147. Pastey MK, Crowe Jr JE, Graham BS. (1999) RhoA interacts with the fusion glycoprotein of respiratory syncytial virus and facilitates virus-induced syncytium formation. *J Virol* **73**: 7262–7270.
148. Pastey MK, Gower TL, Spearman PW, Crowe Jr JE, Graham BS. (2000) A RhoA-derived peptide inhibits syncytium formation induced by respiratory syncytial virus and parainfluenza virus type 3. *Nat Med* **6**: 35–40.
149. McGinnes LW, Reitter JN, Gravel K, Morrison TG. (2003) Evidence for mixed membrane topology of the Newcastle disease virus fusion protein. *J Virol* **77**: 1951–1963.
150. Pantua H, McGinnes LW, Leszyk J, Morrison TG. (2005) Characterization of an alternate form of Newcastle disease virus fusion protein. *J Virol* **79**: 11660–11670.
151. Schowalter RM, Wurth MA, Aguilar HC, *et al.* (2006) Rho GTPase activity modulates paramyxovirus fusion protein-mediated cell–cell fusion. *Virology* **350**: 323–334.
152. Pontow SE, Heyden NV, Wei S, Ratner L. (2004) Actin cytoskeletal reorganizations and coreceptor-mediated activation of Rac during human immunodeficiency virus-induced cell fusion. *J Virol* **78**: 7138–7147.
153. von Messling V, Milosevic D, Devaux P, Cattaneo R. (2004) Canine distemper virus and measles virus fusion glycoprotein trimers: Partial membrane-proximal ectodomain cleavage enhances function. *J Virol* **78**: 7894–7903.
154. Jain S, McGinnes LW, Morrison TG. (2007) Thiol/disulfide exchange is required for membrane fusion directed by the Newcastle disease virus fusion protein. *J Virol* **81**: 2328–2339.
155. Jain S, McGinnes LW, Morrison TG. (2008) Overexpression of thiol/disulfide isomerases enhances membrane fusion directed by the Newcastle disease virus fusion protein. *J Virol* **82**: 12039–12048.
156. Jain S, McGinnes LW, Morrison TG. (2009) Role of thiol/disulfide exchange in Newcastle disease virus entry. *J Virol* **83**: 241–249.

157. Pickles RJ, McCarty D, Matsui H, Hart PJ, Randell SH, Boucher RC. (1998) Limited entry of adenovirus vectors into well-differentiated airway epithelium is responsible for inefficient gene transfer. *J Virol* **72**: 6014–6023.

158. Zhang L, Bukreyev A, Thompson CI, *et al.* (2005) Infection of ciliated cells by human parainfluenza virus type 3 in an *in vitro* model of human airway epithelium. *J Virol* **79**: 1113–1124.

159. Brunner J. (1993) New photolabeling and crosslinking methods. *Annu Rev Biochem* **62**: 483–514.

160. Tanaka Y, Bond MR, Kohler JJ. (2008) Photocrosslinkers illuminate interactions in living cells. *Mol Biosyst* **4**: 473–480.

161. Chan DC, Chutkowski CT, Kim PS. (1998) Evidence that a prominent cavity in the coiled coil of HIV type 1 gp41 is an attractive drug target. *Proc Natl Acad Sci U S A* **95**: 15613–15617.

162. Weissenhorn W, Dessen A, Harrison SC, Skehel JJ, Wiley DC. (1997) Atomic structure of the ectodomain from HIV-1 gp41. *Nature* **387**: 426–430.

163. Bullough PA, Hughson FM, Skehel JJ, Wiley DC. (1994) Structure of influenza haemagglutinin at the pH of membrane fusion. *Nature* **371**: 37–43.

164. Skehel JJ, Wiley DC. (2000) Receptor binding and membrane fusion in virus entry: The influenza hemagglutinin. *Annu Rev Biochem* **69**: 531–569.

165. Bour S, Geleziunas R, Wainberg MA. (1995) The human immunodeficiency virus type 1 (HIV-1) CD4 receptor and its central role in promotion of HIV-1 infection. *Microbiol Rev* **59**: 63–93.

166. Broder CC, Collman RG. (1997) Chemokine receptors and HIV. *J Leukoc Biol* **62**: 20–29.

Chapter 4

What Controls the Distinct VSV RNA Synthetic Processes of Replication and Transcription?

Gail Williams Wertz[*,†,‡], Summer E. Galloway[§] and Djamila Harouaka[†,‡,¶]

1. Introduction

For the non-segmented negative-stranded (NNS) RNA viruses, genomic RNA replication and mRNA transcription are both catalyzed by the virally encoded RNA-dependent RNA polymerase (RdRp), which consists of the large (L) protein in complex with the phosphoprotein (P) co-factor.[1,2] Despite the use of the RdRp for both synthetic processes, transcription and replication are distinct processes, distinguished primarily by the fact that replication requires ongoing protein synthesis[3] and that the polymerase must ignore the *cis*-acting signals at each gene junction that signal initiation and termination of transcripts and modification of the mRNA 5′ and 3′ ends. Vesicular stomatitis virus (VSV) has been utilized as a model for studying various aspects of the NNS RNA virus life cycle. As such, many of the central tenets of the control of NNS RNA virus gene expression and genome replication have been elucidated through studies with VSV. Table 1 highlights key differences between transcription and replication. While our understanding of the characteristics of transcription and replication has improved in recent years, the mechanisms by which these two remarkably distinct processes are controlled remain unresolved and progress in this area has been limited.

---

[*]Corresponding author.

[†]Department of Pathology, The University of Virginia, Charlottesville, VA, USA.

[‡]Department of Microbiology, The University of Virginia, Charlottesville, VA, USA.

[§]Department of Microbiology and Immunology, Emory University, Atlanta, GA, USA.

[¶]Department of Microbiology, University of Alabama at Birmingham, Birmingham, AL, USA.

Table 1. Key differences between VSV transcription and replication.

S.No	Transcription	Replication
1	Protein synthesis is not required; mRNAs are not encapsidated.	Requires concomitant *de novo* synthesis of N protein and appropriate molar ratio of N to P protein.[4] The genomic and antigenomic RNAs are always encapsidated with N protein.
2	Transcription is obligatorily sequential with a decreasing polar gradient of monocistronic mRNA product abundance from 3′ to 5′ on the genome.[8,45–47]	Replication is a two-step process: The full-length positive sense *antigenome* is synthesized first; the *antigenome* is then used as template for negative sense genomic RNA replication in an asymmetric process.[48,49]
3	Transcription initiates at the first gene start sequence in infected cells; at the genomic 3′ end in *in vitro* systems.[8,9,11]	Replication initiates at the 3′ end of the genome.[11]
4	The promoter for transcription includes sequence elements in the leader (Le) at nts 19–29 and 34–46, and nts 47–50 at the Le/N junction.[42] The terminal 15–18 nts of the Le are essential for replication and encapsidation of genomes, thus their role in transcription is difficult to test.	The terminal 15 nts of Le or TrC are the minimal *cis*-acting signal for replication.[38–40,44]
5	The polymerase responds to *cis*-acting signals at the beginning and end of each gene. Conserved sequences at the 3′ end of each gene direct initiation of transcription and modification of the 5′ end by capping.[50] Conserved sequences at the 5′ end of each gene, 3′…AUACUUUUUUU…5′, are critical to termination and polyadenylation.[51,52] The AUAC directs termination and its base composition is critical for slippage and realignment on the U7 tract to generate the mRNA 3′ polyA tail.[51,53] The length of the U tract is critical for termination, polyadenylation, and downstream initiation.[51,54]	During replication, the polymerase ignores the *cis*-acting signals at the gene junctions that direct transcription and modification of mRNAs transcripts.
6	The polymerase is able to move both forward and backward on the template during transcription to search (scan) for an initiation signal as shown by the finding that if a proper initiation signal does not follow a termination signal then the RdRp can scan both forward and backward to locate one.[55]	The process of replication is highly processive. Prematurely terminated products are not found in the presence of adequate amounts of nucleocapsid protein.

Rather than providing an extensive review of advances in the field of RNA synthesis, this brief comment will be limited to some recent advances that point to key, as yet largely unexplored areas that merit further investigation to gain insight into how the two distinct RNA synthetic processes, replication and transcription, are controlled, and what factors are involved in their regulation. The primary focus will be on the role of *trans*-acting factors in regulating transcription and RNA replication, as this encompasses the majority of recent advances in the field of RNA replication. However, brief mention will be made regarding the role of genomic termini, as these represent a critical component by virtue of their sequence and/or their interaction with *trans*-acting viral proteins in regulating transcription and RNA replication.

The active template for both RNA replication and transcription is the genomic RNA encapsidated with the nucleocapsid (N) protein.[1] A major distinction between the two RNA synthetic events is the requirement for *de novo* protein synthesis for RNA replication, whereas transcription can occur in the absence of protein synthesis or *in vitro* requiring only the appropriate salts and ribonucleotides.[1,3] The requirement for protein synthesis during RNA replication is fulfilled by the synthesis of the N protein in a concentration-dependent manner.[4] The N protein is needed in stoichiometric amounts to encapsidate newly replicated RNAs. Transcription is therefore the first RNA synthetic activity in the infected cell, as it must precede replication in order to generate N mRNA and allow for translation of sufficient amounts of the N protein to support replication.

The demonstration that the N protein fulfilled the replication requirement for protein synthesis led to the hypothesis that the synthesis and availability of N protein might be involved in the decision to replicate, perhaps by binding to the leader (Le+) RNA synthesized following initiation by the RdRp and thereby allowing the polymerase to continue through the Le/N junction to generate the antigenomic RNA, or to transcribe, by failure of the N protein to bind the Le+ thereby releasing free Le, followed by polymerase initiation at the N mRNA start site.[5–7] While the requirement for N protein synthesis in replication is undisputed, there is no clear evidence that N protein production alone is the key discriminator between the two RNA synthetic processes.

2. Regulation of RNA Synthesis by the Use of Different Initiation Sites

The initial UV mapping experiments showing that VSV transcription is obligatorily sequential and polar,[8] could not discriminate whether a target as small as the 3′ leader affected downstream mRNA synthesis. Thus, delineating whether transcription initiated directly at the 3' terminus with leader (Le+) RNA synthesis as the first step in the obligatorily sequential transcription of mRNA was not possible. Therefore, the question of where the polymerase initiates transcription, whether at the 3' terminus with synthesis of Le+ or at the first gene start site, was addressed by biochemical analyses.

In vitro reconstitution studies examining transcription in the presence of limiting nucleotides (nts) showed that the RdRp accessed the template at the extreme 3' end and that transcription of N mRNA occurred through prior transcription of Le+.[9] However, studies with the polR1 mutant of VSV, which has a single amino acid mutation in the template-associated N protein, showed that the polR1 virus synthesized N mRNA in excess over Le+.[10] This finding was incompatible with a single initiation site and suggested distinct initiation sites for transcription and replication: at the extreme 3' terminus for synthesis of the Le+ and antigenome and at the first gene start site for mRNA transcription.

A reinvestigation of the site of polymerase initiation using UV transcriptional mapping of recombinant engineered viruses indicated that, *in the infected cell,* transcription and replication initiated at different sites on the genome.[11] This study examined the site of RdRp initiation using recombinant viruses engineered to allow UV target sizes as small as that of the leader RNA to be discriminated. These experiments showed that during primary transcription, the UV target size of the leader RNA region of the genome did *not* impact initiation of the first mRNA, prompting the *prima facie* interpretation that transcription initiated at the first gene start site in the infected cell, without prior synthesis of the leader RNA.[11]

Importantly, the reinvestigation of initiation sites compared the results of the UV mapping experiments in infected cells with those obtained *in vitro* using the same viruses and the same techniques.[11] Surprisingly, it was found that initiation of transcription in infected cells differed from that in a cell-free system. In infected cells, transcription initiated directly at the first gene start site (as described above), whereas *in vitro* the leader RNA was necessarily transcribed prior to initiation of the first mRNA. Both sets of experiments used UV-irradiated purified virions, either to infect cells or to direct cell-free transcription following detergent activation.[11] The contrasting results suggest either that nucleocapsids released by detergent activation differ somehow from those released into the cytoplasm during infection, perhaps by retaining their association with the viral matrix (M) protein, a known inhibitor of viral transcription, or that infected cells provide a factor absent from the cell-free system that allows transcription to initiate internally. It would be of interest to repeat the *in vitro* experiments using as templates both nucleocapsids purified from cells infected with the engineered viruses and nucleocapsids from which the M protein had been stripped.

3. What Controls Transcription and RNA Replication?

The finding that transcription and replication initiate at different sites *in vivo* suggests that the two RNA synthetic processes may be regulated through the use of separate initiation sites. The question now is what controls the selection of initiation site? Three viral proteins are required for replication of the VSV genome: N, P, and L. Studies of the role of these proteins in RNA synthesis have shown that all three proteins can influence the processes of transcription and RNA replication, and have revealed the potential complexity involved in the regulation of the two processes. Mutations in the core polymerase, L protein, and in the polymerase cofactor, P protein, have been identified that affect the ability of the RdRp to transcribe or replicate. Mutations in the N protein have also been shown to affect the balance between RNA replication and transcription. Thus, the polymerase components, the nucleocapsid template or a combination of these factors, could be involved in determining the decision to replicate or transcribe. The remainder of this commentary will give examples of mediators of these processes with focus on addressing gaps in our current knowledge.

3.1. *The RdRp complex*

The RdRp complex is one possibility for regulating the selection of different initiation sites. It is possible that two putative forms of the polymerase exist that can distinguish the two sites of initiation.[12] Mutants of the core polymerase, L protein, and the polymerase cofactor, P protein, have been found that affect replication and transcription differentially. In addition, as yet unidentified host factors may also be involved in modifying the polymerase complex.

3.1.1. *The core polymerase, L*

The large (L) proteins of the NNS RNA viruses are, as their name implies, large, multifunctional proteins that are responsible for catalyzing the RNA synthetic activities of transcription and replication, as well as the activities required for modification of the 5′ and 3′ ends of the viral mRNAs.[2,13–15] Extensive sequence and phylogenetic analyses have subdivided the NNS RNA virus L proteins into six domains of conservation linked by variable regions.[16–18] There have been numerous investigations of the potential functions of the various domains using both forward and reverse genetic approaches. These studies have provided valuable insight into the role of the L protein in regulating transcription and replication. A few of the key examples are discussed below.

3.1.2. Polymerase mutants that differentially affect transcription and RNA replication

Temperature-sensitive (ts) mutations in the L protein have been observed that have differential effects on transcription and replication.[19,20] One mutant virus, in particular, ts(G)114 has the phenotype of inhibition of transcription following shift to non-permissive temperature while RNA replication is unaffected or increased.[20–22] Sequence analysis of the L gene of tsG114 identified three amino acid changes that were shown to be associated with its ts and RNA defective phenotypes: D575G, E1117G, and I1937T.[22] By inserting these mutations individually or in concert into the L gene of a VSV cDNA and recovering the virus, a specific amino acid substitution in domain II (D575G) of the VSV L protein was identified that significantly affected total RNA synthesis, but when in combination with the two additional amino acid substitutions identified in the ts(G)114 L protein, leads to a specific reduction in mRNA transcription, but not replication.[22]

Additionally, studies to examine the role of residues involved in mediating 5' mRNA modifications revealed that, in general, mutation of specific residues involved in these processes negatively affected viral mRNA transcription, whereas the levels of RNA replication were unaffected or increased in some cases.[23–25] These and numerous other analyses of ts mutants have shown that the polymerase has domains that are involved in determining whether the enzyme efficiently catalyzes replication or transcription. There have been substantial recent advances in our understanding of the domains involved in transcription and modification of mRNA. However, the functions of all six domains of the L protein as well as the coordination of the various domains during transcription versus replication have yet to be determined. Determination of the structure and function of all domains of the L protein will be the key to understanding how this large, multifunctional polymerase catalyzes two such distinct reactions (Table 1) as replication and transcription.

3.1.3. Two complexes of polymerase

Using immunoaffinity chromatography to mammalian translation elongation factor-1α (EF-1α), two complexes with the L protein were isolated from VSV infected cells.[12] One complex, which eluted with EF-1α contained, in addition to EF-1α, VSV L and P proteins, heat shock protein 60, and a sub-molar amount of cellular guanylyltransferase (GT). The other complex, found in the flow-through of the column, contained the VSV proteins N, P, and L as identified by western blotting with antibodies specific for those proteins. When an N-RNA template was added to the fraction eluting with EF-1α, capped mRNAs but no leader were synthesized, indicating it was a preparation of transcriptase. When an N-RNA template was added to the flow-through fraction *in vitro*, polydisperse RNA larger than mRNAs, but smaller than genome, and leader RNA were synthesized. This fraction was proposed as a replicase, and by further anti-P immunoaffinity chromatography it was reported to contain a complex of L–(N–P) and an N–P complex.[12]

The above work fractionating infected cell extracts using antibody to a host protein, separated two putative polymerase complexes as described. The important question, as yet unanswered by these studies, is what controls the interaction of the polymerase with these different host and viral factors? If the core polymerase is the same, there must be additional factors that regulate whether the polymerase binds different host factors or the N protein. These might be differentially modified viral proteins, such as the phosphorylated P protein, as described below, or unidentified host factors.

A further key question is what other host proteins are associated with the different complexes? The proteins associated with the two polymerase complexes separated by EF-1α affinity chromatography were characterized by western blotting.[12] However, only four antibodies were tested. Because of this limited analysis, it is not known what other proteins might also be associated with the two forms of polymerase. In more recent work by Grdzelishvili and his colleagues, the proteins from highly purified virions isolated from various cell types were subjected to mass spectrometry and have revealed a plethora of host factors potentially involved in the VSV life cycle.[26] Some of these proteins had been previously identified for VSV, such as EF-1α, but others represent newly identified host factors whose contribution to the VSV life cycle and/or viral RNA synthesis is unknown.[26] While many proteins found in the virion are likely to be involved in assembly and exit from the cell, it is quite possible that host factors required for early biosynthetic processes, such as transcription, are co-packaged in the virion. Nevertheless, more studies of this nature as well as follow-up with functional studies will be crucial to understanding and appreciating the complexity of the interactions between virus and host, and given the large number of cofactors and interacting proteins characterized for RNA polymerases, it is probable that numerous other host factors interact with the polymerase and have a role in regulating transcription and RNA replication.

3.1.4. *The polymerase cofactor: Phosphoprotein, P*

The primary structure of the P protein has been divided functionally into four domains: domains, I, II, III, and a hinge region.[27] The P protein of VSV Indiana is phosphorylated within domains I and II. Mutational analysis of the phosphorylated residues Ser 60, Thr 62, and Ser 64 in domain I showed that when all three residues were substituted with alanine, the ability of the RdRp to transcribe was inhibited, whereas genome replication was not affected.[28,29] However, alteration of these residues to glutamic acid[28] or aspartic acid[29] did not affect transcription, suggesting a net negative charge as a result of phosphorylation was important for transcription. The effect of these P protein phosphorylation sites on transcription and replication was examined further in studies using recombinant viruses.[30] It was found that a virus having all three residues substituted with alanine was not recoverable, but a virus having Ser 60, Thr 62, and Ser 64 substituted with glutamic acid was readily recoverable and displayed similar growth kinetics and levels of mRNA synthesis as wild-type and viruses containing single alanine substitutions of each residue.[30] In addition to their role in transcription, it has also been shown that the phosphorylation of domain I residues by CKII is required for multimerization of P[29] and its interaction with L,[31] both of which are necessary functions for transcription and replication.

In contrast, phosphorylation of both Ser 226 and Ser 227 in domain II was shown to be important for replication, but did not affect transcription using a minigenome system.[32] The L protein has been shown to interact with serines 226 and 227[33] and this interaction may occur in a phosphorylation-independent manner.[31] Analysis of mutations of these residues in the background of recombinant virus showed that virus containing alanine substitution at both sites was not recoverable, but viruses containing single alanine substitutions were readily recoverable. However, unlike the virus having the S226A substitution in the P protein, for which wild-type levels of virus growth and RNA synthesis were observed, the virus having the S227A mutation displayed reduced virus growth kinetics, but increased viral mRNA synthesis.[30] Unlike the observation for residues 60, 62, and 64, substitution of serines 226 and 227 with glutamic acid did not appreciably rescue the defect in replication.[32] Importantly, substitution of targeted phosphorylation sites with phosphomimetic amino acids is static, whereas phosphorylation is usually a dynamic process, which may be a critical point when considering the formation of replication complexes. The temporal nature of transcription and RNA replication during VSV infection may be highly sensitive to the formation of polymerase complexes that favor one synthetic process over another. Early in infection when the primary biosynthetic event is transcription, due to the requirement for synthesis of encapsidation-competent N protein, having complexes that prefer to replicate may negatively impact the balance between these two processes or may affect another function of P, thus providing a rationale for why replication-dependent P protein phosphorylation has evolved to require more refined regulation than merely being dependent on the presence of a net negative charge in this region.

After the structure of the VSV N protein in contact with RNA was determined and showed the RNA is sequestered between the N- and C-terminal lobes of the

N protein,[34] the structure of the C-terminal domain of the P protein bound to the N-RNA complex was solved.[35] This latter structure showed that neither Ser 226 nor Ser 227 makes direct contact with the N protein, but rather each sits exposed directly above the entrance to the RNA cavity aligned with the interior face of the C-lobe of the N protein.[35] The authors propose that the P_{CTD} could bring L in close contact with the RNA because the two serine residues at 226 and 227 might represent a boundary of contact between L and N proteins.

The available data indicate that phosphorylation of the P protein can affect the ability of the polymerase to replicate or transcribe. Further, the sites of phosphorylation in domain II are located at a key position with respect to the interacting N and L proteins, which may have significance for the regulation of replication. Despite the new insights revealed by these studies, they highlight the importance of understanding the critical question of how and what controls the dynamic phosphorylation of the P protein?

3.2. The nucleocapsid protein affects RNA synthetic activities

The structure of the VSV N protein bound to RNA shows that the RNA is tightly sequestered between two lobes of the N protein[34] (For structure description of VSV proteins, see chapter by Luo *et al.* in this book). Based on this structure, it is unclear how the polymerase can access the genomic RNA encapsidated by the N protein. Transient domain movement to open the lobes of N has been proposed to allow the polymerase access to the RNA.[34] This places the N protein in a key position to influence the use of the template for transcription or RNA replication since it is the guardian of access to the genome. Consistent with this, as mentioned above, the polR1 mutation in the N protein template was one of the first alterations in the N protein shown to affect the activity of the polymerase.[10]

Following the determination of the structure of the N protein in complex with RNA, structure-based investigations of key regions of N protein were carried out and identified several mutations in the N protein that differentially affected transcription and RNA replication. Mutations in the C-terminal loop of the N protein (Q346A, Q347A, and C349A) that are predicted to break contacts within the C-terminal loop, thus providing greater flexibility, were shown to cause up to 10 fold increase in RNA replication without an equivalent effect on transcription.[36] Conversely, mutation of a different residue within the C-terminal loop, F348A, predicted to break contacts with the N-terminal arm and the underlying C-terminal lobe of the N protein, allowed for wild-type levels of replication but almost completely abrogated transcription. Interestingly, genomic ribonucleoprotein complexes (RNPs) formed by the F348A N protein were nuclease-sensitive, whereas the antigenomic RNPs were nuclease-resistant, thus providing a rationale for the defect in transcription.[36] Perhaps more intriguing is why a nuclease-sensitive genomic RNP would affect transcription, but not anti-genomic RNA replication? An alternative explanation might be that the F348A N protein interacts differentially with the various phosphorylated forms of the P protein, thus the F348A N protein might have a defect in association with the transcriptionally active P protein. While the interaction of the F348A N protein with the P protein was examined by co-immunoprecipitation and found to be only slightly reduced compared with wild-type,[36] the phosphorylated form of P that was pulled down was not examined. Similar to the phenotype of the F348A N protein mutant, several mutations in a highly conserved central region of the N protein (282-GLSSKSPYSS-291) decreased transcription without affecting replication; however, neither the nuclease sensitivities of the RNPs formed by these mutant N proteins nor their interaction with the various forms of the P protein have been examined.[37]

3.3. cis-Acting sequences in the genomic termini can affect RNA replication and transcription

The genomic termini of VSV have been shown to regulate RNA synthesis. The 3′ end of the genomic RNA (i.e., the leader region) acts as a promoter for the two distinct RNA synthetic processes, transcription and replication, whereas the 3′ end of the antigenomic RNA (i.e., the trailer complement, TrC, region) acts exclusively as a promoter for replication of genomic RNA. Numerous studies examining the role of the termini in the two RNA synthetic processes have shown that the genomic termini are multifunctional, signaling encapsidation, assembly into virus particles, transcription, and replication.[38–44] Studies examining the role of nts in the leader region have shown that nts 19–29 and 34–46 of the leader region are essential to the initiation of transcription.[42] In a separate study in which deletions of the leader region were examined for their effects on RNA synthesis, it was shown that deletion of nts 1–6, 7–12, or 13–18 resulted in completely defective templates that did not support any (+) sense RNA synthesis.[38] Additionally, this study confirmed the aforementioned study in showing that deletion of clusters of 6 nts between 19 and 47 did not appreciably affect RNA replication, but significantly decreased transcription.[38]

The initial study examining the VSV genomic termini focused on the feature of terminal complementarity.[41] It was found for VSV that the extent of complementarity at the extreme termini of the genomic RNA template has a profound effect on the balance of transcription and replication.[41] Studies using VSV sub-genomic replicons engineered to increase the wild-type VSV Indiana 8 nts of complementarity to 18, 22, 28, or 51 nts have shown that increasing terminal complementarity results in progressively increasing levels of RNA replication.[41] Concomitant with this observed increase in RNA replication was a decrease in the levels of transcription. It is not known what advantages increased terminal complementarity would provide the virus or through what mechanism it acts; it may be associated with efficiency of initiation of encapsidation or polymerase binding. Nevertheless, the observation merits further investigation. Highly robust RNA replication may not provide an evolutionary advantage to a virus population, as defective interfering particles may be generated or the error rate in the population may become too high to sustain a viable population of virus. Thus, modulating a delicate balance between these two biosynthetic processes of RNA replication and transcription is likely a key component of the virus life cycle.

4. Summary

All of the viral proteins involved in transcription and RNA replication have been shown to have potential to affect these biosynthetic processes differentially. The available evidence indicates that there are different entry sites for transcription and replication,[11] but it is not known what controls the choice of initiation site. There are different polymerase complexes[12] that may be involved in initiation site selection, and the phosphorylation states of the P protein are known to affect the ability of the polymerase to replicate or transcribe.[28–32] Further, there are regions of the L and N proteins that, when altered, affect transcription and RNA replication differentially, suggesting that different regions/domains of these proteins may have relevance for regulating each biosynthetic event, either through their interactions with another viral component or an unidentified host factor.[20–25] How each of these critical factors is controlled and, importantly, how they affect the interaction of the polymerase complex with the template are not known.

Major advances in understanding how the VSV N protein encapsidated template might function were made with the determination of the structure of the nucleo-capsid protein in complex with RNA and by the subsequent determination of the structure on the N:RNA complex in contact with the C-terminal domain of the P protein.[34,35] These studies have provided insight into how the P protein associates with the N:RNA template and allowed proposals of how the N:RNA template and the P protein may be poised to interact with the polymerase. Work is underway on the structure of the L protein and a first look at the L protein by negative stain electron microscopy has been produced by the Whelan laboratory. It is likely that studies of a P:L complex will be necessary to understand the complex interaction possibilities for access of the polymerase to the N-bound RNA template and further, how the polymerase complex(es) influences the two distinct RNA synthetic processes.

Acknowledgments

The authors thank L. Andrew Ball and Deena Jacob for their constructive comments. Research in the Wertz laboratory was supported by NIH grant AI12464 from the NIAID.

References

1. Emerson, S.U. and R.R. Wagner, Dissociation and reconstitution of the transcriptase and template activities of vesicular stomatitis B and T virions. *J Virol*, 1972. **10**(2): 297–309.
2. Emerson, S.U. and Y. Yu, Both NS and L proteins are required for *in vitro* RNA synthesis by vesicular stomatitis virus. *J Virol*, 1975. **15**(6): 1348–56.
3. Davis, N.L. and G.W. Wertz, Synthesis of vesicular stomatitis virus negative-strand RNA *in vitro*: Dependence on viral protein synthesis. *J Virol*, 1982. **41**(3): 821–32.
4. Patton, J.T., N.L. Davis and G.W. Wertz, N protein alone satisfies the requirement for protein synthesis during RNA replication of vesicular stomatitis virus. *J Virol*, 1984. **49**(2): 303–9.
5. Leppert, M., *et al.*, Plus and minus strand leader RNAs in negative strand virus-infected cells. *Cell*, 1979. **18**(3): 735–47.
6. Blumberg, B.M., C. Giorgi and D. Kolakofsky, N protein of vesicular stomatitis virus selectively encapsidates leader RNA *in vitro*. *Cell*, 1983. **32**(2): 559–67.
7. Blumberg, B.M., M. Leppert and D. Kolakofsky, Interaction of VSV leader RNA and nucleocapsid protein may control VSV genome replication. *Cell*, 1981. **23**(3): 837–45.
8. Ball, L.A. and C.N. White, Order of transcription of genes of vesicular stomatitis virus. *Proc Natl Acad Sci U S A*, 1976. **73**(2): 442–6.
9. Emerson, S.U., Reconstitution studies detect a single polymerase entry site on the vesicular stomatitis virus genome. *Cell*, 1982. **31**(3 Pt 2): 635–42.
10. Chuang, J.L. and J. Perrault, Initiation of vesicular stomatitis virus mutant polR1 transcription internally at the N gene *in vitro*. *J Virol*, 1997. **71**(2): 1466–75.
11. Whelan, S.P. and G.W. Wertz, Transcription and replication initiate at separate sites on the vesicular stomatitis virus genome. *Proc Natl Acad Sci U S A*, 2002. **99**(14): 9178–83.
12. Qanungo, K.R., *et al.*, Two RNA polymerase complexes from vesicular stomatitis virus-infected cells that carry out transcription and replication of genome RNA. *Proc Natl Acad Sci U S A*, 2004. **101**(16): 5952–7.
13. Gupta, A.K., M. Mathur and A.K. Banerjee, Unique capping activity of the recombinant RNA polymerase (L) of vesicular stomatitis virus: Association of cellular capping enzyme with the L protein. *Biochem Biophys Res Commun*, 2002. **293**(1): 264–8.
14. Rhodes, D.P., S.A. Moyer and A.K. Banerjee, *In vitro* synthesis of methylated messenger RNA by the virion-associated RNA polymerase of vesicular stomatitis virus. *Cell*, 1974. **3**: 327–33.
15. Testa, D. and A.K. Banerjee, Two methyltransferase activities in the purified virions of vesicular stomatitis virus. *J Virol*, 1977. **24**(3): 786–93.

16. Poch, O., *et al.*, Sequence comparison of five polymerases (L proteins) of unsegmented negative-strand RNA viruses: Theoretical assignment of functional domains. *J Gen Virol*, 1990. **71**(Pt 5): 1153–62.

17. Poch, O., *et al.*, Identification of four conserved motifs among the RNA-dependent polymerase encoding elements. *Embo J*, 1989. **8**(12): 3867–74.

18. Svenda, M., *et al.*, Analysis of the large (L) protein gene of the porcine rubulavirus LPMV: Identification of possible functional domains. *Virus Res*, 1997. **48**(1): 57–70.

19. Pringle, C.R., Genetic characteristics of conditional lethal mutants of vesicular stomatitis virus induced by 5-fluorouracil, 5-azacytidine, and ethyl methane sulfonate. *J Virol*, 1970. **5**(5): 559–67.

20. Perlman, S.M. and A.S. Huang, RNA synthesis of vesicular stomatitis virus V. Interactions between transcription and replication. *J Virol*, 1973. **12**(6): 1395–400.

21. Wertz, G.W., Isolation of possible replicative intermediate structures from vesicular stomatitis virus-infected cells. *Virology*, 1978. **85**(1): 271–85.

22. Galloway, S.E. and G.W. Wertz, A temperature sensitive VSV identifies L protein residues that affect transcription but not replication. *Virology*, 2009. **388**(2): 286–93.

23. Galloway, S.E., P.E. Richardson and G.W. Wertz, Analysis of a structural homology model of the 2′-O-ribose methyltransferase domain within the vesicular stomatitis virus L protein. *Virology*, 2008. **382**(1): 69–82.

24. Li, J., J.S. Chorba and S.P. Whelan, Vesicular stomatitis viruses resistant to the methylase inhibitor sinefungin upregulate RNA synthesis and reveal mutations that affect mRNA cap methylation. *J Virol*, 2007. **81**(8): 4104–15.

25. Li, J., *et al.*, A conserved motif in region v of the large polymerase proteins of nonsegmented negative-sense RNA viruses that is essential for mRNA capping. *J Virol*, 2008. **82**(2): 775–84.

26. Moerdyk-Schauwecker, M., S.I. Hwang and V.Z. Grdzelishvili, Analysis of virion associated host proteins in vesicular stomatitis virus using a proteomics approach. *Virol J*, 2009. **6**: 166.

27. Paul, P.R., D. Chattopadhyay and A.K. Banerjee, The functional domains of the phosphoprotein (NS) of vesicular stomatitis virus (Indiana serotype). *Virology*, 1988. **166**(2): 350–7.

28. Pattnaik, A.K., *et al.*, Phosphorylation within the amino-terminal acidic domain I of the phosphoprotein of vesicular stomatitis virus is required for transcription but not for replication. *J Virol*, 1997. **71**(11): 8167–75.

29. Gao, Y. and J. Lenard, Multimerization and transcriptional activation of the phosphoprotein (P) of vesicular stomatitis virus by casein kinase-II. *Embo J*, 1995. **14**(6): 1240–7.

30. Das, S.C. and A.K. Pattnaik, Phosphorylation of vesicular stomatitis virus phosphoprotein P is indispensable for virus growth. *J Virol*, 2004. **78**(12): 6420–30.

31. Gao, Y. and J. Lenard, Cooperative binding of multimeric phosphoprotein (P) of vesicular stomatitis virus to polymerase (L) and template: Pathways of assembly. *J Virol*, 1995. **69**(12): 7718–23.

32. Hwang, L.N., *et al.*, Optimal replication activity of vesicular stomatitis virus RNA polymerase requires phosphorylation of a residue(s) at carboxy-terminal domain II of its accessory subunit, phosphoprotein P. *J Virol*, 1999. **73**(7): 5613–20.

33. Chen, J.L., T. Das and A.K. Banerjee, Phosphorylated states of vesicular stomatitis virus P protein *in vitro* and *in vivo*. *Virology*, 1997. **228**(2): 200–12.

34. Green, T.J., *et al.*, Structure of the vesicular stomatitis virus nucleoprotein-RNA complex. *Science*, 2006. **313**(5785): 357–60.

35. Green, T.J. and M. Luo, Structure of the vesicular stomatitis virus nucleocapsid in complex with the nucleocapsid-binding domain of the small polymerase cofactor, P. *Proc Natl Acad Sci U S A*, 2009. **106**(28): 11713–8.

36. Harouaka, D. and G.W. Wertz, Mutations in the C-terminal loop of the nucleocapsid protein affect vesicular stomatitis virus RNA replication and transcription differentially. *J Virol*, 2009. **83**(22): 11429–39.

37. Nayak, D., *et al.*, Single-amino-acid alterations in a highly conserved central region of vesicular stomatitis virus N protein differentially affect the viral nucleocapsid template functions. *J Virol*, 2009. **83**(11): 5525–34.

38. Li, T. and A.K. Pattnaik, Overlapping signals for transcription and replication at the 3′ terminus of the vesicular stomatitis virus genome. *J Virol*, 1999. **73**(1): 444–52.

39. Pattnaik, A.K., *et al.*, The termini of VSV DI particle RNAs are sufficient to signal RNA encapsidation, replication, and budding to generate infectious particles. *Virology*, 1995. **206**(1): 760–4.

40. Pattnaik, A.K., *et al.*, Infectious defective interfering particles of VSV from transcripts of a cDNA clone. *Cell*, 1992. **69**(6): 1011–20.

41. Wertz, G.W., *et al.*, Extent of terminal complementarity modulates the balance between transcription and replication of vesicular stomatitis virus RNA. *Proc Natl Acad Sci U S A*, 1994. **91**(18): 8587–91.

42. Whelan, S.P. and G.W. Wertz, Regulation of RNA synthesis by the genomic termini of vesicular stomatitis virus: Identification of distinct sequences essential for transcription but not replication. *J Virol*, 1999. **73**(1): 297–306.

43. Whelan, S.P. and G.W. Wertz, The 5′ terminal trailer region of vesicular stomatitis virus contains a position-dependent *cis*-acting signal for assembly of RNA into infectious particles. *J Virol*, 1999. **73**(1): 307–15.

44. Li, T. and A.K. Pattnaik, Replication signals in the genome of vesicular stomatitis virus and its defective interfering particles: Identification of a sequence element that enhances DI RNA replication. *Virology*, 1997. **232**(2): 248–59.

45. Ball, L.A., Transcriptional mapping of vesicular stomatitis virus *in vivo*. *J Virol*, 1977. **21**(1): 411–4.

46. Abraham, G. and A.K. Banerjee, Sequential transcription of the genes of vesicular stomatitis virus. *Proc Natl Acad Sci U S A*, 1976. **73**(5): 1504–8.

47. Villarreal, L.P., M. Breindl and J.J. Holland, Determination of molar ratios of vesicular stomatitis virus induced RNA species in BHK21 cells. *Biochemistry*, 1976. **15**(8): 1663–7.

48. Green, T.J., *et al.*, Study of the assembly of vesicular stomatitis virus N protein: Role of the P protein. *J Virol*, 2000. **74**(20): 9515–24.

49. Davis, N.L., H. Arnheiter and G.W. Wertz, Vesicular stomatitis virus N and NS proteins form multiple complexes. *J Virol*, 1986. **59**(3): 751–4.

50. Wang, J.T., L.E. McElvain and S.P. Whelan, Vesicular stomatitis virus mRNA capping machinery requires specific *cis*-acting signals in the RNA. *J Virol*, 2007. **81**(20): 11499–506.

51. Barr, J.N., S.P. Whelan and G.W. Wertz, *cis*-Acting signals involved in termination of vesicular stomatitis virus mRNA synthesis include the conserved AUAC and the U7 signal for polyadenylation. *J Virol*, 1997. **71**(11): 8718–25.

52. Hwang, L.N., N. Englund and A.K. Pattnaik, Polyadenylation of vesicular stomatitis virus mRNA dictates efficient transcription termination at the intercistronic gene junctions. *J Virol*, 1998. **72**: 477–84.

53. Barr, J.N. and G.W. Wertz, Polymerase slippage at vesicular stomatitis virus gene junctions to generate poly(A) is regulated by the upstream 3′-AUAC-5′ tetranucleotide: Implications for the mechanism of transcription termination. *J Virol*, 2001. **75**(15): 6901–13.

54. Hinzman, E.E., J.N. Barr and G.W. Wertz, Identification of an upstream sequence element required for vesicular stomatitis virus mRNA transcription. *J Virol*, 2002. **76**(15): 7632–41.

55. Barr, J.N., *et al.*, The VSV polymerase can initiate at mRNA start sites located either up or downstream of a transcription termination signal but size of the intervening intergenic region affects efficiency of initiation. *Virology*, 2008. **374**(2): 361–70.

Chapter 5

mRNA Capping by Vesicular Stomatitis Virus and Other Related Viruses

Tomoaki Ogino* and Amiya K. Banerjee*

1. Introduction

Vesicular stomatitis virus (VSV) is a member of the *Vesiculovirus* genus belonging to the *Rhabdoviridae* family in the *Mononegavirales* order, which includes important human and animal pathogens, such as rabies (*Rhabdoviridae*), measles (*Paramyxoviridae*), mumps (*Paramyxoviridae*), Ebola (*Filoviridae*), and Borna disease (*Bornaviridae*) viruses. VSV serves as the paradigm of the non-segmented negative strand (NNS) RNA viruses especially in studies of virus gene expression. VSV contains a single negative strand RNA genome of approximately 11 kilobases, which consists of five genes encoding nucleocapsid (N), phospho- (P), matrix (M), glyco- (G), and large (L) proteins (reviewed in Ref. 1). The N proteins enwrap the genomic RNA to generate a helical nucleocapsid (designated as N-RNA), which serves as a template for transcription as well as replication. A viral RNA-dependent RNA polymerase (RdRp) is composed of the catalytic L protein and its cognate P protein and associates the N-RNA template to form a ribonucleoprotein (RNP). The RNP complex transcribes the genomic RNA both *in vivo* and *in vitro* into five monocistronic mRNAs that contain a methylated cap structure [in the presence of *S*-adenosyl-L-methionine (AdoMet) *in vitro*] and poly(A) tail at the 5′- and 3′-ends, respectively. The observed mRNA modifications are thought to be co-tanscriptionally carried out by the multifunctional L protein. However, unlike eukaryotic and other viral systems, it had been difficult to demonstrate any mRNA processing activities associated with VSV by using exogenous RNAs as substrates, suggesting that mRNA processing events are tightly coupled to mRNA

*Department of Molecular Genetics, Section of Virology, Lerner Research Institute, Cleveland Clinic, Cleveland, OH 44195, USA.

synthesis.[2] Presently, the development of a new *in vitro* assay system allowed solving the mechanism of the mRNA capping catalyzed by the VSV L protein.[3–5] This chapter describes the unconventional mechanism of the cap formation of VSV mRNAs.

2.   Cap Structure in mRNA and Conventional mRNA Capping

The 5′-terminal cap structure [$m^7G(5′)ppp(5′)N$-], in which N7-methylguanosine (m^7G) is linked to the initial nucleoside of mRNA through the inverted 5′-5′ triphosphate bridge (Fig. 1), was initially discovered in dsRNA viral[6,7] and DNA viral[8] mRNAs in 1975. As suggested earlier,[9] the cap structure was subsequently identified in eukaryotic mRNAs.[10–14] Eukaryotic mRNAs have the general 5′-terminal structure shown as $m^7G(5′)ppp(5′)N(m)pN(m)$-, where the capping guanine base is universally methylated at the N7 position and the first and second nucleosides of mRNA are methylated at the 2′-O positions to varying degrees to form cap 0 (m^7GpppN-), cap 1 ($m^7GpppNm$-), and cap 2 ($m^7GpppNmpNm$-) structures (see Fig. 1) (reviewed in Refs. 2,15). When the first nucleoside of mRNA is adenosine, the cap structure is frequently methylated at the adenine-N6 position to generate $m^7G(5′)ppp(5′)m^6Am$- (m^6Am indicates N6,2′-O-dimethyladenosine).[16] Currently, 5′-terminal capping is known to occur as one of major RNA processing events during mRNA biogenesis of all known eukaryotes and many eukaryotic viruses, and is required for various aspects of mRNA metabolism including mRNA splicing, transport, translation, and stability.[2,15,17] As the discovery of the cap structure, a great deal of research has been gained not only to solve the molecular mechanisms of mRNA cap formation but also to identify eukaryotic and viral mRNA capping enzymes (CEs). For eukaryotes, nucleocytoplasmic large DNA viruses (e.g., vaccinia virus) and dsRNA viruses (e.g., reovirus), the cap structure is formed by a common mechanism involving three enzymes, RNA 5′-triphosphatase (RTPase), GTP:RNA guanylyltransferase (GTase), and mRNA cap (guanine-N7)-methyltransferase (GN7-MTase) (see Fig. 2A) (reviewed in Refs. 2,15, and 18). First, RTPase hydrolyzes a 5′-triphosphate end ($\overset{\gamma'\beta'\alpha'}{pppN}$-) of nascent mRNA into a diphosphate end ($\overset{\beta'\alpha'}{ppN}$-) and inorganic phosphate (P_i). Second, GTase transfers the GMP moiety of GTP ($\overset{\gamma\beta\alpha}{pppG}$) to $\overset{\beta'\alpha'}{ppN}$- through a covalent enzyme-GMP intermediate ($E\text{-}\overset{\alpha}{p}G$) to form the cap core structure ($G\overset{\alpha}{p}\text{-}\overset{\beta'\alpha'}{pp}N$-) with the release of inorganic pyrophosphate (PP_i). Eukaryotic and DNA viral GTases belong to a mononucleotidyltransferase superfamily together with ATP/NAD$^+$-dependent

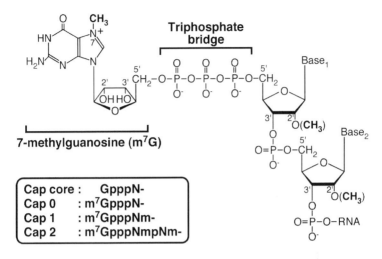

Fig. 1. Structure of the 5'-terminal cap in mRNA.

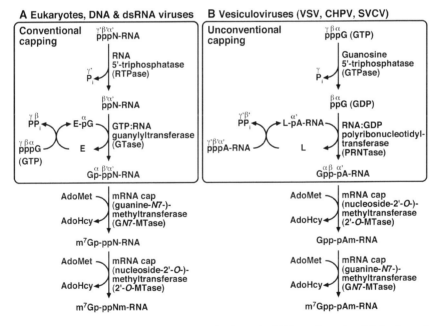

Fig. 2. Conventional and unconventional mechanisms of mRNA cap formation. Mechanisms of mRNA capping (upper reactions) and cap methylation (lower reactions) for eukaryotes, some DNA viruses, and dsRNA viruses (A) are compared with those for vesiculoviruses (B).

DNA ligases and ATP-dependent RNA ligases, and contain the conserved active site motif KxDG, in which the lysine (K) residue is responsible for the formation of the covalent enzyme-(lysyl-N^ε)-GMP intermediate with a phosphoamide bond (reviewed in Ref. 19). Third, cap GN7-MTase methylates the cap core structure at the guanine-N7 position by using AdoMet as a methyl donor to generate the cap 0 structure (m7GpppN). Furthermore, in higher eukaryotes and some viruses including nucleocytoplasmic large DNA viruses and dsRNA viruses, the cap 0 structure is methylated at the ribose-$2'$-O position in the initial nucleoside of mRNA by AdoMet-dependent mRNA cap (nucleoside-$2'$-O-)-methyltransferase ($2'$-O-MTase) to produce the cap 1 structure (m^{7}GpppNm-).

3.   Unique Capping and Cap Methylation Activities Associated with VSV

Shortly after the discovery of the cap structure, detergent-disrupted VSV was found to synthesize the cap core (GpppA-) and cap 1 (m^{7}GpppAm-) structures at the $5'$-termini of *in vitro* mRNAs in the absence and presence of AdoMet, respectively,[20,21] suggesting that all enzymes required for the cap 1 formation are packaged in the VSV virion. Interestingly, *in vitro* synthesized mRNA species (12–18 S) of VSV including N, P, M, and G mRNAs were shown to contain a common $5'$-terminal mRNA-start sequence, (m^7)GpppA(m)pApApCpApG-.[22] Furthermore, all VSV mRNAs synthesized in infected cells were demonstrated to be blocked with the cap structures methylated to different degrees as m^{7}Gppp(m^6)Amp(m^6)A(m)-.[23,24] N6-methylation of the first and second adenosine residues and $2'$-O-methylation of the second adenosine residue were suggested to be catalyzed by cellular enzymes, since these enzymatic activities could not be detected in the VSV virions.

The mechanism of the cap formation by the VSV-associated enzymes appeared to be strikingly different from that by the conventional CEs mentioned above. When *in vitro* transcription was performed in the presence of [β, γ-^{32}P]GTP and other three nucleotides with detergent-disrupted VSV, the [β-^{32}P]phosphate group was found to be incorporated into the $5'$-terminal cap core structure (GpppA) of mRNA,[20] as follows (asterisks indicate ^{32}P-labeled phosphates):

$$\overset{**}{\text{pp}}\,\text{pG} + \text{pppA} + \text{pppC} + \text{pppU} \rightarrow \text{Gp}\overset{*}{\text{pp}}\text{ApApCpApG} \cdots + \overset{*}{\text{P}}_i + \overset{(**)}{\text{PP}}_i(n).$$

Furthermore, by using various NTPs labeled with ^{32}P at different positions (e.g., [α-^{32}P]GTP, [β, γ-^{32}P]ATP, [α-^{32}P]ATP) as substrates for RNA synthesis, the cap core structure of *in vitro* VSV mRNAs was conclusively demonstrated to be composed of the GDP moiety of GTP ($\overset{\gamma\beta\alpha}{\text{pppG}}$) and the AMP moiety of ATP ($\overset{\gamma'\beta'\alpha'}{\text{pppA}}$) as $\overset{\alpha\beta\ \ \alpha'}{\text{Gpp-pA}}$-.[20] Initially, a novel VSV GTase was suggested to transfer the GDP moiety of GTP to a $5'$-monophosphate end of mRNA generated by internal cleavage of a precursor RNA,[25] as shown below:

(1) $\quad \cdots \text{pNpN}\overset{\alpha'}{\text{p}}\text{ApApCpApG}\cdots \rightarrow \cdots \text{pNpN}_{\text{OH}} + \overset{\alpha'}{\text{p}}\text{ApApCpApG}\cdots,$

(2) $\quad \overset{\gamma\beta\alpha}{\text{pppG}} + \overset{\alpha'}{\text{p}}\text{ApApCpApG}\cdots \rightarrow \text{G}\overset{\alpha\beta}{\text{pp}} - \overset{\alpha'}{\text{p}}\text{ApApCpApG}\cdots + \overset{\gamma}{\text{P}}_i$

However, unlike eukaryotic and other viral GTases, any protein in VSV would not form a covalent complex (e.g., enzyme-GDP intermediate) when incubated with [α-^{32}P]GTP. Thus, the precise mechanism by which the VSV RdRp carries out the unique cap formation remained elusive, although several other mechanisms were proposed.[26,27] It is important to note that spring viremia of carp virus (SVCV), a fish vesiculovirus related to VSV, was also shown to incorporate the GDP moiety of GTP into the cap core structure as $\overset{\alpha\beta\ \ \alpha'}{\text{Gpp-pA}}$- during *in vitro* transcription with purified virions or RNPs.[28] However, in contrast to VSV, SVCV was found to carry out an alternative capping mechanism, by which the GMP moiety of GMP-PNP (a non-hydrolyzable analog of GTP) is incorporated into the cap core structure as $\overset{\alpha}{\text{Gp}}$-$\overset{\beta'\alpha'}{\text{ppA}}$- during *in vitro* transcription in the presence of GMP-PNP instead of GTP.[28] Although there are some reports that detergent-disrupted virions and/or purified RNPs of paramyxoviruses (NNS RNA viruses belonging to the *Paramyxoviridae* family) such as Newcastle disease virus, human respiratory syncytial virus (HRSV), and Sendai virus (SeV) synthesize mRNAs containing methylated cap structures *in vitro*,[29–32] origins of phosphate groups forming the $5'$-$5'$ triphosphate linkages in the cap structures of paramyxoviral mRNAs remain unknown. For HRSV, the GDP and GMP moieties of GTP were shown to be incorporated into the cap 0 structure (m^{7}GpppG-) during transcription,[30] invoking two possible combinations of GDP and GMP to form the cap 0 structure such as m^7$\overset{\alpha\beta\ \ \alpha'}{\text{Gpp-p}}$G- and m^7$\overset{\alpha\ \ \beta'\alpha'}{\text{Gp-pp}}$G-.

Another interesting finding is that the VSV-associated MTase methylates the cap core structure only at the ribose-$2'$-O position in the presence of lower concentrations ($<0.1\,\mu$M) of AdoMet to produce a unique GpppAm cap structure on *in vitro* transcripts.[33] In the presence of higher concentrations ($>5\,\mu$M) of AdoMet, the fully methylated cap 1 structure (m^{7}GpppAm) is formed on *in vitro* VSV transcripts.[33] These observations suggested that two VSV MTases having different K_m values for AdoMet sequentially methylate the cap core structure at the $2'$-O position followed by the guanine-$N7$ position. This order of cap methylation is in direct contrast to that of higher eukaryotes[34–36] and other viruses including vaccinia virus (a DNA virus),[37–40] reovirus (a dsRNA virus),[41] paramyxoviruses such as HRSV and SeV,[30,32,42,43] and West Nile virus (a positive strand RNA virus)[44] (see Fig. 2).

4. Establishment of an *in vitro* Assay System to Detect VSV Cap-forming Activity

Recently, to explore the unconventional capping mechanism, an *in vitro* assay system was developed to measure the cap-forming activity associated with the VSV RNP containing the N, P, and L proteins.[3] As described above, all VSV mRNAs have the common mRNA-start sequence (AACGA-) and their 5′-termini are capped,[22] whereas the leader RNA (47 nucleotides) synthesized from the 3′-terminus of the genomic RNA starts with an uncapped sequence [(p)ppAACAG)] that is different from the mRNA-start sequence.[45,46] In addition, several studies have shown that 5′-triphosphorylated short RNAs starting with the AACAG sequence (11–42 nucleotide long of the 5′-end sequences of the N and P mRNAs) are abortively synthesized during *in vitro* VSV transcription.[26,47–49] These observations suggested that 5′-triphosphorylated nascent mRNAs, which are internally initiated at the beginning of each of the genes, specifically serve as substrates for the unique VSV CE. Based on these findings, 5′-triphosphorylated oligo-RNAs corresponding to the VSV mRNA-start and leader RNA-start sequences (pppAACAG and pppACGAA, respectively) were synthesized by T7 RNA polymerase using unique templates composed of oligo-DNAs.[3] By using these oligo-RNAs as substrates, it was demonstrated that the VSV RNP selectively caps the 5′-end of the mRNA-start sequence in the presence of GTP to produce GpppAACAG.[3] Subsequently, in a capping reaction with GTP labeled with ^{32}P at different positions ([α-^{32}P]GTP, [β-^{32}P]GTP, and [γ-^{32}P]GTP) as substrates, it was confirmed that the RNP indeed incorporates the GDP moiety of GTP into the 5′-terminal cap structure as $\overset{\alpha\ \beta\ \ \alpha'}{\text{Gpp-pA-}}$ formed on the exogenous oligo-RNA,[3] as normally formed in *in vitro* VSV transcripts.[20] The establishment of such an *in vitro* capping system led to the demonstration of the following critical findings[3]: (1) The VSV CE could cap pppRNA, but not ppRNA or pRNA, in contrast to the conventional CEs that specifically use ppRNA, generated from pppRNA, as a substrate.[2,15,18] The RNA substrate specificity for the VSV CE, thus, ruled out the previous capping models, in which pRNA or long RNA precursor was suggested to act as a substrate,[25,27] and also indicated that the VSV pre-mRNA initiated with pppAACAG is the genuine substrate for the VSV CE in the VSV mRNA-synthesizing machinery. (2) The VSV CE could use GDP, instead of GTP, as a substrate. This observation led to the discovery of a guanosine $5'$-triphosphatase (GTPase) activity associated with the VSV RNP, which hydrolyzes GTP to GDP. (3) The recombinant VSV L protein alone as well as the native RNP could catalyze the cap formation using pppAACAG and GTP (or GDP) as substrates, and importantly the recombinant VSV L protein by itself exhibited the GTPase activity.

The VSV L protein was found to efficiently form the cap structure on RNAs with the ARCNG (R = A/G) sequence, in which the first A and third pyrimidine residues are essential for the cap formation.[3,4] In an earlier study, Stillman and Whitt[50] showed that introduction of mutations into the first three nucleotides of the conserved gene-start sequence ($3'$-UUGUCNNUAG-) in a model VSV genome abolished $5'$-end mRNA processing and production of full-length mRNA and suggested that the $5'$-ARC sequence of mRNA generated by transcription of the $3'$-UYG (Y = U/C) gene-start sequence is required for capping (and/or cap methylation) of the nascent transcripts and subsequent processive RNA chain elongation. Recently, a similar mutagenesis study suggested the importance of the $3'$-UYG gene-start sequence in mRNA capping.[51] It is interesting to note that the SeV L protein specifically methylates the guanine-$N7$ position of the cap structure on oligo-RNAs with the SeV mRNA-start sequence ($5'$-AGG-).[43] Taken together, these findings suggest that NNS RNA viral CEs and cap MTases specifically recognize $5'$-end sequences of their mRNAs.

It was subsequently confirmed that the VSV L protein efficiently forms an SDS-resistant protein-RNA complex when incubated with pppAACAG, but not with ppAACAG or pppACGAA.[3] In contrast to mononucleotidyltransferases, such as GTases of the conventional CEs and DNA/RNA ligases, the VSV L protein could not form any covalent protein-NMP complex when incubated with GTP or ATP.[3] Further analyses of the L protein-RNA complex (referred to as L-pRNA) revealed that the $5'$-end phosphate of the RNA appeared to be covalently linked to the L protein through a phosphoamide bond, possibly to a histidine residue.[3] Interestingly, the L-pRNA complex formation was found to be specifically decreased in the presence of GDP.[3] Based on these findings and coupled with the analogy to the enzyme-GMP complex formation of GTases,[52–54] it was speculated that the L-pRNA complex is a covalent intermediate in an unconventional capping reaction (RNA transfer to GDP) catalyzed by a new enzyme, designated as RNA:GDP polyribonucleotidyltransferase (PRNTase).[3] Recently, the L-pRNA complex was purified and directly shown to transfer pRNA to GDP ($0.25\,\mu M$), but not to other NDPs, to generate the capped RNA (Gpp-pRNA).[5] Furthermore, the L-pRNA complex could transfer pRNA to PP$_i$ ($5\,\mu M$) to regenerate pp-pRNA,[5] indicating that the formation of the L-pRNA complex is reversible.

These findings indicate that the L-pRNA complex is a *bona fide* enzyme-pRNA intermediate in the RNA transfer reaction. Interestingly, dGDP was shown to function as an RNA acceptor, to a similar extent as GDP, to generate a unique $2'$-deoxyguanosine($5'$)triphospho($5'$)adenosine (dGpppA) cap structure.[5] It is not surprising that the VSV L protein produces the dGpppA cap, because an earlier study[55] indeed showed that detergent-disrupted VSV is able to synthesize dGpppA-capped mRNAs containing internal dGMP residues (DNA–RNA chimeras) in the presence of dGTP instead of GTP. Thus, these findings indicate that the $2'$-hydroxyl group of guanosine is not recognized by the putative PRNTase domain as well as the putative RdRp domain in the VSV L protein. The VSV L protein produces an unusual cap structure with a -4 net charge, guanosine($5'$)tetraphospho($5'$)adenosine (G$\overset{\alpha\beta\gamma}{ppp}$-$\overset{\alpha'}{p}$A-), as a minor product, which is formed by the transfer of pRNA to GTP as an RNA acceptor by the PRNTase activity prior to hydrolysis of GTP to GDP by the GTPase activity.[4] It should be noted that GTases of the conventional CEs can also synthesize tetraphosphate-containing cap-like structures as G$\overset{\alpha}{p}$-$\overset{\gamma'\beta'\alpha'}{ppp}$N- by a different mechanism, i.e., the guanylyl transfer to NTP or pppRNA.[56–59]

5.   Mechanism of Unconventional mRNA Capping by the VSV L Protein

Based on the findings described above, the unconventional mechanism of VSV mRNA capping was proposed as shown in Fig. 2B. At the first step, the GTPase activity of the L protein removes the γ-phosphate group of GTP ($\overset{\gamma\beta\alpha}{ppp}$G) to generate GDP ($\overset{\beta\alpha}{pp}$G), which is in turn used as an RNA acceptor. Subsequently, the putative PRNTase domain in the VSV L protein transfers the $\overset{\alpha'}{p}$RNA moiety of $\overset{\gamma'\beta'\alpha'}{ppp}$pRNA with the conserved VSV mRNA-start sequence (AACAG) to GDP to produce G$\overset{\alpha\beta}{pp}$-$\overset{\alpha'}{p}$RNA via a covalent enzyme–$\overset{\alpha'}{p}$RNA intermediate (L-$\overset{\alpha'}{p}$RNA complex). Therefore, the RNA capping reaction of VSV is directly opposite to the conventional capping reaction (Fig. 2A), in which GTase transfers the GMP moiety of GTP to $\overset{\beta'\alpha'}{pp}$RNA, generated from $\overset{\gamma'\beta'\alpha'}{ppp}$RNA by RTPase, to form G$\overset{\alpha}{p}$-$\overset{\beta'\alpha'}{pp}$RNA. After the formation of the cap core structure (GpppA) by VSV PRNTase, it is suggested to be sequentially methylated at the $2'$-O position of the first adenosine residue followed by the guanine-$N7$ position by the VSV cap MTases to produce the cap 1 structure (m^{7}GpppAm-).[33]

6. Active Site of VSV PRNTase

The VSV L protein (2,109 amino acids) carries five amino acid sequence blocks (I–VI, see Fig. 3), which are conserved in the L proteins of the NNS RNA viruses.[60] The N-terminal region including the block III was suggested to function as the RdRp domain,[60] and the putative divalent metal ion coordination motif (GDN) in the block III was demonstrated to be required for the transcriptase activity of the VSV L protein.[61] Furthermore, the C-terminal region including the block VI was identified as the cap MTase domain (for details see Chapter 7, this book).[43,62–65] In

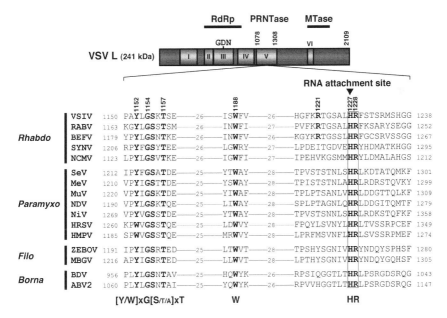

Fig. 3. Active site of the PRNTase domain in the VSV L protein. The VSV L protein is shown as a schematic at the top of the figure. The positions of the six conserved amino acid sequence blocks (I–VI) and the putative RdRp and cap MTase domains are indicated. The local sequence of the VSV L protein including the active site of the PRNTase domain is aligned with those of other NNS RNA viral L proteins. The numbers indicate the amino acid positions in the proteins. The H1227 residue of the VSV L protein was identified as the covalent RNA attachment in the PRNTase domain. The conserved motifs are shown at the bottom of the figure. Virus names and GenBank accession numbers are as follows: VSIV, vesicular stomatitis Indiana virus (K02378); RABV, rabies virus (M13215); BEFV, bovine ephemeral fever virus (AF234533); SYNV, sonchus yellow net virus (L32603); NCMV, northern cereal mosaic virus (AB030277); SeV, Sendai virus (X03614); MeV, measles virus (M20865); MuV, mumps virus (D10575); NDV, Newcastle disease virus (X05399); NiV, Nipah virus (AF212302); HRSV, human respiratory syncytial virus (M75730); HMPV, human metapneumovirus (AF371337); ZEBOV, Zaire ebolavirus (AF086833); MBGV, Marburg virus; BDV (Z29337), Borna disease virus (U04608); ABV2, avian bornavirus 2 (FJ620690).

order to localize the active site of the PRNTase domain in the VSV L protein, the covalent RNA attachment site was mapped by biochemical and mass spectrometric analyses.[5] As shown in Fig. 3, the histidine residue at position 1,227 (H1227) in the conserved histidine–arginine (HR) motif (H1227–R1228) of the VSV L protein was found to be covalently linked to the 5′-monophosphate end of the RNA via a phosphoamide bond.[5] Importantly, the HR motif is remarkably conserved in the L proteins of more than 80 NNS RNA viruses.[5] It is noteworthy that the L proteins of four known fish rhabdoviruses (e.g., infectious hematopoietic necrosis virus) belonging to the *Novirhabdovirus* genus in the family *Rhabdoviridae* contain a similar histidine–lysine sequence instead of the HR motif.[5] By mutational analyses of the VSV L protein, the HR motif and a basic amino acid residue (R1221) in the vicinity of the HR motif were shown to be essential for the PRNTase activity at the step of the covalent L-pRNA intermediate formation, but not for the GTPase activity.[5] For the PRNTase activity of the VSV L protein, the H1227 residue in the HR motif could not be replaced with other basic amino acids, while histidine could be substituted for the R1228 residue although to a lesser extent.[5] Thus, it seems that the H1227 residue in the VSV L protein plays a critical role in the L-pRNA intermediate formation, and the R1228 residue probably regulates this reaction. For the L protein of Chandipura virus (CHPV, a vesiculovirus closely related to VSV), the R1211, H1217, and R1218 residues (the counterparts of the R1221, H1227, and R1228 residues of the VSV L protein) were demonstrated to be crucial for the PRNTase activity at the step of the L-pRNA formation.[66] As the arginine residue (R1221 for VSV) located six residues upstream of the HR motif is conserved only in the L proteins of vesiculoviruses (e.g., VSV, CHPV), lyssaviruses (e.g., rabies virus), and ephemerovirus (bovine ephemeral fever virus) belonging to the *Rhabdoviridae* family, this residue was suggested to contribute to some step of the L-pRNA intermediate formation specific to above rhabdoviruses.[5,66] Li *et al.*[67] recently identified the G1154, T1157, H1227, and R1228 residues of the VSV L protein that are required for the formation of a cap-like structure sensitive to tobacco acid pyrophosphatase. Interestingly, the G1154 and T1157 residues are found within the [Y/W]xG[S/T/A]xT motif that is conserved in the L proteins of all known NNS RNA viruses (data not shown) and located ~75 residues upstream of the HR motif (see Fig. 3). The W1188 residue is another conserved residue located between the [Y/W]xG[S/T/A]xT and HR motifs. Currently, precise roles of these conserved amino acid residues, if any, in mRNA capping and the exact boundary of the putative PRNTase domain remain unknown.

As shown in Fig. 4, an electron lone pair on the ε2-nitrogen in the H1227 residue of the VSV L protein was suggested to nucleophilically attack the α-phosphorus in the 5′-triphosphate end of the RNA to form the enzyme-(histidyl-$N^{\varepsilon 2}$)-pRNA

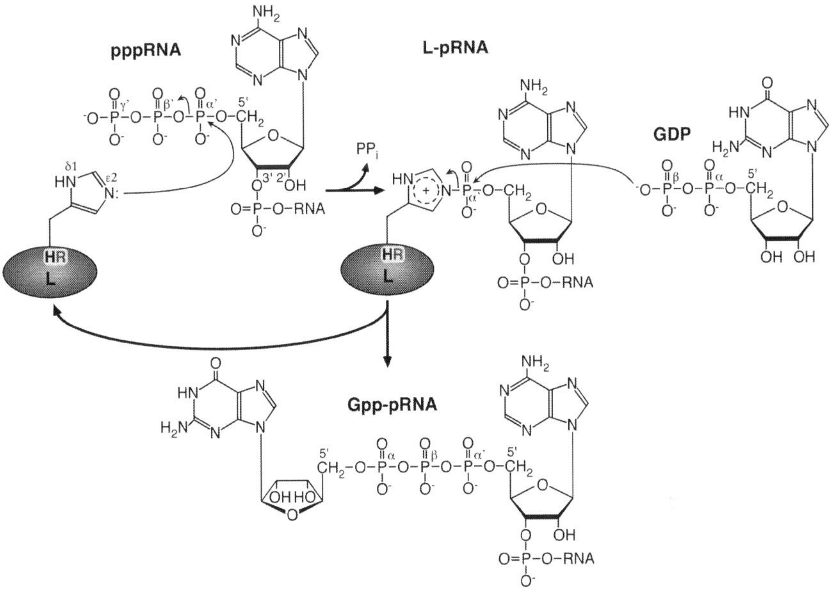

Fig. 4. Proposed chemical reactions for the unconventional mRNA capping by the VSV L protein.

intermediate (L-pRNA complex) with the concomitant release of PP$_i$.[5] To release the RNA linked to the L protein as the capped RNA, the β-phosphoryl group of GDP may nucleophilically attack the 5′-terminal α-phosphorus of the RNA in the L-pRNA intermediate (see Fig. 4).

7.   Concluding Remarks

Over the three decades, the mechanism of the VSV mRNA capping had remained elusive. The development of the new *in vitro* assay system to detect the mRNA cap-forming activity of the VSV L protein led to our understanding of how VSV mRNAs acquire the cap structure identical to that in eukaryotic mRNAs using a widely disparate mechanism. The striking conservation of the amino acid residues required for the PRNTase activity of the vesiculovirus L proteins among other NNS RNA viral L proteins strongly suggested that the PRNTase domains in the L proteins of vesiculoviruses and, by extension, other NNS RNA viruses have evolved from a common ancestor independently of the eukaryotic and other viral capping systems. Thus, PRNTase of the NNS RNA viruses unequivocally becomes a rational target for future development of specific antiviral agents. The precise mechanism by which the mRNAs are co-transcriptionally capped still remains poorly understood. It can be envisaged that, during mRNA synthesis, the 5′-pppAACAG end of nascent VSV mRNA, extruded from the RdRp domain of the VSV L protein, may react with the catalytic H1227 residue within the PRNTase domain to form the L-pRNA intermediate. GDP, generated from GTP, may subsequently react with the L-pRNA, resulting in capping of the elongating nascent mRNA. The minimum length of capped nascent mRNAs produced during *in vitro* transcription was reported to be 23 nucleotides,[68] thus, suggesting that RNA capping occurs at an early stage of mRNA chain elongation. After RNA capping, the GpppA cap core structure on the nascent mRNA elongating from the RdRp domain of the L protein may gain access to the carboxy-terminal MTase domain in the same polypeptide to produce the cap 1 structure. Detail investigation along this line would certainly provide deeper insight into the mechanism of co-transcriptional capping and methylation of NNS RNA virus mRNAs.

References

1. Lyles DS, Rupprecht CE. (2007) Rhabdoviridae. In: Knipe DM, Howley PM (eds), *Fields Virology*, pp. 1363–1408. Lippincott Williams & Wilkins, Philadelphia.
2. Banerjee AK. (1980) 5′-terminal cap strucyture in eucaryotic messenger ribonucleic acids. *Microbiol Rev* **44**: 175–205.
3. Ogino T, Banerjee AK. (2007) Unconventional mechanism of mRNA capping by the RNA-dependent RNA polymerase of vesicular stomatitis virus. *Mol Cell* **25**: 85–97.
4. Ogino T, Banerjee AK. (2008) Formation of guanosine(5′)tetraphospho(5′)adenosine cap structure by an unconventional mRNA capping enzyme of vesicular stomatitis virus. *J Virol* **82**: 7729–7734.
5. Ogino T, Yadav SP, Banerjee AK. (2010) Histidine-mediated RNA transfer to GDP for unique mRNA capping by vesicular stomatitis virus RNA polymerase. *Proc Natl Acad Sci U S A*, **107**: 3463–3468.
6. Furuichi Y, Miura K. (1975) A blocked structure at the 5′ terminus of mRNA from cytoplasmic polyhedrosis virus. *Nature* **253**: 374–375.
7. Furuichi Y, Morgan M, Muthukrishnan S, *et al.* (1975) Reovirus messenger RNA contains a methylated, blocked 5′-terminal structure: m-7G(5′)ppp(5′)G-MpCp. *Proc Natl Acad Sci U S A* **72**: 362–366.
8. Wei CM, Moss B. (1975) Methylated nucleotides block 5′-terminus of vaccinia virus messenger RNA. *Proc Natl Acad Sci U S A* **72**: 318–322.
9. Rottman F, Shatkin AJ, Perry RP. (1974) Sequences containing methylated nucleotides at the 5′ termini of messenger RNAs: Possible implications for processing. *Cell* **3**: 197–199.
10. Wei CM, Gershowitz A, Moss B. (1975) Methylated nucleotides block 5′ terminus of HeLa cell messenger RNA. *Cell* **4**: 379–386.

11. Perry RP, Kelley DE. (1975) Methylated constituents of heterogeneous nuclear RNA: Presence in blocked 5′ terminal structures. *Cell* **6**: 13–19.

12. Furuichi Y, Morgan M, Shatkin AJ, *et al.* (1975) Methylated, blocked 5′ termini in HeLa cell mRNA. *Proc Natl Acad Sci U S A* **72**: 1904–1908.

13. Adams JM, Cory S. (1975) Modified nucleosides and bizarre 5′-termini in mouse myeloma mRNA. *Nature* **255**: 28–33.

14. Desrosiers RC, Friderici KH, Rottman FM. (1975) Characterization of Novikoff hepatoma mRNA methylation and heterogeneity in the methylated 5′ terminus. *Biochemistry* **14**: 4367–4374.

15. Furuichi Y, Shatkin AJ. (2000) Viral and cellular mRNA capping: Past and prospects. *Adv Virus Res* **55**: 135–184.

16. Wei C, Gershowitz A, Moss B. (1975) N6, O2′-dimethyladenosine a novel methylated ribonucleoside next to the 5′ terminal of animal cell and virus mRNAs. *Nature* **257**: 251–253.

17. Cougot N, van Dijk E, Babajko S, *et al.* (2004) Cap-tabolism. *Trends Biochem Sci* **29**: 436–444.

18. Shuman S. (2001) Structure, mechanism, and evolution of the mRNA capping apparatus. *Prog Nucleic Acid Res Mol Biol* **66**: 1–40.

19. Shuman S, Lima CD. (2004) The polynucleotide ligase and RNA capping enzyme superfamily of covalent nucleotidyltransferases. *Curr Opin Struct Biol* **14**: 757–764.

20. Abraham G, Rhodes DP, Banerjee AK. (1975) Novel initiation of RNA synthesis *in vitro* by vesicular stomatitis virus. *Nature* **255**: 37–40.

21. Abraham G, Rhodes DP, Banerjee AK. (1975) The 5′ terminal structure of the methylated mRNA synthesized *in vitro* by vesicular stomatitis virus. *Cell* **5**: 51–58.

22. Rhodes DP, Banerjee AK. (1976) 5′-terminal sequence of vesicular stomatitis virus mRNA's synthesized *in vitro*. *J Virol* **17**: 33–42.

23. Moyer SA, Abraham G, Adler R, *et al.* (1975) Methylated and blocked 5′ termini in vesicular stomatitis virus *in vivo* mRNAs. *Cell* **5**: 59–67.

24. Moyer SA, Banerjee AK. (1976) *In vivo* methylation of vesicular stomatitis virus and its host-cell messenger RNA species. *Virology* **70**: 339–351.

25. Banerjee AK, Abraham G, Colonno RJ. (1977) Vesicular stomatitis virus: Mode of transcription. *J Gen Virol* **34**: 1–8.

26. Testa D, Chanda PK, Banerjee AK. (1980) Unique mode of transcription *in vitro* by vesicular stomatitis virus. *Cell* **21**: 267–275.

27. Shuman S. (1997) A proposed mechanism of mRNA synthesis and capping by vesicular stomatitis virus. *Virology* **227**: 1–6.

28. Gupta KC, Roy P. (1980) Alternate capping mechanisms for transcription of spring viremia of carp virus: Evidence for independent mRNA initiation. *J Virol* **33**: 292–303.

29. Colonno RJ, Stone HO. (1976) Newcastle disease virus mRNA lacks 2′-O-methylated nucleotides. *Nature* **261**: 611–614.

30. Barik S. (1993) The structure of the 5′ terminal cap of the respiratory syncytial virus mRNA. *J Gen Virol* **74**(Pt 3): 485–490.

31. Mizumoto K, Muroya K, Takagi T, *et al.* (1995) Protein factors required for *in vitro* transcription of Sendai virus genome. *J Biochem* **117**: 527–534.

32. Takagi T, Muroya K, Iwama M, *et al.* (1995) *In vitro* mRNA synthesis by Sendai virus: Isolation and characterization of the transcription initiation complex. *J Biochem* **118**: 390–396.

33. Testa D, Banerjee AK. (1977) Two methyltransferase activities in the purified virions of vesicular stomatitis virus. *J Virol* **24**: 786–793.

34. Wei C, Moss B. (1977) 5′-Terminal capping of RNA by guanylyltransferase from HeLa cell nuclei. *Proc Natl Acad Sci U S A* **74**: 3758–3761.

35. Mizumoto K, Lipmann F. (1979) Transmethylation and transguanylylation in 5′-RNA capping system isolated from rat liver nuclei. *Proc Natl Acad Sci U S A* **76**: 4961–4965.

36. Langberg SR, Moss B. (1981) Post-transcriptional modifications of mRNA. Purification and characterization of cap I and cap II RNA (nucleoside-2′-)-methyltransferases from HeLa cells. *J Biol Chem* **256**: 10054–10060.

37. Ensinger MJ, Martin SA, Paoletti E, *et al.* (1975) Modification of the 5′-terminus of mRNA by soluble guanylyl and methyl transferases from vaccinia virus. *Proc Natl Acad Sci U S A* **72**: 2525–2529.

38. Martin SA, Moss B. (1975) Modification of RNA by mRNA guanylyltransferase and mRNA (guanine-7-)methyltransferase from vaccinia virions. *J Biol Chem* **250**: 9330–9335.

39. Martin SA, Moss B. (1976) mRNA guanylyltransferase and mRNA (guanine-7-)-methyltransferase from vaccinia virions. Donor and acceptor substrate specificites. *J Biol Chem* **251**: 7313–7321.

40. Barbosa E, Moss B. (1978) mRNA(nucleoside-2′-)-methyltransferase from vaccinia virus. Characteristics and substrate specificity. *J Biol Chem* **253**: 7698–7702.

41. Furuichi Y, Muthukrishnan S, Tomasz J, *et al.* (1976) Mechanism of formation of reovirus mRNA 5′-terminal blocked and methylated sequence, m7GpppGmpC. *J Biol Chem* **251**: 5043–5053.

42. Liuzzi M, Mason SW, Cartier M, *et al.* (2005) Inhibitors of respiratory syncytial virus replication target cotranscriptional mRNA guanylylation by viral RNA-dependent RNA polymerase. *J Virol* **79**: 13105–13115.

43. Ogino T, Kobayashi M, Iwama M, *et al.* (2005) Sendai virus RNA-dependent RNA polymerase L protein catalyzes cap methylation of virus-specific mRNA. *J Biol Chem* **280**: 4429–4435.

44. Ray D, Shah A, Tilgner M, *et al.* (2006) West Nile virus 5′-cap structure is formed by sequential guanine N-7 and ribose 2′-O methylations by nonstructural protein 5. *J Virol* **80**: 8362–8370.

45. Colonno RJ, Banerjee AK. (1976) A unique RNA species involved in initiation of vesicular stomatitis virus RNA transcription *in vitro*. *Cell* **8**: 197–204.

46. Colonno RJ, Banerjee AK. (1978) Complete nucleotide sequence of the leader RNA synthesized *in vitro* by vesicular stomatitis virus. *Cell* **15**: 93–101.

47. Lazzarini RA, Chien I, Yang F, *et al.* (1982) The metabolic fate of independently initiated VSV mRNA transcripts. *J Gen Virol* **58**: 429–441.

48. Pinney DF, Emerson SU. (1982) Identification and characterization of a group of discrete initiated oligonucleotides transcribed *in vitro* from the 3′ terminus of the N-gene of vesicular stomatitis virus. *J Virol* **42**: 889–896.

49. Piwnica-Worms H, Keene JD. (1983) Sequential synthesis of small capped RNA transcripts *in vitro* by vesicular stomatitis virus. *Virology* **125**: 206–218.

50. Stillman EA, Whitt MA. (1999) Transcript initiation and 5′-end modifications are separable events during vesicular stomatitis virus transcription. *J Virol* **73**: 7199–7209.

51. Wang JT, McElvain LE, Whelan SP. (2007) Vesicular stomatitis virus mRNA capping machinery requires specific *cis*-acting signals in the RNA. *J Virol* **81**: 11499–11506.

52. Shuman S, Hurwitz J. (1981) Mechanism of mRNA capping by vaccinia virus guanylyltransferase: Characterization of an enzyme–guanylate intermediate. *Proc Natl Acad Sci U S A* **78**: 187–191.

53. Mizumoto K, Kaziro Y, Lipmann F. (1982) Reaction mechanism of mRNA guanylyltransferase from rat liver: Isolation and characterization of a guanylyl–enzyme intermediate. *Proc Natl Acad Sci U S A* **79**: 1693–1697.

54. Venkatesan S, Moss B. (1982) Eukaryotic mRNA capping enzyme–guanylate covalent intermediate. *Proc Natl Acad Sci U S A* **79**: 340–344.

55. Schubert M, Lazzarini RA. (1982) *In vitro* transcription of vesicular stomatitis virus. Incorporation of deoxyguanosine and deoxycytidine, and formation of deoxyguanosine caps. *J Biol Chem* **257**: 2968–2973.

56. Smith RE, Furuichi Y. (1982) A unique class of compound, guanosine-nucleoside tetraphosphate G(5′)pppp(5′)N, synthesized during the *in vitro* transcription of cytoplasmic polyhedrosis virus of Bombyx mori. Structural determination and mechanism of formation. *J Biol Chem* **257**: 485–494.

57. Wang D, Shatkin AJ. (1984) Synthesis of Gp4N and Gp3N compounds by guanylyltransferase purified from yeast. *Nucleic Acids Res* **12**: 2303–2315.

58. Cleveland DR, Zarbl H, Millward S. (1986) Reovirus guanylyltransferase is L2 gene product lambda 2. *J Virol* **60**: 307–311.

59. Yu L, Shuman S. (1996) Mutational analysis of the RNA triphosphatase component of vaccinia virus mRNA capping enzyme. *J Virol* **70**: 6162–6168.

60. Poch O, Blumberg BM, Bougueleret L, *et al.* (1990) Sequence comparison of five polymerases (L proteins) of unsegmented negative-strand RNA viruses: Theoretical assignment of functional domains. *J Gen Virol* **71**: 1153–1162.

61. Sleat DE, Banerjee AK. (1993) Transcriptional activity and mutational analysis of recombinant vesicular stomatitis virus RNA polymerase. *J Virol* **67**: 1334–1339.

62. Bujnicki JM, Rychlewski L. (2002) In silico identification, structure prediction and phylogenetic analysis of the 2′-O-ribose (cap 1) methyltransferase domain in the large structural protein of ssRNA negative-strand viruses. *Protein Eng* **15**: 101–108.

63. Ferron F, Longhi S, Henrissat B, *et al.* (2002) Viral RNA-polymerases — a predicted 2′-O-ribose methyltransferase domain shared by all Mononegavirales. *Trends Biochem Sci* **27**: 222–224.

64. Grdzelishvili VZ, Smallwood S, Tower D, *et al.* (2005) A single amino acid change in the L-polymerase protein of vesicular stomatitis virus completely abolishes viral mRNA cap methylation. *J Virol* **79**: 7327–7337.

65. Li J, Fontaine-Rodriguez EC, Whelan SP. (2005) Amino acid residues within conserved domain VI of the vesicular stomatitis virus large polymerase protein essential for mRNA cap methyltransferase activity. *J Virol* **79**: 13373–13384.

66. Ogino T, Banerjee AK. (2010) The HR motif in the RNA-dependent RNA polymerase L protein of Chandipura virus is required for unconventional mRNA capping activity *J Gen Virol*, **91**: 1311–1314.

67. Li J, Rahmeh A, Morelli M, *et al.* (2008) A conserved motif in region v of the large polymerase proteins of nonsegmented negative-sense RNA viruses that is essential for mRNA capping. *J Virol* **82**: 775–784.

68. Schubert M, Harmison GG, Sprague J, *et al.* (1982) *In vitro* transcription of vesicular stomatitis virus: Initiation with GTP at a specific site within the N cistron *J Virol* **43**: 166–173.

Chapter 6

Structural Disorder within the Measles Virus Nucleoprotein and Phosphoprotein: Functional Implications for Transcription and Replication

Sonia Longhi*

1. The Replicative Complex of Measles Virus

Measles virus (MeV) is an important human pathogen within the *Paramyxoviridae* family in the *Mononegavirales* order, which also includes the *Rhabdoviridae, Bornaviridae*, and *Filoviridae* families. MeV is further classified in the *Morbillivirus* genus within the *Paramyxovirinae* subfamily of the *Paramyxoviridae*. *Mononegavirales* members are characterized by a non-segmented, single-stranded RNA genome of negative polarity that is encapsidated by the nucleoprotein (N) to form a helical nucleocapsid. The viral RNA is tightly bound within the nucleocapsid and does not dissociate during RNA synthesis, as shown by resistance of the MeV genome to silencing by siRNA.[1] Hence, this ribonucleoprotein complex, rather than naked RNA, is the template for both transcription and replication. These latter activities are carried out by the RNA-dependent RNA polymerase that is composed of the large (L) protein and the phosphoprotein (P). The P protein is an essential polymerase co-factor in that it tethers the L protein onto the nucleocapsid template. This ribonucleoprotein complex made of RNA, N, P, and L constitutes the basic elements of the viral transcriptase and replicase (i.e., the viral replicative unit) (Fig. 1A).

---
*Architecture et Fonction des Macromolécules Biologiques, UMR 6098 CNRS et Universités d'Aix-Marseille I et II, 163, Avenue de Luminy, Case 932, 13288 Marseille Cedex 09, France. E-mail: Sonia.Longhi@afmb.univ-mrs.fr

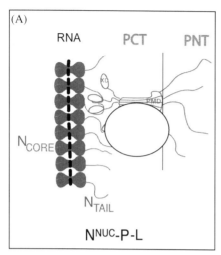
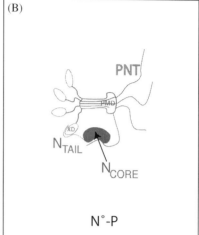

Fig. 1. Schematic representation of the N^{NUC}–P–L (A) and N°–P (B) complexes of MeV (modified from Refs. 18 and 29). The disordered N_{TAIL} (aa 401–525) and PNT (aa 1–230) regions are represented by lines. The encapsidated RNA is shown as a dotted line embedded in the middle of N by analogy with RV, VSV, and RSV N:RNA complexes.[74,75,77] The multimerization domain of P (aa 304–375, PMD) is represented with a dumb-bell shape according to Tarbouriech *et al.*[39] P is depicted as tetrameric by analogy with SeV P.[39] The polymerase complex is formed by L and by a tetramer of P. The tetrameric P is shown bound to N^{NUC} through 3 of its 4 C-terminal XD (aa 459–507) "arms," as in the model of Curran and Kolakofsky.[158] The segment connecting PMD and XD is represented as disordered according to Refs. 26 and 25, respectively. The L protein is shown as an oval contacting P through both PMD, by analogy with SeV,[47] and PNT, by analogy with Rinderpest virus.[46]

Once the viral ribonucleoprotein complexes are released in the cytoplasm of infected cells, transcription of viral genes occurs using endogenous NTPs as substrate. Following primary transcription, the polymerase switches to a processive mode and ignores the gene junctions to synthesize a full, complementary strand of genome length. This positive-stranded RNA (antigenome) does not serve as a template for transcription and its unique role is to provide an intermediate in genome replication. The intracellular concentration of the N protein is the main element controlling the relative level of transcription *versus* replication. When N is limiting, the polymerase functions preferentially as a transcriptase, thus leading to an increase in the intracellular concentration of viral proteins, including N. When N levels are high enough to allow encapsidation of the nascent RNA chain, the polymerase functions preferentially as a replicase[2] (see Refs. 3–6 for reviews on transcription and replication). Studies on Sendai virus (SeV, a paramyxovirus)[7] and on the vesicular stomatitis virus (VSV, a rhabdovirus)[8] have shown that the polymerase in replication mode consists of an L–P–N complex, whereas in transcription mode it is a complex of L, P, and cellular proteins.

The nucleoprotein is the most abundant viral structural protein. It consists of a globular N-terminal domain (aa 1–400, N_{CORE}) and of a flexible C-terminal domain (aa 401–525, N_{TAIL}). The combination of these two domains supports numerous N protein functions that go far beyond that of a mere structural component of the viral core particle. Within infected cells, N is found in a soluble, monomeric form (referred to as $N°$) and in a nucleocapsid assembled form (referred to as N^{NUC}). The soluble form of N is localized in both the cytosol and the nucleus.[9,10] Sato and co-workers have recently identified the determinants of the cytoplasmic to nuclear trafficking within the nucleoprotein sequence of MeV and canine distemper virus (CDV), a closely related Morbillivirus. They both possess a novel nuclear localization signal (NLS) at positions 70–77 and a nuclear export signal (NES). The NLS has a novel leucine/isoleucine-rich motif (TGILISIL), whereas the NES is composed of a leucine-rich motif (LLRSLTLF). While in CDV the NES occurs at positions 4–11, in MeV it is located in the C-terminus.[11] In both viruses, the nuclear export of N is CRM1-independent. At present, the intranuclear function(s) of N is unknown.

Following synthesis of the N protein, a chaperone is required to maintain this latter protein in a soluble and monomeric form in the cytoplasm. This role is played by the P protein, whose association simultaneously prevents illegitimate self-assembly of N and retains N in the cytoplasm.[12,13] This soluble $N°$–P complex (Fig. 1B) is used as the substrate for the encapsidation of the nascent genomic RNA chain during replication. It has recently been suggested that the $N°$–P complex has a very short half-life.[2] The assembled form of N also forms complexes with P, either isolated (N^{NUC}–P) or bound to L (N^{NUC}–P–L), which are essential to RNA synthesis by the viral polymerase.[14,15]

The viral polymerase, which is responsible for both transcription and replication, is poorly characterized. It is thought to carry out most (if not all) enzymatic activities required for transcription and replication, including nucleotide polymerization, mRNA capping, and polyadenylation. As no functional Paramyxoviridae polymerase has been biochemically characterized so far, most of our present knowledge arises from bioinformatics studies. Notably, using bioinformatics approaches, a ribose-2'-O-methyltransferase domain, possibly involved in capping of viral mRNAs, was identified within the C-terminal region of Mononegavirales polymerases (with the exception of *Bornaviridae* and *Nucleorhabdoviruses*).[16] Consistent with this prediction, the methyltransferase activity has been demonstrated biochemically within the C-terminal region of the closely related SeV polymerase (aa 1756–2228).[17]

For all Mononegavirales members, the viral genomic RNA is always encapsidated by the N protein, and genomic replication does not occur in the absence of N°
and without concurrent encapsidation of the nascent genomic RNA chain. Therefore, during RNA synthesis the viral polymerase has to interact with the N:RNA
complex and to use the N°–P complex as a substrate for encapsidation of nascent
genomic RNA. Hence, the components of the viral replication machinery, namely
P, N, and L, engage in a complex macromolecular ballet.

Although the understanding of the precise role(s) of N, P, and L within the
replicative complex of MeV has benefited of significant breakthroughs in recent
years (see Refs. 18–21 for reviews), rather limited three-dimensional information
on the replicative machinery is available. The scarcity of high-resolution structural
data stems from several facts: (i) the difficulty of obtaining homogenous polymers
of N suitable for X-ray analysis,[22,23] (ii) the low abundance of L in virions and its
very large size that is a challenge to heterologous expression, and (iii) the structural
flexibility of N and P.[18,20,24–28]

2. Structural Disorder within the N and P Proteins

In the course of the structural and functional characterization of MeV replicative
complex proteins, my group discovered that the N and P proteins contain disordered
regions up to 230 residues in length that possess the sequence and biochemical
features that typify intrinsically disordered proteins (IDPs).[18,20,24–29]

IDPs are functional proteins (with peculiar sequence properties) that lack highly
populated secondary and tertiary structure under physiological conditions of pH and
salinity in the absence of a partner, existing as dynamic ensembles of conformers
(for recent reviews see Refs. 30 and 31). Such proteins are highly abundant in
nature, which led to coining the term of "unfoldome[32]" to designate the portion of
the proteome that embraces IDPs. The functional repertoire of IDPs complements
the functions of ordered proteins, with IDPs being often associated with various
human diseases and often involved in regulation, signaling, and control, where
binding to multiple partners and high-specificity/low-affinity interactions play a
crucial role.

Using bioinformatics approaches (as described in Refs. 33 and 34), my group
further showed that the abundance of structural disorder is a conserved feature of
the N and P proteins of viruses of the Paramyxovirinae subfamily.[25] By combining
computational and experimental approaches (as described in Ref. 35), large disordered regions were also shown to occur within the P protein of rabies virus (RV, a
Rhabdoviridae member)[36] and of respiratory syncytial virus (RSV, a *Pneumovirinae* member within the *Paramyxoviridae* family).[37]

2.1. *Structural organization of the phosphoprotein*

Similar to the nucleoprotein, the P protein provides several functions in transcription and replication. Beyond serving as a chaperone for N, P binds to the nucleocapsid, thus tethering the polymerase onto the nucleocapsid template. The actual oligomeric state of MeV P is unknown. However, by analogy with the closely related SeV,[38,39] it is thought to be tetrameric (Fig. 1).

In MeV, the P gene encodes multiple proteins, including P, V, and C (for reviews see Refs. 3 and 4). While the C protein is encoded by an alternate open reading frame (ORF) within the P gene through ribosome initiation at an alternative translation codon, the V protein is translated from a P messenger obtained upon co-transcriptional insertion of a G at position 753 of the P mRNA. The V protein thus shares with the P protein in the N-terminal module (PNT, aa 1–230) and possesses a unique C-terminal, zinc-binding domain (ZnBD). The P gene organization suggests that the P protein is a modular protein, consisting of at least two domains: an N-terminal domain (aa 1–230, PNT) common to both P and V, and a C-terminal domain (aa 231–507, PCT) unique to the P protein (Fig. 2A). Transcription requires only the PCT domain, whereas genome replication also requires PNT. Within Paramyxovirinae, PNT plays the role of a chaperone for newly-synthesized N (N°) and this interaction leads to the formation of the encapsidation complex (N°–P) that is used as a substrate by the polymerase during RNA replication (for reviews see Refs. 3 and 4). Within the N°–P complex, P to N binding is mediated by the dual PNT–N_{CORE} and PCT–N_{TAIL} interaction[40] (Fig. 1B).

2.1.1. *The intrinsically disordered PNT domain*

Analysis of the hydrodynamic properties of recombinant PNT shows that this domain has a very extended shape in solution, and possesses a Stokes radius much larger (41 Å) than expected for a globular form and rather consistent with an extended (disordered) conformation.[24] The absence of a globular core was thereafter confirmed by limited proteolysis experiments and by far-UV circular dichroism (CD) and NMR studies.[24]

Noteworthy, PNT is consistently predicted to be disordered by several disorder predictors (see Refs. 21 and 24), including PONDR,[41] the method of the hydrophobicity/mean charge ratio,[42] as well as most disorder predictors implemented within the MeDor metaserver for the prediction of disorder.[43] These results indicate that the lack of stable secondary structure in PNT does not arise from a purification artifact, being rather an intrinsic property encoded in its primary structure. These results have been further extended to the W protein of SeV (the counterpart of MeV PNT) and to the PNT domains from other Morbillivirus members.[24]

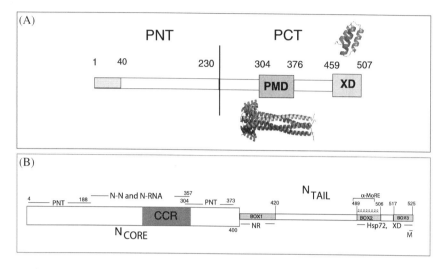

Fig. 2. (A) Schematic representation of the modular organization of MeV P, where globular and disordered regions are represented by large and narrow boxes, respectively. The vertical line separating PNT and PCT is located at the border between the region shared by P and V and the region unique to P. The hydrophobic, α-helical region at the N-terminus of P is highlighted. The crystal structures of MeV XD (pdb code 1OKS)[49] and of SeV PMD (pdb code 1EZJ)[39] are also shown. Structures were drawn using Pymol.[159] (B) Schematic representation of the modular organization of MeV N, where the disordered N$_{TAIL}$ domain is represented by a narrow box. The location of N–N, N–PNT, and RNA-binding sites is indicated. The CCR (aa 258–357) involved in oligomerization and in RNA binding is depicted in dark gray. The three N$_{TAIL}$ boxes conserved among *Morbillivirus* members[88] are shown (light gray), as is the location of the NR, hsp70, M, and XD binding sites.

What are the functional implications of the intrinsic disorder of PNT? PNT is reminiscent of transcriptional acidic activator domains (AADs). AADs play a role in recruiting the transcriptional machinery via protein–protein interactions and their function does not rely on a precise tertiary structure. Recently, several studies have shown that AADs act through bulky hydrophobic residues scattered within acidic residues. The numerous, charged residues would help to keep these hydrophobic residues in an aqueous environment, allowing them to establish weak, short-distance contacts with hydrophobic patches in their targets. According to this model, specificity of AADs for their physiological partners is determined by other factors dictating their co-localization with those partners (on DNA), and affinity is ensured by the intervention of multiple activation domains that strengthen the interaction.[44] AADs undergo induced folding (i.e., a disorder-to-order transition) in the presence of their physiological partner.[44] In the case of SeV, PNT was shown to be required for the synthesis of genomic RNA.[45] This activity is related to the involvement of PNT in the formation of the N°–P complex that is the substrate for the encapsidation of the nascent RNA chain, but could also be attributed to a weak PNT–L interaction, as shown in the case of Rinderpest virus.[46] In contrast, a stable L–P interaction site has been mapped within the PCT domain.[47] Given the similarity of AADs to PNT, the co-localization of PNT and L on the N:RNA complex would be ensured by the presence of a stable, independent L–P interaction site. The co-localization, together with the presence of multiple PNT domains within the P tetramer, would strengthen the PNT–L interaction.

The incubation of PNT in the presence of increasing concentrations of 2,2,2-trifluoroethanol (TFE) induces a pronounced gain of α-helicity. TFE is an organic solvent that mimics the hydrophobic environment experienced by proteins during protein–protein interactions. It is widely used as a probe for regions that have a propensity to undergo induced folding. Limited proteolysis experiments of PNT in the presence of TFE led to the identification of a thermolysin-resistant fragment. This fragment, spanning residues 27–99, contains a protein region (aa 27–38) with a strong propensity to fold as an α-helix. This α-helix may represent one of the secondary structure elements involved in the possible disorder-to-order transition of PNT upon binding to a partner.[24]

Does PNT actually undergo induced folding upon interaction with its physiological partner(s)? It is conceivable that the N-terminus of P folds upon binding to N°. This gain of structure may favor recognition of the N°–P complex by the polymerase and the proper positioning of N monomers on the nascent RNA chain. However, direct answers to this question await the availability of the two potential physiological PNT partners, namely N° and the L protein.

2.1.2. *The partly disordered PCT domain*

Beyond PNT, other disordered regions have been identified within P. Indeed, PCT has a modular organization, being composed of alternating disordered and structured regions.[25] (see Fig. 2A). Among these latter, the region spanning residues 304–376 P (referred to as PMD for P multimerization domain) is responsible for the oligomerization of P,[48] whereas the X domain (XD) is responsible for binding to both N_{TAIL}[49] and the cellular ubiquitin E3 ligase Pirh2.[48] In Morbilliviruses, sequence analysis predicts a coiled-coil region within the PMD. The coiled-coil organization has been experimentally confirmed in the case of SeV[39] and Rinderpest virus[50] PMDs. My group has previously reported the crystal structure of MeV XD and shown that it consists of a triple α-helical bundle[49] (see Fig. 2A). High-resolution structural data are also available for the XDs of the closely related mumps virus (MuV) and SeV, the structures of which have been solved by X-ray crystallography and NMR, respectively.[51,52] The MuV XD (aa 343–391 of P) has a few notable distinguishing properties with respect to MeV and SeV XD. Indeed, MuV XD has been shown to exist as a molten globule in solution, being loosely packed and devoid of stable tertiary structure.[51,53] In addition, contrary to the MeV and SeV XDs, MuV XD does not interact with N_{TAIL} and rather establishes contacts with the structured N_{CORE} region of N.[53]

In all Paramyxovirinae, PMD and XD are separated by a flexible linker region predicted to be poorly ordered.[25] Indeed, in the case of SeV, NMR studies carried out on the 474–568 region of P (referred to as PX) showed that the region upstream XD (aa 474–515 of P) is disordered.[54,55] The flexibility and solvent exposure of this linker region have also been experimentally determined in the case of MeV, where recombinant PCT was shown to undergo spontaneous cleavage at position 436.[26] Finally, in Morbilliviruses, an additional flexible region (referred to as "spacer") occurs upstream PMD.[25]

The disordered nature of PNT and of the "spacer" region connecting PNT to PMD likely reflects a way of alleviating evolutionary constraints within overlapping ORFs, in agreement with previous reports that pointed out a relationship between overlapping genes and structural disorder.[56–58] Indeed, PNT partially overlaps the C protein (being encoded by the same RNA region) and the spacer region partially overlaps the ORF encoding the ZnBD of the V protein (Fig. 2A).[4] Disorder, which is encoded by a much wider portion of sequence space as compared with order, can indeed represent a strategy by which genes encoding overlapping reading frames can lessen evolutionary constraints imposed on their sequence by the overlap, allowing the encoded overlapping protein products to sample a wider sequence space without losing function.

2.2. Structural organization of the nucleoprotein

Deletion analyses and electron microscopy studies have shown that Paramyxoviridae nucleoproteins are divided into two regions: a structured N-terminal moiety, N_{CORE} (aa 1–400 in MeV), which contains all the regions necessary for self-assembly and RNA-binding,[23,53,59–64] and a C-terminal domain, N_{TAIL} (aa 401–525 in MeV), which is intrinsically disordered[26] and cannot be visualized in cryo-electron microscopy reconstructions of nucleocapsids[65] (Fig. 2B). N_{TAIL} protrudes from the globular body of N_{CORE} and is exposed at the surface of the viral nucleocapsid.[23,66,67] N_{TAIL} contains the regions responsible for binding to P in both $N°$–P and N^{NUC}–P complexes (see Fig. 1).[26,53,61,62]

When expressed in heterologous systems, N self-assembles to form large helical nucleocapsid-like particles with a broad size distribution.[68–70] MeV nucleocapsids, as visualized by negative stain transmission electron microscopy, have a typical herringbone-like appearance.[23,26,65,70,71] The nucleocapsid of all Paramyxoviridae has a considerable conformational flexibility and can adopt different helical pitches (the axial rise *per* turn) and twists (the number of subunits per turn) resulting in conformations differing in their extent of compactness.[65–67,70–72] Because of this property, Paramyxoviridae nucleoproteins are poorly amenable to high-resolution structural characterizations. Owing to variable helical parameters, recombinant or viral nucleocapsids are also difficult to analyze using electron microscopy coupled with image analysis. Despite these technical drawbacks, elegant electron microscopy studies by two independent groups led to real-space helical reconstruction of MeV nucleocapsids.[65,71] These studies showed that the most extended nucleocapsid conformation has a helical pitch of 66 Å, whereas twist varies from 13.04 to 13.44.[65] Notably, these studies also highlighted a cross-talk between N_{CORE} and N_{TAIL}, as judged based on the observation that the removal of the disordered N_{TAIL} domain leads to increased nucleocapsid rigidity, with significant changes in both pitch and twist (see Refs. 26, 65, and 71). Distinct nucleocapsid morphologies have been associated with either high or low levels of viral transcriptional activity, making variations in N_{TAIL}–N_{CORE} interaction a possible determinant of viral gene expression.[73]

Conversely, high-resolution structural data are available for the nucleoproteins from two Rhabdoviridae members, namely RV and VSV,[74–76] and for RSV,[77] for which the crystal structures of N:RNA rings have been solved. The nucleoprotein of these viruses consists of two lobes and possesses an extended terminal arm that makes contacts with a neighboring N monomer. The RNA is tightly packed between the two N lobes and, in the case of RV and VSV, points towards the inner cavity of the N:RNA rings. Conversely, in the case of RSV the RNA is located on the external face of the N:RNA rings.[77] Thus, in all nucleoproteins, the RNA is not accessible to the solvent and has to be partially released from N to become accessible to the polymerase. In addition, the crystal structure of the nucleoprotein of the Borna disease virus has also been solved. Although it also shares a bilobal morphology with the N proteins from the above-mentioned viruses, its structure was not solved from N:RNA rings but from the N protein alone and was shown to consist of a tightly packed homotetramer centered at the four-fold crystallographic axis of the crystal.[78]

Functional and structural similarities between the nucleoproteins of *Rhabdoviridae* and of *Paramyxovirinae* are well established. In particular, they share the same organization in two well-defined regions, N_{CORE} and N_{TAIL}, and in both families a central conserved region (CCR) is involved in RNA-binding and self-assembly of N.[64,79] Incubation of the RV nucleocapsid with trypsin results in the removal of the C-terminal region (aa 377–450).[79] This N_{TAIL}-free nucleocapsid is no longer able to bind to P, thus suggesting that in *Rhabdoviridae*, N_{TAIL} plays a role in the recruitment of P as in the case of Paramyxoviridae.[22] However, in contrast to MeV, the RV N_{TAIL} domain is structured, as is that of VSV.[74,75] In the case of RSV, the last 12–20 residues of N are disordered in the crystal.[77] Although VSV, RV, and RSV nucleoproteins share the same bilobed morphology, presently it is not known whether this morphology is also conserved in MeV. Bioinformatics analyses predict that MeV N_{CORE} is organized into two subdomains (aa 1–130 and aa 145–400; see Refs. 33 and 80) separated by a variable, antigenic loop (aa 131–149),[81] that might fold cooperatively into a bilobed morphology.

As indicated previously, the N region responsible for binding to P within the MeV N^{NUC}–P complex is located within N_{TAIL}, whereas the N°–P complex involves an additional interaction between N_{CORE} and the disordered N-terminal domain of P (PNT) (Fig. 1). In other words, the N°–P complex reflects dual PNT–N_{CORE} and PCT–N_{TAIL} interactions[40] (Fig. 1). Studies on SeV suggested that an N°–P complex is absolutely necessary for the polymerase to initiate encapsidation.[82] Therefore, formation of the N°–P complex would have at least two separate functions: (i) prevent illegitimate self-assembly of N, and (ii) allow the polymerase to deliver N to the nascent RNA to initiate replication.

The regions within N_{CORE} responsible for binding to PNT within the N°–P complex have been mapped to residues 4–188 and 304–373, with the latter region being not strictly required for binding and rather favoring it[61] (Fig. 2B). However, precise mapping of such regions is hard because N_{CORE} does not have a modular structure, and consequently it is difficult to distinguish between gross structural defects and specific effects of deletions.

2.2.1.   The intrinsically unstructured N_{TAIL} domain

In Morbilliviruses, N_{TAIL} is responsible for binding to P in both N°–P and N^{NUC}–P complexes.[26,53,61,62] Within the N^{NUC}–P complex, N_{TAIL} is also responsible for the interaction with the polymerase (L–P) complex[26,53,61,62] (see Fig. 1).

Computational, biochemical, hydrodynamic, and spectroscopic analyses similar to those carried out during the characterization of PNT confirmed that N_{TAIL} belongs to the family of IDPs.[26] Although N_{TAIL} is mostly unfolded in solution, it nevertheless retains a certain degree of compactness based on its Stokes radius (27 Å) and ellipticity values at 200 and 222 nm.[26] Altogether, these characteristics indicate that N_{TAIL} is a pre-molten globule,[26,27] i.e., it has a conformational state intermediate between a random coil and a molten globule and possesses a certain degree of residual compactness due to the presence of residual and fluctuating secondary and/or tertiary structures.[83,84] It has been proposed that the residual intramolecular interactions that typify the pre-molten globule state may enable a more efficient start of the folding process induced by a partner by lowering the entropic cost of the folding coupled with binding process.[85–87]

That N_{TAIL} does indeed undergo induced folding was documented by CD studies, where N_{TAIL} was shown to undergo an α-helical transition in the presence of to the C-terminal X domain (XD, aa 459–507) of P.[49] Using computational and experimental approaches,[27] an α-helical molecular recognition element (α-MoRE, aa 488–499 of N) involved in binding to P has been identified within one (namely Box2) of three N_{TAIL} regions (referred to as Box1, Box2, and Box3) that are conserved within Morbillivirus members[88] (see Fig. 2B). MoREs are short, order-prone regions within IDPs that have a certain propensity to bind to a partner and thereby to undergo induced folding.[89–92]

A model of the interaction between XD and the α-MoRE of N_{TAIL} has been proposed[49] and then experimentally confirmed by X-ray crystallography studies.[93] Those studies showed that the α-MoRE of N_{TAIL} is embedded in a large hydrophobic cleft delimited by helices 2 and 3 of XD to form a pseudo-four helix arrangement (see Fig. 3).

Small-angle X-ray scattering (SAXS) studies provided a low-resolution model of the N_{TAIL}–XD complex, which showed that most of N_{TAIL} (aa 401–488) remains disordered within the complex (Fig. 3).[28] The lack of a protruding appendage corresponding to the extreme C-terminus of N_{TAIL} suggests that, beyond Box2, Box3 could also be involved in binding to XD.[28] That XD does indeed affect the conformation of Box3 has been confirmed by heteronuclear NMR (HN-NMR) studies.[94] In further support for a role of Box3 in binding, surface plasmon resonance studies showed that the removal of Box3 results in a strong increase in the equilibrium dissociation constant, with the K_D increasing from 80 nM to 12 μM.[28] Surprisingly, however, a synthetic Box3 peptide (aa 505–525) was found to exhibit an insignificant affinity for XD (K_D of approximately 1 mM) (see Ref. 21). In the same vein, HN-NMR experiments using ^{15}N-labeled XD pointed out the lack of magnetic perturbation within this latter upon addition of unlabeled Box3 peptide, consistent with the lack of stable contacts between XD and Box3.[95] The discrepancy between the data obtained with N_{TAIL} truncated proteins and with peptides could be accounted for by assuming that Box3 would act only in the context of N_{TAIL} and not in isolation. Thus, according to this model, Box3 and Box2 would be functionally coupled in the binding of N_{TAIL} to XD: the burying of the hydrophobic side of the α-MoRE in the hydrophobic cleft formed by helices α2 and α3 of XD would provide the primary driving force in the N_{TAIL}–XD interaction and then Box3 would act to further stabilize the bound conformation.

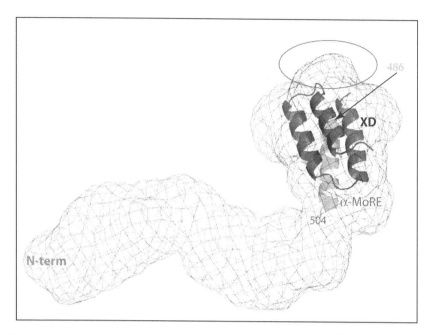

Fig. 3. Global shape of the N_{TAIL}–XD complex as derived by SAXS studies.[28] The circle points to the lack of a protruding shape from the globular body of the model. The crystal structure of the complex between XD and the N_{TAIL} region encompassing residues 486–504 (pdb code 1T6O)[93] is shown. The picture was drawn using Pymol.[159] Data were taken from Ref. 28.

In the view of unraveling the impact of XD binding on Box3, the N_{TAIL}–XD interaction has also been investigated by using site-directed spin-labeling electron paramagnetic resonance (EPR) spectroscopy, a technique based on the analysis of the mobility of spin labels covalently grafted to selected protein sites. Altogether, these studies allowed mapping the region of N_{TAIL} undergoing XD-induced α-helical folding to residues 488–502, while the downstream region does not adopt a helical conformation albeit it is significantly affected by the addition of XD.[96,97] Notably, XD and 20% TFE have a comparable impact on the 505–522 region, thus suggesting that the reduced mobility that the Box3 region experiences upon binding to XD is due neither to a steric hindrance exerted by XD nor to a direct interaction with XD, and rather arises only from α-helical folding of the neighboring Box2.[97] The reduced conformational freedom of Box3 may favor the establishment of weak, non-specific contacts with XD. At present, the exact role that Box3 plays in the stabilization of the N_{TAIL}–XD complex remains to be unraveled. Indeed, if recent data clearly indicate that transient long-range tertiary contacts between Box2 and Box3 are unlikely,[94] it is still unclear whether Box3 contributes to binding to XD through weak (transient) non-specific contacts with this latter or rather through another unknown mechanism.

Strikingly, the mobility of the 488–502 region was found to be restrained even in the absence of the partner,[97,98] a behavior that could be accounted for by the existence of a transiently populated folded state. The N_{TAIL} region spanning residues 491–499 adopts an α-helical conformation in about 50% of the conformers sampled by unbound N_{TAIL}, which has been recently experimentally confirmed by HN-NMR.[94] These findings are in agreement with previous reports which showed that the conformational space of MoREs[90] in the unbound state is often restricted by their inherent conformational propensities, thereby reducing the entropic cost of binding.[85–87,99] The occurrence of a transiently populated α-helix even in the absence of the partner suggests that the molecular mechanism governing the folding of N_{TAIL} induced by XD could rely on conformer selection (i.e., selection by the partner of a pre-existing conformation).[100,101] Recent data based on a quantitative analysis of NMR titration studies[94] suggest, however, that the binding reaction may also imply a binding intermediate in the form of a weak, non-specific encounter complex and hence may also occur through a "fly casting" mechanism.[102]

Finally, EPR equilibrium displacement experiments showed that the XD-induced folding of N_{TAIL} is a reversible phenomenon.[96,97] These results represent the first experimental evidence indicating that N_{TAIL} adopts its original pre-molten globule conformation after dissociation of XD. This latter point is particularly relevant taking into consideration that the contact between XD and N_{TAIL} within the replicative complex has to be dynamically made and broken to allow the polymerase to progress along the nucleocapsid template during both transcription and replication. Hence, the complex cannot be excessively stable for this transition to occur efficiently at a high rate as described in the following section.

3. Functional Role of Structural Disorder within N and P in Terms of Transcription and Replication

The K_D value between N_{TAIL} and XD is in the 100 nM range.[28] This affinity is considerably higher than that derived from isothermal titration calorimetry studies (K_D of 13 μM).[53] It should be pointed out, however, that these latter studies were carried out using an N_{TAIL} peptide encompassing residues 477–505. A weak binding affinity, implying fast association and dissociations rates, would ideally fulfill the requirements of a polymerase complex that has to cartwheel on the nucleocapsid template during both transcription and replication. However, a K_D in the μM range would not seem to be physiologically relevant considering the low intracellular concentrations of P in the early phases of infection, and the relatively long half-life of active P–L transcriptase complex tethered on the nucleocapsid template, which has been determined to be well over 6 h.[2] Moreover, such a weak affinity is not consistent with the ability to readily purify nucleocapsid–P complexes using rather stringent techniques such as CsCl isopycnic density centrifugation.[73,103–105] On the other hand, a more stable XD–N_{TAIL} complex would be expected to hinder the processive movement of P along the nucleocapsid template. In agreement with this model, the elongation rate of MeV polymerase was found to be rather slow (three nucleotides/s)[2] and N_{TAIL} amino acid substitutions that lower the affinity towards XD result in enhanced transcription and replication levels, as well as in increased polymerase rate (Oglesbee *et al.*, unpublished data). In addition, the C-terminus of N_{TAIL} has been shown to have an inhibitory role upon transcription and genome replication, as indicated by minireplicon experiments, where deletion of the C-terminus of N enhances basal reporter gene expression.[106] Deletion of the C-terminus of N also reduces the affinity of XD for N_{TAIL}, providing further support for modulation of XD–N_{TAIL} binding affinity as a basis for polymerase elongation rate. Thus, Box3 would dynamically control the strength of the N_{TAIL}–XD interaction, by stabilizing the complex probably through several weak, non-specific contacts with XD. Removal of Box3 or interaction of Box3 with other partners as described in the following paragraph would reduce the affinity of N_{TAIL} for XD thus stimulating transcription and replication. Modulation of XD–N_{TAIL} binding affinity could be dictated by interactions between N_{TAIL} and cellular and/or viral co-factors. Indeed, the requirement for cellular or viral co-factors in both transcription and replication has been already documented in the case of MeV,[107] and of other Mononegavirales members.[108,109] Furthermore, in both CDV and MeV, viral transcription and replication are enhanced by the major inducible heat shock protein (hsp70), and this stimulation relies on interaction with N_{TAIL}.[106,110–116] These co-factors may serve as elongation factors and could act by modulating the strength of the interaction between the polymerase complex and the nucleocapsid template as described below.

N_{TAIL} also influences the physical properties of the nucleocapsid helix that is formed by N_{CORE}.[26,71] Electron microscopic analysis of nucleocapsids formed by either N or N_{CORE} indicates that the presence of N_{TAIL} is associated with a greater degree of fragility, evidenced by the tendency of helices to break into individual ring structures. This fragility is associated with the evidence of increased nucleocapsid flexibility, with helices formed by N_{CORE} alone forming rods.[26,71] It is therefore conceivable that the induced folding of N_{TAIL} resulting from the interaction with P (and/or other physiological partners) could also exert an impact on the nucleocapsid conformation in such a way as to affect the structure of the replication

Fig. 4. Schematic representation of the nucleocapsid (left) highlighting the structure of the replication promoter composed of two discontinuous units juxtaposed on successive helical turns (see regions wrapped by the N monomers highlighted by an asterisk) and cryo-electron microscopy reconstructions of MeV nucleocapsid (right). Courtesy of D. Bhella, MRC, Glasgow, UK. Modified from Refs. 19 and 20.

promoter. Indeed, the replication promoter, located at the 3′ end of the viral genome, is composed of two discontinuous elements that form a functional unit when juxtaposed on two successive helical turns[117] (Fig. 4). The switch between transcription and replication could be dictated by variations in the helical conformation of the nucleocapsid, which would result in a modification in the number of N monomers (and thus of nucleotides) *per* turn, thereby disrupting the replication promoter in favor of the transcription promoter (or *vice versa*). Morphological analyses, showing the occurrence of a large conformational flexibility within Paramyxoviridae nucleocapsids,[65,70,103,118] tend to corroborate this hypothesis. In further support for a possible role of cellular co-factors in affecting the nucleocapsid conformation thereby favoring transcription and replication, hsp70–nucleocapsid complexes of CDV exhibit an expanded helical diameter, an increased fragility (as judged by tendency to fragment into rings), and an enhanced exposure of the genomic RNA to nuclease degradation.[103]

In the same vein, preliminary data indicate that the incubation of MeV nucleo-capsids in the presence of XD triggers unwinding of the nucleocapsid, thus possibly enhancing the accessibility of genomic RNA to the polymerase complex (Bhella and Longhi, unpublished data). Hence, it is tempting to propose that the XD-induced α-helical folding of N_{TAIL} could trigger the opening of the two lobes of N_{CORE} thus rendering the genomic RNA accessible to the solvent. Altogether, these data estab-lish a relationship between N_{TAIL} binding partners, nucleocapsid conformation, and exposure of genomic RNA.

The presence of unstructured domains on both N and P would allow for coor-dinated interactions between the polymerase complex and a large surface area of the nucleocapsid template, including successive turns of the helix. Unstructured regions are in fact considerably more extended in solution than globular ones. For instance, MeV PNT has a Stokes radius of 4 nm.[24] However, the Stokes radius only reflects a mean dimension. Indeed, the maximal extension of PNT as measured by SAXS (Longhi and Receveur-Bréchot, unpublished data) is considerably larger (>40 nm). In comparison, one turn of the nucleocapsid is 18 nm in diameter and 6 nm high.[70] Thus, PNT could easily stretch over several turns of the nucleocapsid, and since P is multimeric, N°–P might have a considerable extension (Fig. 5). In the same vein, it is striking that SeV and MeV PCT, which interact with the intrin-sically disordered N_{TAIL} domain, is composed of a flexible linker.[26,52,54,119] This certainly suggests the need for a great structural flexibility. This flexibility could be necessary for the tetrameric P to bind several turns of the helical nucleocapsid. Indeed, the promoter signals for the polymerase are located on the first and the second turn of the SeV nucleocapsid (see Fig. 4).[117]

Similarly, the maximal extension of N_{TAIL} in solution is of 13 nm.[26] The very long reach of disordered regions could enable them to act as linkers and to tether partners on large macromolecular assemblies, thereby acting as scaffolding engines as already described for intrinsically disordered scaffold proteins.[120,121] Accord-ingly, one role of the tentacle-like N_{TAIL} projections in actively replicating nucleo-capsids could be to put into contact several proteins within the replicative complex, such as the N°–P and the P–L complexes (see Fig. 5).

In Respiroviruses and Morbilliviruses, PNT contains binding sites for N°[46,122,123] and for L.[45,46,124] This pattern of interactions among N°, P, and L, mediated by unstructured regions of either P or N, suggests that N°, P, and L might interact simultaneously at some point during replication. Notably, the existence of a N–P–L tripartite complex has been recently proven by co-immunoprecipitation studies in the case of VSV, where this tripartite complex constitutes the replicase complex, as opposed to the transcriptase complex that consists of L, P, and host cellular proteins, such as hsp60.[8,125]

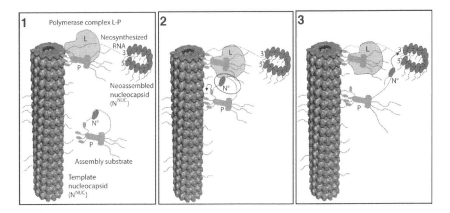

Fig. 5. (A) Model of the polymerase complex actively replicating genomic RNA. Disordered regions are represented by lines. Within the N^{NUC}–P complex, P is represented as bound to N^{NUC} through three of its four terminal XD arms according to the model of Ref. 158. For the sake of clarity, only a few N_{TAIL} regions are drawn. The numbering of the different panels indicates the chronology of events. (1) L is bound to a P tetramer. The newly synthesized RNA is shown as already partially encapsidated. (2) The encapsidation complex, N°–P, binds to the nucleocapsid template through three of its four XD arms. The extended conformation of N_{TAIL} and PNT would allow the formation of a tripartite complex between N°, P, and the polymerase (circled). It is tempting to imagine that the proximity of the polymerase (or an unknown signal from this latter) may promote the release of N° by XD, thus leading to N° incorporation within the assembling nucleocapsid. The N° release would also lead to cartwheeling of the L–P complex through binding of the free XD arm onto the nucleocapsid template (see arrow) as in the model of Ref. 158. (3) PNT delivers N° to the newly assembled nucleocapsid (see arrow).

A model can be proposed, where during replication, the extended conformation of PNT and N_{TAIL} would be key to allowing contact between the assembly substrate (N°–P) and the polymerase complex (L–P), thus leading to a tripartite N°–P–L complex (Fig. 5). This model emphasizes the plasticity of intrinsically disordered regions, which might give a considerable reach to the elements of the replicative machinery.

Interestingly, there is a striking parallel between the N_{TAIL}–XD interaction and the PNT–N_{CORE} interaction (Fig. 1B). Both interactions are not stable by themselves and must be strengthened by the combination of other interactions. This might ensure easy breaking and reforming of interactions and could result in transient, easily modulated interactions. One can speculate that the gain of structure of N_{TAIL} upon binding to XD could result in stabilization of the N–P complex. At the same time, folding of N_{TAIL} would result in a modification in the pattern of solvent-accessible regions resulting in the shielding of specific regions of interaction. As a result, N_{TAIL} would no longer be available for binding to its other partners. Although induced folding likely stabilizes the N–P complex, the dynamic nature of these interactions could rely on (i) the intervention of viral and/or cellular co-factors modulating the strength of such interactions and (ii) the ability of the IDP to establish weak affinity interactions through residual disordered regions.

4. Structural Disorder and Molecular Partnership

One of the functional advantages of disorder is related to an increased plasticity that enables disordered regions to bind to numerous structurally distinct targets.[126–128] In agreement, intrinsic disorder is a distinctive and common feature of "hub" proteins, with disorder serving as a determinant of protein interactivity.[126–128] The lack of a rigid 3D structure allows IDPs to establish interactions that are characterized by a high specificity and a low affinity: while the former is ensured by the very large surface area that is generally buried in complexes involving IDPs,[129] the low affinity arises from an unfavorable entropic contribution associated with the disorder-to-order transition.[84,130–137] The extent of the entropic penalty is, however, tightly related to the extent of conformational sampling of the pre-recognition state, i.e., on the degree to which MoREs are preconfigured in solution prior to binding: the occurrence of a partly preconfigured MoREs in the unbound state in fact reduces the entropic cost of binding thereby enhancing affinity.[85–87,91,92,99] As such, IDPs exhibit a wide binding diversity, with some of them binding their partners with strong affinities, as is the case of the N_{TAIL}–XD and N_{TAIL}–hsp70 complexes (see below).

The disordered nature of N_{TAIL} confers to this N domain a large malleability enabling it to adapt to various partners and to form complexes that are critical for both transcription and replication. Hence, thanks to its exposure at the surface of the viral nucleocapsid, N_{TAIL} establishes numerous interactions with various viral partners, including P, the P–L complex, and the matrix protein.[138] Beyond viral partners, N_{TAIL} also interacts with cellular proteins, including hsp70,[106,114] the interferon regulatory factor 3,[139,140] the cell protein responsible for the nuclear export of N,[11] and possible components of the cell cytoskeleton.[141,142] Moreover, N_{TAIL} within viral nucleocapsids released from infected cells also binds to cell receptors involved in MeV-induced immunosuppression.[143–145] While N_{CORE} specifically interacts with FcγRII,[145] N_{TAIL} interacts with a yet uncharacterized nucleoprotein receptor (NR) that is expressed at the surface of dendritic cells of lymphoid origin (both normal and tumoral),[144] and of T and B lymphocytes.[145]

Similarly, both MeV and SeV PNT domains have been reported to interact with multiple partners, with the former interacting with N[40] and cellular proteins,[146] and the latter interacting with the unassembled form of N (N°) and the L protein.[45,122]

4.1. *Interaction between N_{TAIL} and hsp70*

In both CDV and MeV, the interaction between N_{TAIL} and hsp70 stimulates both transcription and genome replication.[106,110–116] In MeV, two binding sites for hsp70 have been identified.[106,114] High-affinity binding (K_D of 10 nM) is supported by the α-MoRE, and hsp70 was shown to competitively inhibit binding of XD to N_{TAIL}.[114] A second low-affinity binding site is present within Box3.[106,115] Variability in Box3 sequence does not influence the binding affinity of N_{TAIL} for hsp70, since this reaction is primarily determined by the α-MoRE, yet the ability of hsp70 to directly bind Box3 does have functional consequences.[114] Analysis of infectious virus containing the N522D substitution within Box3 shows loss of hsp70-dependent stimulation of transcription but not of genome replication.[114] These findings suggest that hsp70 could enhance transcription and genome replication by reducing the stability of P–N_{TAIL} complexes, thereby promoting successive cycles of binding and release, which are essential to polymerase movement along the nucleocapsid template.[28,114] Hsp70-dependent reduction of the stability of P–N_{TAIL} complexes would rely on competition between hsp70 and XD for binding to N_{TAIL} through (i) competition for binding to the α-MoRE (and this would occur at low hsp70 concentrations) and (ii) neutralization of the contribution of the C-terminus of N_{TAIL} to the formation of a stable P–N_{TAIL} complex (and this would occur in the context of elevated cellular levels of hsp70 and only for MeV strains that support hsp70 binding in this region).[114] Another possible mechanism for the hsp70-dependent enhancement of viral transcription and replication is related to the ability of hsp70 to maintain the nucleocapsid in a transcription-competent form (Oglesbee and Gerlier, unpublished data).

The basis for the separable effects of hsp70 on genome replication *versus* transcription remains to be shown, with template changes unique to a replicase *versus* transcriptase being a primary candidate. The latter could involve unique nucleocapsid ultrastructural morphologies, with hsp70-dependent morphologies being well documented for CDV (see Refs. 103 and 118). Another possibility is that the transcriptase and replicase complexes may be different, with the latter possibly including hsp70 by analogy to the replicase complex of VSV that was shown to incorporate hsp60.[8]

As for the functional role of hsp70 in the context of MeV infection, it has been proposed that the elevation in hsp70 levels could contribute to virus clearance.[147,148]

Indeed, the stimulation of viral transcription and replication by hsp70 is also associated with cytopathic effects leading to apoptosis and release of viral proteins in the extracellular compartment.[110–112] Once released in the extracellular environment, both viral proteins and heat shock proteins have the potential to promote innate and adaptive antiviral immune responses, thereby leading to virus clearance (see Refs. 116 and 147 and references cited therein).

Many heat shock proteins function as chaperone complexes, and high-affinity N_{TAIL}–hsp70 binding has been recently shown to require an hsp40 co-chaperone that interacts primarily with the hsp70 nucleotide-binding domain and displays no significant affinity for N_{TAIL}. Hsp40 directly enhances hsp70 ATPase activity in an N_{TAIL}-dependent manner, and formation of hsp40–hsp70–N_{TAIL} intracellular complexes requires the presence of N_{TAIL} Box2 and Box3. These results are consistent with the functional interplay between the hsp70 nucleotide and substrate binding domains, where ATP hydrolysis is rate limiting to high-affinity binding to client proteins and is enhanced by hsp40.[149] Further investigation is required to determine whether limitations in intracellular levels of hsp40 may influence permissiveness to viral gene expression or development of antiviral immune responses.

5. Conclusions

As thoroughly discussed in this chapter, N and P are multifunctional proteins exerting multiple biological functions. The molecular bases for this property reside in the ability of their disordered regions to establish interactions with various partners. As many, if not all, of these interactions are critical for transcription and genome replication, they provide excellent targets for antiviral agents. In this context, the discovery that the N and P domains supporting these multiple protein interactions are intrinsically disordered is particularly relevant: indeed, protein–protein interactions mediated by disordered regions provide interesting drug discovery targets with the potential to increase significantly the discovery rate for new compounds.[150]

Although it would be unjustified to formally rule out the possibility that MeV PNT and N_{TAIL} domains may undergo some degree of folding in the context of the entire P and N proteins, a few studies directly support the conclusion that disorder is retained in the context of the entire proteins. Indeed, in SeV, the PNT region has been shown to be disordered not only in isolation but also in the context of the full-length protein,[151,152] as were the linker region between PMD and XD in the context of MeV PCT[38] and the region upstream XD in both SeV and MeV.[26,54,55] Similarly, the MeV N_{TAIL} region is also disordered within the entire nucleoprotein, being equally accessible to monoclonal antibodies in isolation and within recombinant nucleocapsids, and being not visible by electron microscopy.[26] As for the possibility that PNT and N_{TAIL} are effectively disordered *in vivo*, further experimental work is required to address this question. In this regard, it is noteworthy that available data addressing the disordered state of IDPs *in vivo* are mixed. On the one hand, in-cell NMR experiments on the natively unfolded FlgM protein suggest a more folded conformation in the cellular environment of live bacteria.[153] In contrast, a few reports support a disordered state for other IDPs either in living cells or in the presence of crowding agents that mimic the crowded environment of cells (for examples, see Refs. 26, 154 and 155).

Recent studies showed that viruses and eukaryota have 10 times more conserved disorder (roughly 1%) than archaea and bacteria (0.1%),[156] and also pointed out that viral proteins, and in particular proteins from RNA viruses, are enriched in short disordered regions.[157] In this latter study, the authors propose that beyond affording a broad partnership, the wide occurrence of disordered regions in viral proteins could also be related to the typical high mutation rates of RNA viruses, representing a strategy for buffering the deleterious effects of mutations.[157]

Taking into account these considerations, as well as the correlation between overlapping genes and disorder,[56–58] I propose that the main advantage of the abundance of disorder within viruses would reside in pleiotropy and genetic compaction. Indeed, disorder provides a solution to reduce both genome size and molecular crowding, where a single gene would (i) encode a single (regulatory) protein product that can establish multiple interactions via its disordered regions and hence exerts multiple concomitant biological effects, and/or (ii) would encode more than one product by means of overlapping reading frames. In fact, since disordered regions are less sensitive to structural constraints than ordered ones, the occurrence of disorder within one or both protein products encoded by an overlapping ORF can represent a strategy to alleviate evolutionary constraints imposed by the overlap. As such, disorder would confer to viruses the ability to "handle" overlaps, thus further expanding the coding potential of viral genomes.

Acknowledgments

The author would like to thank all the persons who contributed to the works herein described. In particular, within the AFMB laboratory, the author would like to thank Jean-Marie Bourhis, Benjamin Morin, Stéphanie Costanzo, Matteo Colombo, Marie Couturier, Sabrina Rouger, Elodie Liquière, Bruno Canard, Kenth Johansson, David Karlin, François Ferron, Véronique Receveur-Brechot, Hervé

Darbon, Cédric Bernard, Valérie Campanacci, and Christian Cambillau. The author would also like to thank his co-workers Keith Dunker (Indiana University, USA), David Bhella (MRC, Glasgow, UK), Denis Gerlier (LabVirPAth, Lyon, France), Michael Oglesbee (Ohio State University, USA) Hélène Valentin and Chantal Rabourdin-Combe (INSERM, Lyon, France), Gary Daughdrill (University of South Florida, USA), Martin Blackledge and Malene Ringkjobing-Jensen (IBS, Grenoble, France), André Fournel, Valérie Belle and Bruno Guigliarelli (BIP, CNRS, Marseille, France). The author wishes to express her gratitude to Frédéric Carrière (EIPL, CNRS, Marseille, France) for having introduced her to EPR spectroscopy and for constant support. The author is also grateful to David Bhella who is the author of Fig. 4. The author also wants to thank Vladimir Uversky and Denis Gerlier for stimulating discussions. The studies mentioned in this chapter were carried out with the financial support of the European Commission, program RTD, QLK2-CT2001-01225, "Towards the design of new potent antiviral drugs: Structure-function analysis of *Paramyxoviridae* polymerase," of the Agence Nationale de la Recherche, specific program "Microbiologie et Immunologie," ANR-05-MIIM-035-02, "Structure and disorder of measles virus nucleoprotein: Molecular partnership and functional impact," and of the National Institute of Neurological Disorders and Stroke, specific program "The cellular stress response in viral encephalitis," R01 NS031693-11A2.

References

1. Bitko, V. and S. Barik, Phenotypic silencing of cytoplasmic genes using sequence-specific double-stranded short interfering RNA and its application in the reverse genetics of wild type negative-strand RNA viruses. *BMC Microbiol*, 2001. **1**: 34.

2. Plumet, S., W.P. Duprex and D. Gerlier, Dynamics of viral RNA synthesis during measles virus infection. *J Virol*, 2005. **79**(11): 6900–8.

3. Longhi, S. and B. Canard, Mécanismes de transcription et de réplication des Paramyxoviridae. *Virologie*, 1999. **3**(3): 227–40.

4. Lamb, R.A. and D. Kolakofsky, Paramyxoviridae: The Viruses and Their Replication, in *Fields Virology*, B.N. Fields, D.M. Knipe and P.M. Howley, Editors. Lippincott-Raven: Philadelphia, PA. 2001, pp. 1305–1340.

5. Albertini, A.A.V., G. Schoehn and R.W. Ruigrok, Structures impliquées dans la réplication et la transcription des virus à ARN non segmentés de sens négatif. *Virologie*, 2005. **9**(2): 83–92.

6. Roux, L., Dans le génome des Paramyxovirinae, les promoteurs et leurs activités sont façonnés par la règle de six. *Virologie*, 2005. **9**(1): 19–34.

7. Horikami, S.M., *et al.*, Complexes of Sendai virus NP–P and P–L proteins are required for defective interfering particle genome replication *in vitro*. *J Virol*, 1992. **66**(8): 4901–8.

8. Qanungo, K.R., *et al.*, Two RNA polymerase complexes from vesicular stomatitis virus-infected cells that carry out transcription and replication of genome RNA. *Proc Natl Acad Sci U S A*, 2004. **101**(16): 5952–7.

9. Gombart, A.F., A. Hirano and T.C. Wong, Conformational maturation of measles virus nucleocapsid protein. *J Virol*, 1993. **67**(7): 4133–41.

10. Horikami, S.M. and S.A. Moyer, Structure, transcription, and replication of measles virus. *Curr Top Microbiol Immunol*, 1995. **191**: 35–50.

11. Sato, H., *et al.*, Morbillivirus nucleoprotein possesses a novel nuclear localization signal and a CRM1-independent nuclear export signal. *Virology*, 2006. **352**(1): 121–30.

12. Huber, M., *et al.*, Measles virus phosphoprotein retains the nucleocapsid protein in the cytoplasm. *Virology*, 1991. **185**(1): 299–308.

13. Spehner, D., R. Drillien and P.M. Howley, The assembly of the measles virus nucleoprotein into nucleocapsid-like particles is modulated by the phosphoprotein. *Virology*, 1997. **232**(2): 260–8.

14. Ryan, K.W. and A. Portner, Separate domains of Sendai virus P protein are required for binding to viral nucleocapsids. *Virology*, 1990. **174**(2): 515–21.

15. Buchholz, C.J., *et al.*, The carboxy-terminal domain of Sendai virus nucleocapsid protein is involved in complex formation between phosphoprotein and nucleocapsid-like particles. *Virology*, 1994. **204**(2): 770–6.

16. Ferron, F., *et al.*, Viral RNA-polymerases — A predicted 2′-*O*-ribose methyltransferase domain shared by all Mononegavirales. *Trends Biochem Sci*, 2002. **27**(5): 222–4.

17. Ogino, T., *et al.*, Sendai virus RNA-dependent RNA polymerase L protein catalyzes cap methylation of virus-specific mRNA. *J Biol Chem*, 2005. **280**(6): 4429–35.

18. Bourhis, J.M., B. Canard and S. Longhi, Structural disorder within the replicative complex of measles virus: Functional implications. *Virology*, 2006. **344**: 94–110.

19. Bourhis, J.M. and S. Longhi, Measles Virus Nucleoprotein: Structural Organization and Functional Role of the Intrinsically Disordered C-terminal Domain, in *Measles Virus Nucleoprotein*, S. Longhi, Editor. Nova Publishers Inc.: Hauppage, NY. 2007, p. 1–35.

20. Longhi, S., Nucleocapsid structure and function. *Curr Top Microbiol Immunol*, 2009. **329**: 103–128.

21. Longhi, S. and M. Oglesbee, Structural disorder within the measles virus nucleoprotein and phosphoprotein. *Protein Pept Lett*, 2010. **17**(8): 961–78.

22. Schoehn, G., *et al.*, Structure of recombinant rabies virus nucleoprotein–RNA complex and identification of the phosphoprotein binding site. *J Virol*, 2001. **75**(1): 490–8.

23. Karlin, D., S. Longhi and B. Canard, Substitution of two residues in the measles virus nucleoprotein results in an impaired self-association. *Virology*, 2002. **302**(2): 420–32.

24. Karlin, D., *et al.*, The N-terminal domain of the phosphoprotein of morbilliviruses belongs to the natively unfolded class of proteins. *Virology*, 2002. **296**(2): 251–62.

25. Karlin, D., *et al.*, Structural disorder and modular organization in Paramyxovirinae N and P. *J Gen Virol*, 2003. **84**(Pt 12): 3239–52.

26. Longhi, S., *et al.*, The C-terminal domain of the measles virus nucleoprotein is intrinsically disordered and folds upon binding to the C-terminal moiety of the phosphoprotein. *J Biol Chem*, 2003. **278**(20): 18638–48.

27. Bourhis, J., *et al.*, The C-terminal domain of measles virus nucleoprotein belongs to the class of intrinsically disordered proteins that fold upon binding to their physiological partner. *Virus Res*, 2004. **99**: 157–67.

28. Bourhis, J.M., *et al.*, The intrinsically disordered C-terminal domain of the measles virus nucleoprotein interacts with the C-terminal domain of the phosphoprotein via two distinct sites and remains predominantly unfolded. *Protein Sci*, 2005. **14**: 1975–92.

29. Bourhis, J.M., B. Canard and S. Longhi, Désordre structural au sein du complexe réplicatif du virus de la rougeole: Implications fonctionnelles. *Virologie*, 2005. **9**: 367–83.

30. Radivojac, P., *et al.*, Intrinsic disorder and functional proteomics. *Biophys J*, 2007. **92**(5): 1439–56.

31. Uversky, V.N., The mysterious unfoldome: Structureless, underappreciated, yet vital part of any given proteome. *J Biomed Biotechnol*, 2009. **2010**: 568068.

32. Cortese, M.S., *et al.*, Uncovering the unfoldome: Enriching cell extracts for unstructured proteins by Acid treatment. *J Proteome Res*, 2005. **4**(5): 1610–8.

33. Ferron, F., *et al.*, A practical overview of protein disorder prediction methods. *Proteins*, 2006. **65**(1): 1–14.

34. Bourhis, J.M., B. Canard and S. Longhi, Predicting protein disorder and induced folding: From theoretical principles to practical applications. *Curr Protein Pept Sci*, 2007. **8**(2): 135–49.

35. Receveur-Bréchot, V., *et al.*, Assessing protein disorder and induced folding. *Proteins Struct Funct Bioinform*, 2006. **62**: 24–45.

36. Gerard, F.C., *et al.*, Modular organization of rabies virus phosphoprotein. *J Mol Biol*, 2009. **388**(5): 978–96.

37. Llorente, M.T., *et al.*, Structural analysis of the human respiratory syncitial virus phosphoprotein: Characterization of an α-helical domain involved in oligomerization. *J Gen Virol*, 2006. **87**: 159–69.

38. Tarbouriech, N., *et al.*, On the domain structure and the polymerization state of the Sendai virus P protein. *Virology*, 2000. **266**(1): 99–109.

39. Tarbouriech, N., *et al.*, Tetrameric coiled coil domain of Sendai virus phosphoprotein. *Nat Struct Biol*, 2000. **7**(9): 777–81.

40. Chen, M., J.C. Cortay and D. Gerlier, Measles virus protein interactions in yeast: New findings and caveats. *Virus Res*, 2003. **98**(2): 123–9.

41. Li, X., *et al.*, Predicting protein disorder for N-, C-, and internal regions. *Genome Inform Ser Workshop Genome Inform*, 1999. **10**: 30–40.

42. Uversky, V.N., J.R. Gillespie and A.L. Fink, Why are "natively unfolded" proteins unstructured under physiologic conditions? *Proteins*, 2000. **41**(3): 415–27.

43. Lieutaud, P., B. Canard and S. Longhi, MeDor: A metaserver for predicting protein disorder. *BMC Genomics*, 2008. **9**(Suppl 2): S25.

44. Melcher, K., The strength of acidic activation domains correlates with their affinity for both transcriptional and non-transcriptional proteins. *J Mol Biol*, 2000. **301**(5): 1097–112.

45. Curran, J., J.B. Marq and D. Kolakofsky, An N-terminal domain of the Sendai paramyxovirus P protein acts as a chaperone for the NP protein during the nascent chain assembly step of genome replication. *J Virol*, 1995. **69**(2): 849–55.

46. Sweetman, D.A., J. Miskin and M.D. Baron, Rinderpest virus C and V proteins interact with the major (L) component of the viral polymerase. *Virology*, 2001. **281**(2): 193–204.

47. Smallwood, S., K.W. Ryan and S.A. Moyer, Deletion analysis defines a carboxylproximal region of Sendai virus P protein that binds to the polymerase L protein. *Virology*, 1994. **202**(1): 154–63.

48. Chen, M., *et al.*, Inhibition of ubiquitination and stabilization of human ubiquitin E3 ligase PIRH2 by measles virus phosphoprotein. *J Virol*, 2005. **79**(18): 11824–36.

49. Johansson, K., *et al.*, Crystal structure of the measles virus phosphoprotein domain responsible for the induced folding of the C-terminal domain of the nucleoprotein. *J Biol Chem*, 2003. **278**(45): 44567–73.

50. Rahaman, A., *et al.*, Phosphoprotein of the rinderpest virus forms a tetramer through a coiled coil region important for biological function. A structural insight. *J Biol Chem*, 2004. **279**(22): 23606–14.

51. Kingston, R.L., *et al.*, Structure of the nucleocapsid-binding domain from the mumps virus polymerase; an example of protein folding induced by crystallization. *J Mol Biol*, 2008. **379**(4): 719–31.

52. Blanchard, L., *et al.*, Structure and dynamics of the nucleocapsid-binding domain of the Sendai virus phosphoprotein in solution. *Virology*, 2004. **319**(2): 201–11.

53. Kingston, R.L., A.B. Walter and L.S. Gay, Characterization of nucleocapsid binding by the measles and the mumps virus phosphoprotein. *J Virol*, 2004. **78**(16): 8615–29.

54. Bernado, P., *et al.*, A structural model for unfolded proteins from residual dipolar couplings and small-angle X-ray scattering. *Proc Natl Acad Sci U S A*, 2005. **102**(47): 17002–7.

55. Houben, K., *et al.*, Intrinsic dynamics of the partly unstructured PX domain from the Sendai virus RNA polymerase cofactor P. *Biophys J*, 2007. **93**(8): 2830–44.

56. Jordan, I.K., B.A. Sutter and M.A. McClure, Molecular evolution of the Paramyxoviridae and Rhabdoviridae multiple-protein-encoding P gene. *Mol Biol Evol*, 2000. **17**(1): 75–86.

57. Narechania, A., M. Terai and R.D. Burk, Overlapping reading frames in closely related human papillomaviruses result in modular rates of selection within E2. *J Gen Virol*, 2005. **86**(Pt 5): 1307–13.

58. Rancurel, C., *et al.*, Overlapping genes produce proteins with unusual sequence properties and offer insight into de novo protein creation. *J Virol*, 2009. **83**(20): 10719–36.

59. Curran, J., *et al.*, The hypervariable C-terminal tail of the Sendai paramyxovirus nucleocapsid protein is required for template function but not for RNA encapsidation. *J Virol*, 1993. **67**(7): 4358–64.

60. Buchholz, C.J., *et al.*, The conserved N-terminal region of Sendai virus nucleocapsid protein NP is required for nucleocapsid assembly. *J Virol*, 1993. **67**(10): 5803–12.

61. Bankamp, B., *et al.*, Domains of the measles virus N protein required for binding to P protein and self-assembly. *Virology*, 1996. **216**(1): 272–7.

62. Liston, P., *et al.*, Protein interaction domains of the measles virus nucleocapsid protein (NP). *Arch Virol*, 1997. **142**(2): 305–21.

63. Myers, T.M., A. Pieters and S.A. Moyer, A highly conserved region of the Sendai virus nucleocapsid protein contributes to the NP–NP binding domain. *Virology*, 1997. **229**(2): 322–35.

64. Myers, T.M., S. Smallwood and S.A. Moyer, Identification of nucleocapsid protein residues required for Sendai virus nucleocapsid formation and genome replication. *J Gen Virol*, 1999. **80**(Pt 6): 1383–91.

65. Bhella, D., A. Ralph and R.P. Yeo, Conformational flexibility in recombinant measles virus nucleocapsids visualised by cryo-negative stain electron microscopy and real-space helical reconstruction. *J Mol Biol*, 2004. **340**(2): 319–31.

66. Heggeness, M.H., A. Scheid and P.W. Choppin, Conformation of the helical nucleocapsids of paramyxoviruses and vesicular stomatitis virus: Reversible coiling and uncoiling induced by changes in salt concentration. *Proc Natl Acad Sci U S A*, 1980. **77**(5): 2631–5.

67. Heggeness, M.H., A. Scheid and P.W. Choppin, The relationship of conformational changes in the Sendai virus nucleocapsid to proteolytic cleavage of the NP polypeptide. *Virology*, 1981. **114**(2): 555–62.

68. Spehner, D., A. Kirn and R. Drillien, Assembly of nucleocapsid-like structures in animal cells infected with a vaccinia virus recombinant encoding the measles virus nucleoprotein. *J Virol*, 1991. **65**(11): 6296–300.

69. Warnes, A., *et al.*, Expression of the measles virus nucleoprotein gene in *Escherichia coli* and assembly of nucleocapsid-like structures. *Gene*, 1995. **160**(2): 173–8.

70. Bhella, D., *et al.*, Significant differences in nucleocapsid morphology within the Paramyxoviridae. *J Gen Virol*, 2002. **83**(Pt 8): 1831–9.

71. Schoehn, G., *et al.*, The 12 A structure of trypsin-treated measles virus N-RNA. *J Mol Biol*, 2004. **339**(2): 301–12.

72. Egelman, E.H., *et al.*, The Sendai virus nucleocapsid exists in at least four different helical states. *J Virol*, 1989. **63**(5): 2233–43.

73. Robbins, S.J., R.H. Bussell and F. Rapp, Isolation and partial characterization of two forms of cytoplasmic nucleocapsids from measles virus-infected cells. *J Gen Virol*, 1980. **47**(2): 301–10.

74. Green, T.J., *et al.*, Structure of the vesicular stomatitis virus nucleoprotein–RNA complex. *Science*, 2006. **313**(5785): 357–60.

75. Albertini, A.A., *et al.*, Crystal structure of the rabies virus nucleoprotein–RNA complex. *Science*, 2006. **313**(5785): 360–3.

76. Luo, M., *et al.*, Conserved characteristics of the rhabdovirus nucleoprotein. *Virus Res*, 2007. **129**(2): 246–51.

77. Tawar, R.G., *et al.*, 3D structure of a nucleocapsid-like nucleoprotein–RNA complex of respiratory syncytial virus. *Science*, 2009. **326**: 1279–83.

78. Rudolph, M.G., *et al.*, Crystal structure of the borna disease virus nucleoprotein. *Structure (Camb)*, 2003. **11**(10): 1219–26.

79. Kouznetzoff, A., M. Buckle and N. Tordo, Identification of a region of the rabies virus N protein involved in direct binding to the viral RNA. *J Gen Virol*, 1998. **79**(Pt 5): 1005–13.

80. Bourhis, J., B. Canard and S. Longhi, Predicting protein disorder and induced folding: From theoretical principles to practical applications. *Curr Protein Pept Sci*, 2007. **8**: 135–49.

81. Giraudon, P., M.F. Jacquier and T.F. Wild, Antigenic analysis of African measles virus field isolates: Identification and localisation of one conserved and two variable epitope sites on the NP protein. *Virus Res*, 1988. **10**(2–3): 137–52.

82. Baker, S.C. and S.A. Moyer, Encapsidation of Sendai virus genome RNAs by purified NP protein during *in vitro* replication. *J Virol*, 1988. **62**(3): 834–8.

83. Uversky, V.N., Natively unfolded proteins: A point where biology waits for physics. *Protein Sci*, 2002. **11**(4): 739–56.

84. Dunker, A.K., *et al.*, Intrinsically disordered protein. *J Mol Graph Model*, 2001. **19**(1): 26–59.

85. Tompa, P., Intrinsically unstructured proteins. *Trends Biochem Sci*, 2002. **27**(10): 527–33.

86. Fuxreiter, M., *et al.*, Preformed structural elements feature in partner recognition by intrinsically unstructured proteins. *J Mol Biol*, 2004. **338**(5): 1015–26.

87. Lacy, E.R., *et al.*, p27 binds cyclin-CDK complexes through a sequential mechanism involving binding-induced protein folding. *Nat Struct Mol Biol*, 2004. **11**(4): 358–64.

88. Diallo, A., *et al.*, Cloning of the nucleocapsid protein gene of peste-des-petits-ruminants virus: Relationship to other morbilliviruses. *J Gen Virol*, 1994. **75**(Pt 1): 233–7.

89. Garner, E., *et al.*, Predicting binding regions within disordered proteins. *Genome Inform Ser Workshop Genome Inform*, 1999. **10**: 41–50.

90. Oldfield, C.J., *et al.*, Coupled folding and binding with alpha-helix-forming molecular recognition elements. *Biochemistry*, 2005. **44**(37): 12454–70.

91. Mohan, A., *et al.*, Analysis of molecular recognition features (MoRFs). *J Mol Biol*, 2006. **362**(5): 1043–59.

92. Vacic, V., *et al.*, Characterization of molecular recognition features, MoRFs, and their binding partners. *J Proteome Res*, 2007. **6**(6): 2351–66.

93. Kingston, R.L., *et al.*, Structural basis for the attachment of a paramyxoviral polymerase to its template. *Proc Natl Acad Sci U S A*, 2004. **101**(22): 8301–6.

94. Gely, S., *et al.*, Solution structure of the C-terminal X domain of the measles virus phosphoprotein and interaction with the intrinsically disordered C-terminal domain of the nucleoprotein. *J Mol Recogn*, 2010. DOI:10.1002/jmr.1010.

95. Bernard, C., *et al.*, Interaction between the C-terminal domains of N and P proteins of measles virus investigated by NMR. *FEBS Lett*, 2009. **583**(7): 1084–9.

96. Morin, B., *et al.*, Assessing induced folding of an intrinsically disordered protein by site-directed spin-labeling EPR spectroscopy. *J Phys Chem B*, 2006. **110**(41): 20596–608.

97. Belle, V., *et al.*, Mapping alpha-helical induced folding within the intrinsically disordered C-terminal domain of the measles virus nucleoprotein by site-directed spin-labeling EPR spectroscopy. *Proteins*, 2008. **73**(4): 973–88.

98. Kavalenka, A., *et al.*, Conformational analysis of the partially disordered measles virus NTAIL–XD complex by SDSL EPR spectroscopy. *Biophys J*, 2010. **98**(6): 1055–64.

99. Sivakolundu, S.G., D. Bashford and R.W. Kriwacki, Disordered p27Kip1 exhibits intrinsic structure resembling the Cdk2/cyclin A-bound conformation. *J Mol Biol*, 2005. **353**(5): 1118–28.

100. Tsai, C.D., *et al.*, Protein folding: Binding of conformationally fluctuating building blocks via population selection. *Crit Rev Biochem Mol Biol*, 2001. **36**(5): 399–433.

101. Tsai, C.J., *et al.*, Structured disorder and conformational selection. *Proteins*, 2001. **44**(4): 418–27.

102. Shoemaker, B.A., J.J. Portman and P.G. Wolynes, Speeding molecular recognition by using the folding funnel: The fly-casting mechanism. *Proc Natl Acad Sci U S A*, 2000. **97**(16): 8868–73.

103. Oglesbee, M., *et al.*, Isolation and characterization of canine distemper virus nucleo-capsid variants. *J Gen Virol*, 1989. **70**(Pt 9): 2409–19.

104. Robbins, S.J. and R.H. Bussell, Structural phosphoproteins associated with puri-fied measles virions and cytoplasmic nucleocapsids. *Intervirology*, 1979. **12**(2): 96–102.

105. Stallcup, K.C., S.L. Wechsler and B.N. Fields, Purification of measles virus and char-acterization of subviral components. *J Virol*, 1979. **30**(1): 166–76.

106. Zhang, X., *et al.*, Identification and characterization of a regulatory domain on the carboxyl terminus of the measles virus nucleocapsid protein. *J Virol*, 2002. **76**(17): 8737–46.

107. Vincent, S., *et al.*, Restriction of measles virus RNA synthesis by a mouse host cell line: Trans-Complementation by polymerase components or a human cellular factor(s). *J Virol*, 2002. **76**(12): 6121–30.

108. Fearns, R. and P.L. Collins, Role of the M2–1 transcription antitermination pro-tein of respiratory syncytial virus in sequential transcription. *J Virol*, 1999. **73**(7): 5852–64.

109. Hartlieb, B., *et al.*, Oligomerization of Ebola virus VP30 is essential for viral tran-scription and can be inhibited by a synthetic peptide. *J Biol Chem*, 2003. **278**(43): 41830–6.

110. Vasconcelos, D., E. Norrby and M. Oglesbee, The cellular stress response increases measles virus-induced cytopathic effect. *J Gen Virol*, 1998. **79**(Pt 7): 1769–73.

111. Vasconcelos, D.Y., X.H. Cai and M.J. Oglesbee, Constitutive overexpression of the major inducible 70 kDa heat shock protein mediates large plaque formation by measles virus. *J Gen Virol*, 1998. **79**(Pt 9): 2239–47.

112. Oglesbee, M.J., *et al.*, Enhanced production of morbillivirus gene-specific RNAs fol-lowing induction of the cellular stress response in stable persistent infection. *Virology*, 1993. **192**(2): 556–67.

113. Oglesbee, M.J., *et al.*, The highly inducible member of the 70 kDa family of heat shock proteins increases canine distemper virus polymerase activity. *J Gen Virol*, 1996. **77**(Pt 9): 2125–35.

114. Zhang, X., *et al.*, Hsp72 recognizes a P binding motif in the measles virus N protein C-terminus. *Virology*, 2005. **337**(1): 162–74.

115. Carsillo, T., *et al.*, A single codon in the nucleocapsid protein C terminus contributes to *in vitro* and *in vivo* fitness of Edmonston measles virus. *J Virol*, 2006. **80**(6): 2904–12.

116. Oglesbee, M., Nucleocapsid Protein Interactions with the Major Inducible 70 kDa Heat Shock Protein, in *Measles Virus Nucleoprotein*, S. Longhi, Editor. Nova Publishers Inc.: Hauppage, NY, 2007.

117. Tapparel, C., D. Maurice and L. Roux, The activity of Sendai virus genomic and antigenomic promoters requires a second element past the leader template regions: A motif (GNNNNN)3 is essential for replication. *J Virol*, 1998. **72**(4): 3117–28.

118. Oglesbee, M., S. Ringler and S. Krakowka, Interaction of canine distemper virus nucleocapsid variants with 70K heat-shock proteins. *J Gen Virol*, 1990. **71**(Pt 7): 1585–90.

119. Marion, D., *et al.*, Assignment of the 1H, 15N and 13C resonances of the nucleocapsid-binding domain of the Sendai virus phosphoprotein. *J Biomol NMR*, 2001. **21**(1): 75–6.

120. Cortese, M.S., V.N. Uversky and A.K. Dunker, Intrinsic disorder in scaffold proteins: Getting more from less. *Prog Biophys Mol Biol*, 2008. **98**(1): 85–106.

121. Balazs, A., *et al.*, High levels of structural disorder in scaffold proteins as exemplified by a novel neuronal protein, CASK-interactive protein1. *FEBs J*, 2009. **276**(14): 3744–56.

122. Curran, J., T. Pelet and D. Kolakofsky, An acidic activation-like domain of the Sendai virus P protein is required for RNA synthesis and encapsidation. *Virology*, 1994. **202**(2): 875–84.

123. Harty, R.N. and P. Palese, Measles virus phosphoprotein (P) requires the NH2- and COOH-terminal domains for interactions with the nucleoprotein (N) but only the COOH terminus for interactions with itself. *J Gen Virol*, 1995. **76**(Pt 11): 2863–7.

124. Curran, J.A. and D. Kolakofsky, Rescue of a Sendai virus DI genome by other parainfluenza viruses: Implications for genome replication. *Virology*, 1991. **182**(1): 168–76.

125. Gupta, A.K., D. Shaji and A.K. Banerjee, Identification of a novel tripartite complex involved in replication of vesicular stomatitis virus genome RNA. *J Virol*, 2003. **77**(1): 732–8.

126. Dunker, A.K., *et al.*, Flexible nets. *FEBs J*, 2005. **272**(20): 5129–48.

127. Uversky, V.N., C.J. Oldfield and A.K. Dunker, Showing your ID: Intrinsic disorder as an ID for recognition, regulation and cell signaling. *J Mol Recognit*, 2005. **18**(5): 343–84.

128. Haynes, C., *et al.*, Intrinsic disorder is a common feature of hub proteins from four eukaryotic interactomes. *PLoS Comput Biol*, 2006. **2**(8): e100.

129. Tompa, P., The functional benefits of disorder. *J Mol Struct (Theochem)*, 2003. 666–67: 361–71.

130. Dunker, A.K., *et al.*, Protein disorder and the evolution of molecular recognition: Theory, predictions and observations. *Pac Symp Biocomput*, 1998. **3**: 473–84.

131. Wright, P.E. and H.J. Dyson, Intrinsically unstructured proteins: Re-assessing the protein structure-function paradigm. *J Mol Biol*, 1999. **293**(2): 321–31.

132. Dunker, A.K. and Z. Obradovic, The protein trinity-linking function and disorder. *Nat Biotechnol*, 2001. **19**(9): 805–6.

133. Dunker, A.K., C.J. Brown and Z. Obradovic, Identification and functions of usefully disordered proteins. *Adv Protein Chem*, 2002. **62**: 25–49.

134. Uversky, V.N., *et al.*, Biophysical properties of the synucleins and their propensities to fibrillate: Inhibition of alpha-synuclein assembly by beta- and gamma-synucleins. *J Biol Chem*, 2002. **25**: 25.

135. Gunasekaran, K., *et al.*, Extended disordered proteins: Targeting function with less scaffold. *Trends Biochem Sci*, 2003. **28**(2): 81–5.

136. Fink, A.L., Natively unfolded proteins. *Curr Opin Struct Biol*, 2005. **15**(1): 35–41.

137. Dyson, H.J. and P.E. Wright, Intrinsically unstructured proteins and their functions. *Nat Rev Mol Cell Biol*, 2005. **6**(3): 197–208.

138. Iwasaki, M., *et al.*, The matrix protein of measles virus regulates viral RNA synthesis and assembly by interacting with the nucleocapsid protein. *J Virol*, 2009. **83**(20): 10374–83.

139. tenOever, B.R., *et al.*, Recognition of the measles virus nucleocapsid as a mechanism of IRF-3 activation. *J Virol*, 2002. **76**(8): 3659–69.

140. Colombo, M., *et al.*, The interaction between the measles virus nucleoprotein and the interferon regulator factor 3 relies on a specific cellular environment. *Virol J*, 2009. **6**(1): 59.

141. De, B.P. and A.K. Banerjee, Involvement of actin microfilaments in the transcription/replication of human parainfluenza virus type 3: Possible role of actin in other viruses. *Microsc Res Tech*, 1999. **47**(2): 114–23.

142. Moyer, S.A., S.C. Baker and S.M. Horikami, Host cell proteins required for measles virus reproduction. *J Gen Virol*, 1990. **71**(Pt 4): 775–83.

143. Marie, J.C., *et al.*, Mechanism of measles virus-induced suppression of inflammatory immune responses. *Immunity*, 2001. **14**(1): 69–79.

144. Laine, D., *et al.*, Measles virus nucleoprotein binds to a novel cell surface receptor distinct from FcγRII via its C-terminal domain: Role in MV-induced immunosuppression. *J Virol*, 2003. **77**(21): 11332–46.

145. Laine, D., *et al.*, Measles virus nucleoprotein induces cell proliferation arrest and apoptosis through NTAIL/NR and NCORE/FcgRIIB1 interactions, respectively. *J Gen Virol*, 2005. **86**(6): 1771–84.

146. Liston, P., C. DiFlumeri and D.J. Briedis, Protein interactions entered into by the measles virus P, V, and C proteins. *Virus Res*, 1995. **38**(2–3): 241–59.

147. Oglesbee, M.J., M. Pratt and T. Carsillo, Role for heat shock proteins in the immune response to measles virus infection. *Viral Immunol*, 2002. **15**(3): 399–416.

148. Carsillo, T., *et al.*, Hyperthermic pre-conditioning promotes measles virus clearance from brain in a mouse model of persistent infection. *Brain Res*, 2004. **1004**(1–2): 73–82.

149. Couturier, M., *et al.*, High affinity binding between Hsp70 and the C-terminal domain of the measles virus nucleoprotein requires an Hsp40 co-chaperone. *J Mol Recognit*, 2009. DOI: 10.1002/jmr.982.

150. Cheng, Y., *et al.*, Rational drug design via intrinsically disordered protein. *Trends Biotechnol*, 2006. **24**(10): 435–42.

151. Chinchar, V.G. and A. Portner, Functions of Sendai virus nucleocapsid polypeptides: Enzymatic activities in nucleocapsids following cleavage of polypeptide P by *Staphylococcus aureus* protease V8. *Virology*, 1981. **109**(1): 59–71.

152. Deshpande, K.L. and A. Portner, Monoclonal antibodies to the P protein of Sendai virus define its structure and role in transcription. *Virology*, 1985. **140**(1): 125–34.

153. Dedmon, M.M., *et al.*, FlgM gains structure in living cells. *Proc Natl Acad Sci U S A*, 2002. **99**(20): 12681–4.

154. McNulty, B.C., G.B. Young and G.J. Pielak, Macromolecular crowding in the *Escherichia coli* periplasm maintains alpha-synuclein disorder. *J Mol Biol*, 2006. **355**(5): 893–7.

155. Bodart, J.F., *et al.*, NMR observation of Tau in *Xenopus oocytes*. *J Magn Reson*, 2008. **192**(2): 252–7.

156. Chen, J.W., *et al.*, Conservation of intrinsic disorder in protein domains and families: I. A database of conserved predicted disordered regions. *J Proteome Res*, 2006. **5**(4): 879–87.

157. Tokuriki, N., *et al.*, Do viral proteins possess unique biophysical features? *Trends Biochem Sci*, 2009. **34**(2): 53–9.
158. Curran, J. and D. Kolakofsky, Replication of paramyxoviruses. *Adv Virus Res*, 1999. **54**: 403–22.
159. DeLano, W.L., The PyMOL molecular graphics system. *Proteins: Structure, Function and Bioinformatics*, 2002. **30**: 442–54.

Chapter 7

Biochemical and Structural Insights into Vesicular Stomatitis Virus Transcription

Amal A. Rahmeh* and Sean P. J. Whelan*

1. Introduction

The genomes of non-segmented negative-strand (NNS) RNA viruses exist uniquely as a protein–RNA complex in which the RNA is completely encapsidated by a viral nucleocapsid (N) protein. The N-RNA is replicated and transcribed by an RNA-dependent RNA polymerase (RdRp) that is composed of a large (L) core catalytic subunit and a phosphoprotein cofactor (P). To synthesize RNA, the polymerase is required to negotiate structural and functional complexities that arise from the usage of a non-segmented encapsidated template. Structurally, the polymerase must gain access to the RNA genome while maintaining the integrity of the N-RNA. Functionally, the polymerase is required to differentially respond to regulatory sequences that direct genome replication or sequential mRNA transcription. In addition to nucleotide polymerization, the enzymatic activities of L include those responsible for mRNA cap formation and polyadenylation. The various catalytic activities are regulated such that cap formation and polyadenylation are suppressed during replication yet coordinated during transcription to ensure that the nascent mRNAs are properly capped, methylated, and polyadenylated. Vesicular stomatitis virus (VSV) has served as an important prototype to study the complex nature of NNS viral RNA synthesis. This reflects the availability of robust systems to study transcription *in vitro* and to generate recombinant viruses. The combination of those systems has led to significant recent progress in understanding the structural and

*Department of Microbiology & Molecular Genetics, Harvard Medical School, 200 Longwood Ave, Boston, MA 02115, USA.

127

functional organization of the RNA synthesis machinery. This chapter specifically focuses on the catalytic core of this machinery, L, reviewing the recent advances in understanding the mechanisms of its various catalytic activities, their coordination during mRNA synthesis, and their overall structural organization into one multifunctional unit. This information is integrated into the framework of elucidating the mechanisms by which the polymerase circumvents the challenges of transcribing an encapsidated negative-sense RNA genome with remarkable precision.

2.   General Overview of RNA Synthesis

The 11,161-nucleotide genome of VSV encodes the N, P, matrix protein (M), glycoprotein (G), and L genes. Those five genes are flanked at the termini by regulatory leader (Le) and trailer (Tr) regions and are arranged in the order 3′ Le-N-P-M-G-L-Tr 5′ (Fig. 1). During transcription, the RdRp engages the N-RNA template and begins to synthesize mRNA by initiating synthesis in response to a promoter element provided by the leader region and the first gene-start (GS) sequence.[1–4] Shortly following initiation, the polymerase caps the nascent mRNA by recognizing a specific *cis*-acting element present at its 5′ end.[4–6] The cap structure is further modified by the polymerase to become methylated at the ribose 2′-O (2′-O) and

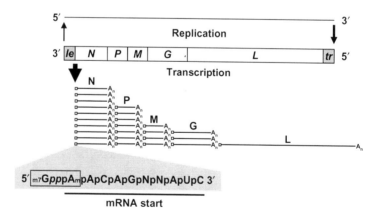

Fig. 1.   Overview of VSV RNA synthesis. The negative-sense RNA genome is depicted with Leader (le), the genes encoding N, P, M, G, and L, and trailer (tr). During replication, the polymerase copies the genome into antigenome and back to genome. During transcription, the polymerase transcribes the five mRNAs encoded by each gene. All mRNAs share a conserved 5′ start sequence (shaded in gray) that is modified by cap addition and cap methylation at the ribose 2′-O and guanine-N-7 positions (boxed). The mRNA 3′end is polyadenylated (An).

guanine-N-7 (G-N-7) positions.[7] At a second *cis*-acting signal on the template, a 23-nt element termed the gene junction, the polymerase terminates and polyadenylates the upstream mRNA and reinitiates and caps the downstream mRNA.[8–11] The events at the gene junction are not 100% efficient such that, for VSV, approximately 30% less of the downstream mRNA is synthesized.[12] This continues along the template producing sequentially less of the five viral mRNAs (Fig. 1). This mode of transcription is referred to as the "start–stop" model of sequential transcription and is well supported by experimental evidence (reviewed in Ref. 13).

During RNA replication, the polymerase initiates at the 3′ end of the template but now ignores all the regulatory signals provided at each gene junction to instead produce a full-length antigenomic RNA. Initiation at the precise 3′ end of the genomic RNA produces a 47-nt leader RNA (Le+) that retains a 5′ ppp terminus.[14,15] This transcript remains uncapped because it lacks the *cis*-acting signals at the 5′ end for mRNA cap addition.[5,6] The finding that Le+ interacts with soluble N protein (termed N°) led to the simple model in which N° controls the template activity of RdRp, switching it from transcription to replication.[16,17] Although the number of genes differs for NNS RNA viruses, the general strategy of transcription and replication is essentially the same.

3. The Machinery of Viral RNA Synthesis

Following the early demonstration that purified VSV retains an active transcriptase capable of viral mRNA synthesis *in vitro*,[18] biochemical experiments were performed to dissect the components of the RNA synthesis machinery. A transcriptionally active ribonucleoprotein (RNP) complex was isolated from virus and separated into the individual components of N-RNA, L, and P. Reconstitution of RNA synthesis *in vitro* from these separated components[19,20] showed that L was required for catalysis of RNA synthesis and that P served as a polymerase cofactor.[21] The utility of the reconstitution system was significantly enhanced by the ability to substitute virus-derived proteins with purified recombinant L and P.[22,23] This has permitted the study of lethal mutations in the L and P genes, providing new insights into the mechanistic role of critical functional residues of these proteins.[23]

The determination of the atomic structure of the N-RNA complex and of domains of P has offered a structural framework to relate the functions of N and P in RNA synthesis. The atomic structures of the N-RNA complexes of VSV and rabies virus show that the RNA is tightly sequestered in a channel formed by the two lobes of each N molecule along adjacent N monomers.[24,25] This implies that N must be removed or substantially remodeled to permit L access to the RNA.

During RNA synthesis, however, the RNA template remains resistant to nuclease degradation, indicating that it remains physically shielded during active copying by the polymerase. This process entails a transient displacement or remodeling of N from the copied stretches of RNA and concomitant physical protection by the polymerase complex. The process whereby L gains access to the RNA template while maintaining the integrity of the N-RNA remains unsolved.

The polymerase cofactor, P, is a 29 kDa highly phosphorylated protein that oligomerizes and binds to the N-RNA template and to L. Biochemical experiments demonstrate that the N-terminus of P contains an L binding domain and that the C-terminus contains an N-RNA binding domain.[26,27] Crystallization of a C-terminal domain of P (P_{CTD}) in complex with N-RNA shows that the P_{CTD} interacts between adjacent lobes of two N molecules.[28] As the N-terminus of P binds L, it seems likely that the interactions between the P_{CTD} and the N-protein facilitate L engagement of the RNA. The atomic structure of a central oligomerization domain of P shows that it is dimeric,[29] though biochemical experiments suggest that P may function as a tetramer.[30] This may reflect a requirement to form a higher order L–P complex for RNA synthesis. The requirements within VSV L for interaction with P remain unknown. However, work with Sendai virus (SeV) implicates the N-terminus of L in this association,[31,32] but a small deletion in the C-terminus of VSV L also inhibits P binding.[33] Thus, major questions remain about how the polymerase complex assembles and gains access to the N-RNA.

Genetic and biochemical evidence have shown that L contains the enzymatic activities required for RNA-dependent RNA polymerization, as well as mRNA cap addition. Sequence alignments among different NNS RNA virus L proteins identified six conserved regions (CR) I–VI[34] (Fig. 2). The CRIII region of L contains

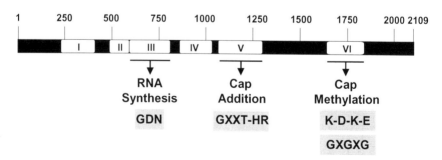

Fig. 2. Functional organization of VSV L. A schematic of VSV L depicting conserved regions (CR) among NNS virus L proteins as white boxes separated by variable regions (VR) marked in black. Signature motifs for RNA polymerization (GDN), cap addition (GXXT-HR) and cap methylation (K-D-K-E and GXGXG) map to CR III, V, and VI respectively.

a set of core motifs that are shared among all template-dependent polynucleotide polymerases. These include a GDN motif where the aspartic acid is equivalent to first aspartic acid in the general XDD motif that coordinates a catalytically essential Mg++ ion. A $D_{714}A$ substitution in the $G_{713}D_{714}N_{715}$ motif of VSV inactivates RNA synthesis and is consistent with the RdRp catalytic domain residing within CRIII.[35] The size of the protein combined with the unique nature of the capping reactions made the identification of the regions of L required for those activities difficult. The cap-forming activities were, however, mapped by engineering extensive single amino acid substitutions to conserved residues within L and testing the activity in RNA synthesis and cap formation *in vitro*. Amino acid substitutions in CRV disrupt cap addition,[23,36] and amino acid substitutions in CRVI disrupt cap methylation[37–39] (Sec. 5). Specific functions have not yet been assigned to CRI, II, and IV of VSV L, but experiments with SeV implicate CRI and II in P interaction and template binding, respectively.[40,41] Between the CRs are variable regions (VR) of limited sequence homology. The VR between CRV and VI has been termed a "hinge region" into which it has been possible to insert GFP while retaining L function.[42,43] Sequence alignments also support the presence of a hinge region between CRII and III for the viruses in the family Paramyxoviridae. The mapping of the different enzymatic activities onto the linear sequence of L is depicted in Fig. 2.

The discrete distribution of motifs that are required for the different activities of L into distinct CRs suggests that the polymerase might function as a series of independent domains. Consistent with this idea of independent functional domains, a C-terminal fragment of SeV L retains G-N-7 methyltransferase (MTase) activity.[44] A recent determination of the general molecular architecture of VSV L provided the first structural evidence for the presence of these motifs in distinct domains of L (Sec. 7). It is important to emphasize, however, that the various activities of L are coordinated so that capping of the nascent mRNA is tightly coupled to its transcription (Sec. 6). Consequently, while the enzymatic activities of the protein may reside in separate regions of the primary amino acid sequence that locates in distinct structural domains, those activities are highly coordinated within the three-dimensional assembly of the RNP complex.

Although the functional polymerase complex contains L and P, the precise identity and stoichiometry of the complex have been difficult to determine unambiguously. Two polymerase complexes that differ in function and composition have been described:[45] a transcriptase comprising L, P, and host factors (translation elongation factor 1A, heat shock protein 60, and substoichiometric amounts of the host capping enzyme) and a replicase comprising the N, P, and L proteins. There is, however, little direct evidence to support that the host components associated with the transcriptase are required for RNA synthesis. For instance, the findings that capping of viral mRNA is catalyzed by L[6,23] and that the leader RNA synthesized *in vitro* remains uncapped, indicate that the reported associated host guanylyltransferase is likely of no functional relevance.

4. Polymerase Initiation Site Choice

The start–stop model of sequential transcription is widely accepted; however, the route by which the polymerase gains access to the first GS sequence remains open to debate. Evidence has accumulated in the support of two possible entry mechanisms: one in which the polymerase initiates at the 3′ end of the genome to transcribe Le+ prior to N mRNA, or the second in which the polymerase initiates transcription internally at the first GS (reviewed in Ref. 13). Evidence for transcription of the Le+ prior to N mRNA comes from reconstituted transcription reactions using purified N-RNA template and separately purified RdRp. When specific NTPs are omitted from the reconstituted reactions, the products of synthesis correspond to the initiation at the 3′ end of the genome and not to initiation at the *N* GS.[46] Application of the technique of ultraviolet (UV) mapping to engineered viral genomes also indicated that *in vitro* the polymerase gains access at the 3′ end of the template. In cells, however, the UV mapping experiments indicated that the polymerase can initiate at the *N* GS independent of prior transcription of the leader RNA.[47] Evidence for internal initiation *in vitro* was first provided by the polR1 mutant of VSV, which contains a single amino acid change in the N protein. This VSV mutant produces Le+ in submolar quantities to the N mRNA, suggesting internal initiation.[48] Further evidence in support of an internal initiation was provided by the purification of two separate polymerase activities from infected cells: one that initiates at the 3′ terminus, and another that initiates internally at the *N* GS site.[45]

One possible explanation to reconcile these seemingly different findings involves the notion of movement of the polymerase along the template without synthesis of a product, termed "polymerase scanning." The concept of scanning comes from observations made predominantly at internal gene junctions in NNS RNA viruses, where following termination of the upstream mRNA the polymerase reinitiates at a downstream gene some distance away (up to 100's of nucleotides).[49,50] Because the product of the intervening region is not found in either of the mRNAs, it is referred to as a non-transcribed intergenic region. Sequence analysis of the *G-L* intergenic region of the New Jersey serotype of VSV showed that in contrast to the 2-nt G/CA intergenic region present between other genes, the *G-L* intergenic region is composed of 21-nts. Remarkably, during transcription reactions performed *in vitro*, the polymerase was shown to initiate at a non-canonical sequence within the intergenic region as well as at the *L* GS.[51] This implies that following termination of synthesis of the G mRNA, the polymerase moves through the intergenic region "scanning" for an initiation sequence. These observations are consistent with mutational analysis of the VSV GS sequence revealing that the requirements for initiation are less fastidious than for 5′ RNA modification.[5,52] As such, the polymerase may actually initiate synthesis on a non-canonical mRNA start sequence during scanning. As the mRNA start sequence would then lack the requirements for cap addition, premature termination would ensue as described below. Such uncapped products would not be detected in cell-based assays, which have provided most of the experimental evidence for scanning, as they would be rapidly degraded by the cellular RNA degradation machinery.

5.   The Unconventional Nature of L-Mediated mRNA Cap Formation

The 5′ m7GpppNm cap structure present at the 5′ ends of mRNAs is added by the sequential action of a series of RNA modifying enzymes.[53] At the time that the mRNA cap structure was discovered, it was quickly realized that VSV capping occurs through a fundamentally distinct mechanism than cellular and most characterized viral systems. Chemical analysis of VSV mRNA synthesized *in vitro* in the presence of $[\beta^{32}]$-GTP and $[^{3}H]$ S-adenosyl methionine (SAM) showed that the 5′ terminus consisted of 5′ m7G*ppp*Am, where the two italicized phosphates are derived from GTP.[4,54] This contrasted with the conventional provenance of one phosphate from GTP. Two RNA MTase activities were proposed since limited concentrations of SAM ($< 0.1\,\mu$M) resulted in a single methylation at the ribose 2′-O position of the first adenosine and chased into a double methylated species at higher concentrations ($>7\,\mu$M).[55] This order of methylation thus appeared to be the opposite to that of conventional cap methylation. Recently, the enzymatic activities that mediate cap addition and methylation were precisely mapped in L and the unconventional mechanisms by which they occur were characterized (Fig. 3).

5.1.   *Mechanism of mRNA cap addition by L*

An L-specific capping activity was demonstrated by the ability of purified recombinant L to cap exogenous 5′-triphosphorylated oligo-RNAs corresponding in sequence to the VSV mRNA start pppApApCpApG. In this reaction, monophosphate RNA is transferred onto a GDP acceptor through a covalent L-RNA intermediate. L was also shown to possess a GTPase activity presumably to

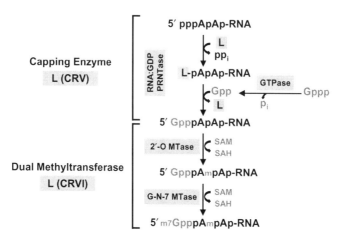

Fig. 3. Cap formation for VSV mRNA. L contains all the enzymatic activities necessary for cap formation; CRV contains the cap addition activity and CRVI contains the dual specificity MTase activity. For cap addition, L forms a covalent intermediate with the 5′ end of the nascent RNA then transfers the RNA onto a GDP acceptor through a PRNTase activity. GTP is hydrolyzed into the GDP acceptor by a GTPase activity. For methylation, L catalyzes two sequential methylation reactions adding a methyl group derived from SAM to the ribose 2′-O position followed by the G-N-7 position.

hydrolyze GTP into the GDP acceptor. Those experiments demonstrated that VSV mRNA cap addition occurs through a GDP:polyribonucleotidyltransferase (PRNTase) activity.[6] The unusual capping activity of L was mapped to CRV, and a signature motif that is essential for mRNA cap formation was defined.[23] The critical residues of this motif were identified by phylogenetic alignment of CRV among NNS virus L proteins, and each conserved residue in this region was systematically evaluated in *in vitro* transcription reconstituted using recombinant L and P with purified N-RNA. Substitutions $G_{1154}A$, $T_{1157}A$, $H_{1227}A$, and $R_{1228}A$ generated truncated transcripts (50–400 nts) with a correctly initiated yet uncapped 5′ triphosphorylated ends. As the chain length at which the mRNA acquires its cap is unknown, it was formally possible that the substitutions in L resulted in defects in polymerase processivity that indirectly led to a failure to cap. This possibility was eliminated by the demonstration that wild type L, but not the CRV mutants, capped a triphosphorylated 5-nt RNA substrate *in trans*. This study thus defined a new motif, $GXXT(X_{68-70})HR$, present in NNS RNA virus L proteins, which plays a direct role in mRNA cap formation. A hidden Markov model search of the protein data bank revealed that this motif was only found in NNS RNA virus L proteins, consistent with the unique chemistry of this capping reaction. As the L-RNA covalent intermediate was shown to be sensitive to hydrolysis by neutral hydroxylamine, this intermediate was suggested to occur via a phosphoamide bond involving a histidine,[6] which explained the identification of H_{1227} as key for mRNA cap formation.[23] Subsequent mass spectrometric analysis confirmed H_{1227} to be the residue involved in the covalent L-pRNA intermediate, in a manner that is influenced by the neighboring R_{1228}.[36]

The role of the GXXT motif has not yet been defined. This motif shares some homology with a GSDS motif present within nucleoside diphosphate kinases that exchange γ phosphate groups from NTP to NDP, but lacks essential catalytic residues needed for that activity.[56] However, many ATP and GTP binding proteins have a phosphate binding loop (p-loop) with a consensus sequence GXXXXGK(S/T).[57] Perhaps the GXXT motif represents a p-loop-like motif that is involved in binding GTP or GDP. A p-loop-like motif has been described for dUTPases from different organisms with a consensus RGXXG(F/Y)GS(S/T)G where the K residue of a canonical P-loop is absent.[58] The L proteins of rhabdoviruses possess a similar RGXXXXYXGSXT motif raising the possibility that the GXXT motif is required for hydrolysis of GTP or binding of the GDP acceptor for the capping reaction.

mRNA cap addition is further distinguished by its requirements for specific *cis*-elements in the GS sequence. For VSV, the conserved GS sequence is composed of 10-nts: 3'-UUGUCNNUAG-5'. Insertion of a 60-nts non-essential gene between Le and *N* allowed the dissection of sequence requirements for initiation of mRNA synthesis and cap addition. Mutations introduced at each position of the GS showed that the transcription of the first gene was detected for all mutants except for U8G, demonstrating that the GS is remarkably tolerant of mutation for initiation of mRNA synthesis. However, changes to positions 1–3 that deviated from the consensus 3'-UYG-5' yielded transcripts that were not capped.[5] Furthermore, recombinant VSV L capped exogenous pppApApCpApG oligos corresponding to the mRNA start but failed to cap pppApCpGpApA oligos corresponding to Le+ start.[6] This stringency in sequence requirement for capping contrasts with the more promiscuous capping apparatus of eukaryotic cells and of the well-characterized vaccinia virus (VV). The sequence specificity of L for capping the mRNA sequences might reflect a regulatory mechanism ensuring that the Le+ and Le− sequences remain uncapped.

5.2. *Mechanism of mRNA cap methylation by L*

The first evidence for an L-associated RNA MTase activity was provided by the complementation of a host range mutant VSV that failed to methylate mRNA *in vitro*.[59] Despite the availability of this mutant, the sequence of its *L* gene was unknown and thus the responsible region of L uncertain. Using sequence and structural data of the *Escherichia coli* ribosomal RNA MTase J (RrmJ) in iterative searches of sequence databases, bioinformatic analyses identified a putative 2'-O MTase domain in CRVI of NNS virus L proteins.[60,61] This prediction identified a set of nine motifs in L corresponding to the highly conserved core structural fold characteristic of SAM-dependent MTases consisting of alternating 7 β strands and 5 α helices. Motifs I–III form an SAM binding cleft with a GXGXG signature, and motifs X, IV, VI, and VIII form a putative substrate binding region and contain an invariant K-D-K-E signature. Biochemical studies of RrmJ and the VV 2'-O MTase (VP39) demonstrated that the K-D-K-E tetrad was critical for catalysis.[62]

Biochemical tests of these predictions in VSV L revealed that CRVI was indeed required for mRNA cap methylation. Recombinant VSV containing alanine substitutions in the conserved $K_{1651}D_{1762}K_{1795}E_{1833}$ of L synthesized unmethylated mRNA *in vitro*.[37] Remarkably, the mRNA lacked both G-N-7 and ribose 2'-O methyl groups. These results suggested that the mRNA cap methylation reactions of VSV occur in the order 2'-O prior to G-N-7, or that this region of L was required for both methylations. Alanine substitutions to the SAM binding motif $G_{1670}DGSGG_{1675}$-D_{1735} provided evidence that the same SAM binding site is used by both the ribose 2'-O and G-N-7 MTase. Alanine substitutions to residues D_{1671} S_{1673} G_{1675} and D_{1735} blocked both 2'-O and G-N-7 methylation, whereas substitutions to G_{1670} and G_{1672} blocked G-N-7 methylation only.[38] These results provided further evidence to support the hypothesis that L is a dual specificity MTase catalyzing cap methylation in the reverse order of the typical cap methylation reactions.

The requirements within CRVI of L for catalysis of the G-N-7 methylation have not yet been defined. Atomic structures of a number of G-N-7 MTases suggest that catalysis is mediated by correct positioning of the substrates and does not depend upon specific conserved residues.[63,64] Similar to the studies with VSV, the flavivirus 2'-O MTase[65] has also been shown to function as a dual specificity MTase.[66–68] Structural and biochemical data support a substrate repositioning model in which the RNA is initially positioned to favor G-N-7 methylation, followed by the relocation of the m7G cap into a GTP binding pocket to reposition the RNA so that the ribose 2'-O position is juxtaposed to the SAM binding site.[69] The lack of atomic structure of VSV L limits our insights into how its dual specificity MTase activity is controlled. However, as discussed below, it seems likely that there will be distinctions to the mechanism involved in flavivirus MTase since 2'-O methylation appears to occur first in the case of VSV.

Using an *in vitro* methylation assay in which purified recombinant VSV L methylates synthetic capped RNA, further mechanistic insights into the coordination of the two MTase activities of L were obtained.[39] Similar to the mRNA capping activity, the MTase activity of L was shown to be sequence specific. At saturating concentrations of SAM, kinetic analysis showed that ribose 2'-O precedes G-N-7 methylation, directly demonstrating that L preferentially methylates mRNA in the opposite order to other mRNA cap MTases. The conventional order of methylation reflects the strong preference for stacking of the G-N-7 methylated mRNA cap structure in a cap-binding pocket that positions the ribose 2'-O position for methylation. The opposite order of methylation in VSV L does not simply reflect a reversal in substrate preference, as pre-2'-O methylated RNA actually slows G-N-7 methylation. In contrast to the dual specificity MTases of flaviviruses where the two activities are readily uncoupled biochemically, those of VSV L mirror one another under most biochemical conditions. When the need for 2'-O methylation was bypassed by using a pre-2'-O methylated RNA, G-N-7 methylation was also diminished by substitutions to the K-D-K-E motif.[39] Collectively those results suggest that the dual specificity L MTase operates *via* a different mechanism to that of the flavivirus MTase. A putative model that accounts for these results is that binding and methylation of the capped RNA first at the 2'-O position induce a conformational change in CRVI, which leads to repositioning of the RNA and activation of the G-N-7 MTase activity.

6. mRNA Cap Formation as a Central Element of Transcriptional Control

The simple fact that L participates in two distinct RNA synthesis reactions: (i) mRNA transcription with the accompanying capping, methylation, and polyadenylation or (ii) genome replication in which those L associated activities are silenced indicates that the activities of L are tightly regulated. A hallmark of replication is the need for an ongoing source of N° protein necessary for the encapsidation of the nascent RNA. Continuing encapsidation is thought to serve a positive role in ensuring replicase processivity, where transient interactions of the replicase with the assembling N° on the growing N-RNA chain lock the polymerase in a highly processive mode on the template.[70]

The transcriptase also uses exquisite mechanisms to fine tune its processivity and prevent premature termination of transcription. The net result is that the transcriptase responds precisely to the initiation and termination elements, negotiates gene junctions, and coordinates transit with the formation of a 5' cap structure and a 3' poly(A) tail (Fig. 4). Such coupling of the various enzymatic activities of L is thought to be achieved through a complex temporal and

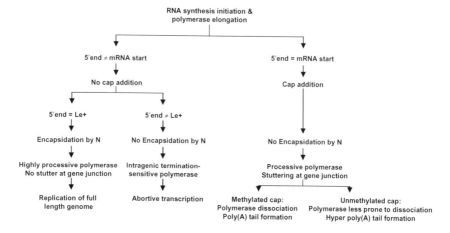

Fig. 4.   Cap formation is a regulatory element for transcription.

spatial organization where substrate binding, catalysis, and/or product release modify the functional interactions between the domains of L. In this section, the importance of capping and methylation in coordination of the activities of L are reviewed.

6.1.   *Cap addition serves as a positive regulator of polymerase elongation*

Several lines of evidence provide compelling support for the idea that cap addition serves as a positive regulatory factor during mRNA transcription. Mutation of the first three positions of the VSV GS 3′-UUGUCNNUAC-5′ to deviate from the consensus 3′-UYG-5′ resulted in the production of transcripts that were correctly initiated, but lacked a cap structure and terminated prematurely.[5,52] Alteration to the capping GXXT-HR motif in CRV of VSV L also resulted in the synthesis of uncapped, prematurely terminated transcripts.[23] Similarly, a chemical inhibitor of respiratory syncytial virus RNA synthesis resulted in the production of short uncapped transcripts.[56] Collectively, these data show that a defect in capping is invariably correlated with a defect in the production of full-length mRNA transcripts. It is notable that all of the above studies were conducted using experimental systems that are based on minigenome assays, non-essential gene insertion in the viral genomic RNA, or reconstituted *in vitro* transcription systems since a defect in capping is deleterious to viral transcription.

Analyses of the uncapped prematurely terminated products of VSV showed that irrespective of the mechanism by which capping is inhibited (polymerase mutation or GS mutation), the resulting transcripts typically terminated between

40 and 400 nts in length. The pattern of the size distribution of these products was specific to the transcribed gene, suggesting that termination did not occur randomly. Subsequent sequence analysis of the prematurely terminated products of the *N* gene revealed that the polymerase typically terminated on A/U-rich elements.[71] This termination is reminiscent of that occurring at an authentic termination sequence 3′-AUACUUUUUUU-5′ where weak base pairing between the template and the nascent strand controls reiterative synthesis on the U tract to generate the polyadenylate tail, eventually resulting in mRNA termination.[8,9] The parallel between the intragenic termination observed following a failure to cap and the termination at an authentic gene junction suggests that cap addition allows the polymerase to bypass intragenic sequence elements that could potentially result in premature termination.

The dependence of the VSV polymerase on capping for processive gene transcription finds a parallel in the eukaryotic RNA polymerase II (Pol II) where capping is an integral component of the maturation of the polymerase into a processive elongation complex.[72] For pol II, the capping enzyme consists of one or two separate polypeptides that are targeted to the polymerase by direct physical interaction with its C-terminal domain. For VSV, both the polymerase and the capping enzyme reside in L. However, the structural organization of L into distinct domains harboring the RdRp and the cap formation signature motifs (Sec. 7) suggests that a similar functional and/or physical interaction could be involved in the regulation of the polymerase properties by the capping enzyme. In fact, mutational analysis of CRV suggests that, in addition to its role in mRNA cap addition, this region may also play a role in maintaining the overall architecture of L since a number of alanine substitutions completely abolished or substantially decreased mRNA synthesis.[23] Conserved clusters of histidines and cysteines in CRV were noted to bear a resemblance to metal binding sites.[34] Such a metal binding site could play a catalytic role, perhaps by coordinating the divalent ions required for the capping reaction. The cysteine residues were suggested to alternatively play a role in the formation of disulfide bridges providing a structural scaffolding for that region of L.[34] Based on these properties of CRV, it is possible that the capping reaction might result in a conformational change in CRV that would affect the neighboring RdRp domain enhancing its processivity, or CRV itself might also influence an RNA exit channel of the polymerase that in turn could impact elongation.

The cap itself might also influence the properties of the polymerase during elongation. One way in which the cap could influence polymerase is by interacting with a binding pocket to help stabilize the nascent mRNA transcript. Although no cap binding pocket has been mapped in L, a likely location could be in the MTase domain. The stage of mRNA synthesis at which the cap structure is methylated is unknown, but experiments with the New Jersey serotype of VSV suggest that methylation occurs when the RNA is several 100 nt long.[73] As capping and methylation occur co-transcriptionally, the cap may remain associated with L for some time during transcription. Such an association may influence the polymerase properties, rendering it less prone to premature termination.

6.2. *Cap methylation impacts polyadenylation*

Initial evidence that the modification of the 5′ end of the mRNA influenced downstream activities of the polymerase actually came from studies of methylation. Addition of S-adenosyl homocysteine (SAH), a competitive inhibitor of SAM, to *in vitro* transcription reactions resulted in the synthesis of full-length transcripts that contained giant poly(A) tails.[74] More insights into the link between methylation and polyadenylation were provided by the analysis of mutants in CRVI of L that were defective in specific aspects of cap methylation. Substitutions to the SAM binding site of L did not result in the synthesis of longer poly(A) tails in the presence of SAH,[75] whereas some substitutions to the catalytic residues resulted in synthesis of long poly(A) tails independent of the presence or absence of exogenous SAH.[71,75] To explain this distinction, it was suggested that SAH binding was responsible for the production of large poly(A) tails and that some substitutions to the catalytic residues result in preferential binding of SAH, which in this case co-purifies with the mutant L proteins.[75]

The link between methylation and polyadenylation in VSV finds a parallel in VV. The ribose 2′-O MTase (VP39) of VV forms a heterodimer with the poly(A) polymerase (VP55) to stimulate polyadenylation by increasing the affinity of VP55 for the poly(A) tail.[76] In the case of VSV, the enzymatic activities reside within a single polypeptide chain, thus if a similar mechanism is operative, complex interactions between the MTase and RdRp domains of L would be anticipated. Consistent with this hypothesis, the finding that VSV mutants resistant to the MTase inhibitor sinefungin contain amino acid substitutions in CRII/III or CRVI of L is suggestive of an interaction between these two distant regions of L.[77] The structural organization of L into a distinct domain harboring the RdRp connected to a separate flexible domain containing the MTase signature motifs (Sec. 7) provides additional evidence in support of such a putative regulatory mechanism.

The coordination between polyadenylation and methylation could provide a further regulatory checkpoint to ensure that the mRNAs are correctly processed. Transcripts that are not authentically methylated at the 5′ end can be hyperpolyadenylated, which can target the resulting mRNA for degradation.[78]

7. First Insights into the Molecular Architecture of VSV L: Functional Implications for an NNS RNA Virus Polymerase

A major gap in our knowledge regarding the RNA synthesis machinery of NNS RNA viruses is the absence of structural data regarding the core of the machine, the L protein. Recently, we have obtained the first structural glimpse at VSV L using negative-stain electron microscopy of highly purified fully functional protein.[79] Class averages of particle images revealed that the overall architecture of L is composed of a core ring-like domain that is decorated with an appendage composing of three globular domains (Fig. 5). The approximate dimensions of the ring are 90–100 Å in diameter with a 23–27 Å stain accessible center, and each of the globular domains of the appendage is 45–50 Å. The globular domains occupy variable spatial positions relative to the ring, suggesting a high degree of conformational mobility in the appendage. Mapping of the conserved regions onto the molecular architecture of L using a deletion mutagenesis approach revealed that the ring domain encompasses the N-terminus through CRIV and the appendage spans CRV through the C-terminus. Thus, the ring domain harbors the signature motifs for the RdRp, whereas the appendage harbors the signature motifs for cap addition and cap methylation.

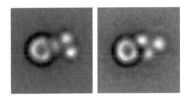

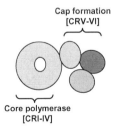

Cap formation
[CRV-VI]

Core polymerase
[CRI-IV]

Fig. 5. Molecular architecture of VSV L. (Top) Two Representative class averages of electron microscopy images of negatively stained VSV L particles. The side length of the individual panels is 30 nm. (Bottom) A schematic of VSV L depicting a general structural organization into a core ring domain spanning the N-terminus through CRIV and an appendage consisting of three globular domains and spanning CRV through the C-terminus.

These first insights into the molecular architecture of L provide a first step for the integration of L into the structural framework provided by the atomic structure of the N-RNA template and portions of P. The center of the L ring is remarkably similar in dimensions to the central opening present in the RdRps of positive and double-strand RNA viruses.[80] By analogy, the RNA template is anticipated to thread through the stain accessible center of L. As the dimensions of the center in L are insufficient to accommodate N protein bound to RNA, this suggests that L displaces N from the template during copying. Such displacement could be mediated by the cofactor P that interacts with L and the N-RNA through its respective N- and C-terminal domains. The molecular architecture of L also offers additional grounds for the interpretation of biochemical and genetic analyses of the functional properties of L. The structural organization of L revealed the location of the RdRp and the cap formation signatures in distinct domains that are demarcated by the ring and the appendage, respectively. Yet, the observed flexibility of the appendage relative to the ring suggests a putative mechanism that could play a role in the differential coordination of the RdRp and the cap forming activities during replication and transcription. For instance, a specific positioning of the capping apparatus relative to the RdRp might permit the suppression of the capping activities during replication, while more dynamic physical and functional interactions could allow the spatial and temporal coordination of polymerase processivity with cap formation during transcription (Sec. 6).

8. Concluding Remarks

As genetic and biochemical analysis of NNS viral RNA synthesis is continually augmented by structural information, a mechanistic picture has begun to emerge. Most recently, the electron microscopic characterization of L provides the first structural insight into the catalytic core. In combination with the atomic level structures of the N-RNA template and portions of P, the EM images of L provide the first complete glimpse of each component of the NNS RNA viral synthesis machine. Future advances mandate the atomic level resolution not only of L but also of the organized N-RNA:L–P complex. These further structural studies will continue to empower genetic and biochemical experiments, yet complete understanding of RNA synthesis will inevitably require more than static images. Understanding the dynamics of this complex machinery will likely entail the application of single molecule techniques. Using such approaches promises to answer the long standing question of where the polymerase enters the template during transcription, how the polymerase gains access to the RNA bases, what happens to the polymerase following termination, and the stages of synthesis at which the various RNA modifying events occur.

Acknowledgments

The authors thank Philip J. Kranzusch for discussion and critical reading of the manuscript.

References

1. Wertz, G.W., *et al.*, Extent of terminal complementarity modulates the balance between transcription and replication of vesicular stomatitis virus RNA. *Proc Natl Acad Sci U S A*, 1994. **91**(18): 8587–91.
2. Whelan, S.P. and G.W. Wertz, The 5′ terminal trailer region of vesicular stomatitis virus contains a position-dependent *cis*-acting signal for assembly of RNA into infectious particles. *J Virol*, 1999. **73**(1): 307–15.
3. Whelan, S.P. and G.W. Wertz, Regulation of RNA synthesis by the genomic termini of vesicular stomatitis virus: Identification of distinct sequences essential for transcription but not replication. *J Virol*, 1999. **73**(1): 297–306.
4. Abraham, G., D.P. Rhodes and A.K. Banerjee, The 5′ terminal structure of the methylated mRNA synthesized *in vitro* by vesicular stomatitis virus. *Cell*, 1975. **5**(1): 51–8.
5. Wang, J.T., L.E. McElvain and S.P. Whelan, The vesicular stomatitis virus mRNA capping machinery requires specific *cis*-acting signals in the RNA. *J Virol*, 2007.
6. Ogino, T. and A.K. Banerjee, Unconventional mechanism of mRNA capping by the RNA-dependent RNA polymerase of vesicular stomatitis virus. *Mol Cell*, 2007. **25**(1): 85–97.
7. Moyer, S.A., *et al.*, Methylated and blocked 5′ termini in vesicular stomatitis virus *in vivo* mRNAs. *Cell*, 1975. **5**(1): 59–67.
8. Barr, J.N., S.P. Whelan and G.W. Wertz, *cis*-Acting signals involved in termination of vesicular stomatitis virus mRNA synthesis include the conserved AUAC and the U7 signal for polyadenylation. *J Virol*, 1997. **71**(11): 8718–25.
9. Barr, J.N. and G.W. Wertz, Polymerase slippage at vesicular stomatitis virus gene junctions to generate poly(A) is regulated by the upstream 3′-AUAC-5′ tetranucleotide: Implications for the mechanism of transcription termination. *J Virol*, 2001. **75**(15): 6901–13.
10. Hwang, L.N., N. Englund and A.K. Pattnaik, Polyadenylation of vesicular stomatitis virus mRNA dictates efficient transcription termination at the intercistronic gene junctions. *J Virol*, 1998. **72**(3): 1805–13.
11. Stillman, E.A. and M.A. Whitt, Mutational analyses of the intergenic dinucleotide and the transcriptional start sequence of vesicular stomatitis virus (VSV) define sequences required for efficient termination and initiation of VSV transcripts. *J Virol*, 1997. **71**(3): 2127–37.

12. Iverson, L.E. and J.K. Rose, Localized attenuation and discontinuous synthesis during vesicular stomatitis virus transcription. *Cell*, 1981. **23**(2): 477–84.

13. Whelan, S.P., J.N. Barr and G.W. Wertz, Transcription and replication of nonsegmented negative-strand RNA viruses. *Curr Top Microbiol Immunol*, 2004. **283**: 61–119.

14. Colonno, R.J., G. Abraham and A.K. Banerjee, Blocked and unblocked 5′termini in vesicular stomatitis virus product RNA *in vitro*: Their possible role in mRNA biosynthesis. *Prog Nucleic Acid Res Mol Biol*, 1976. **19**: 83–7.

15. Colonno, R.J. and A.K. Banerjee, Complete nucleotide sequence of the leader RNA synthesized *in vitro* by vesicular stomatitis virus. *Cell*, 1978. **15**(1): 93–101.

16. Blumberg, B.M., M. Leppert and D. Kolakofsky, Interaction of VSV leader RNA and nucleocapsid protein may control VSV genome replication. *Cell*, 1981. **23**(3): 837–45.

17. Blumberg, B.M., C. Giorgi and D. Kolakofsky, N protein of vesicular stomatitis virus selectively encapsidates leader RNA *in vitro*. *Cell*, 1983. **32**(2): 559–67.

18. Baltimore, D., A.S. Huang and M. Stampfer, Ribonucleic acid synthesis of vesicular stomatitis virus, II. An RNA polymerase in the virion. *Proc Natl Acad Sci U S A*, 1970. **66**(2): 572–6.

19. Banerjee, A.K. and D.P. Rhodes, *In vitro* synthesis of RNA that contains polyadenylate by virion-associated RNA polymerase of vesicular stomatitis virus. *Proc Natl Acad Sci U S A*, 1973. **70**(12): 3566–70.

20. Emerson, S.U. and R.R. Wagner, Dissociation and reconstitution of the transcriptase and template activities of vesicular stomatitis B and T virions. *J Virol*, 1972. **10**(2): 297–309.

21. Emerson, S.U. and R.R. Wagner, L protein requirement for *in vitro* RNA synthesis by vesicular stomatitis virus. *J Virol*, 1973. **12**(6): 1325–35.

22. Mathur, M., T. Das and A.K. Banerjee, Expression of L protein of vesicular stomatitis virus Indiana serotype from recombinant baculovirus in insect cells: Requirement of a host factor(s) for its biological activity *in vitro*. *J Virol*, 1996. **70**(4): 2252–9.

23. Li, J., *et al.*, A conserved motif in region v of the large polymerase proteins of nonsegmented negative-sense RNA viruses that is essential for mRNA capping. *J Virol*, 2008. **82**(2): 775–84.

24. Green, T.J., *et al.*, Structure of the vesicular stomatitis virus nucleoprotein–RNA complex. *Science*, 2006. **313**(5785): 357–60.

25. Albertini, A.A., *et al.*, Crystal structure of the rabies virus nucleoprotein–RNA complex. *Science*, 2006. **313**(5785): 360–3.

26. Ding, H., T.J. Green and M. Luo, Crystallization and preliminary X-ray analysis of a proteinase-K-resistant domain within the phosphoprotein of vesicular stomatitis virus (Indiana). *Acta Crystallogr D Biol Crystallogr*, 2004. **60**(Pt 11): 2087–90.

27. Paul, P.R., D. Chattopadhyay and A.K. Banerjee, The functional domains of the phosphoprotein (NS) of vesicular stomatitis virus (Indiana serotype). *Virology*, 1988. **166**(2): 350–7.

28. Green, T.J. and M. Luo, Structure of the vesicular stomatitis virus nucleocapsid in complex with the nucleocapsid-binding domain of the small polymerase cofactor, P. *Proc Natl Acad Sci U S A*, 2009. **106**(28): 11713–8.

29. Ding, H., *et al.*, Crystal structure of the oligomerization domain of the phosphoprotein of vesicular stomatitis virus. *J Virol*, 2006. **80**(6): 2808–14.

30. Gao, Y. and J. Lenard, Multimerization and transcriptional activation of the phosphoprotein (P) of vesicular stomatitis virus by casein kinase-II. *Embo J*, 1995. **14**(6): 1240–7.

31. Cevik, B., *et al.*, The phosphoprotein (P) and L binding sites reside in the N-terminus of the L subunit of the measles virus RNA polymerase. *Virology*, 2004. **327**(2): 297–306.

32. Holmes, D.E. and S.A. Moyer, The phosphoprotein (P) binding site resides in the N terminus of the L polymerase subunit of sendai virus. *J Virol*, 2002. **76**(6): 3078–83.

33. Canter, D.M. and J. Perrault, Stabilization of vesicular stomatitis virus L polymerase protein by P protein binding: A small deletion in the C-terminal domain of L abrogates binding. *Virology*, 1996. **219**(2): 376–86.

34. Poch, O., *et al.*, Sequence comparison of five polymerases (L proteins) of unsegmented negative-strand RNA viruses: Theoretical assignment of functional domains. *J Gen Virol*, 1990. **71** (Pt 5): 1153–62.

35. Sleat, D.E. and A.K. Banerjee, Transcriptional activity and mutational analysis of recombinant vesicular stomatitis virus RNA polymerase. *J Virol*, 1993. **67**(3): 1334–9.

36. Ogino, T., S.P. Yadav and A.K. Banerjee, Histidine-mediated RNA transfer to GDP for unique mRNA capping by vesicular stomatitis virus RNA polymerase. *Proc Natl Acad Sci U S A*, 2010. **107**(8): 3463–8.

37. Li, J., E.C. Fontaine-Rodriguez and S.P. Whelan, Amino acid residues within conserved domain VI of the vesicular stomatitis virus large polymerase protein essential for mRNA cap methyltransferase activity. *J Virol*, 2005. **79**(21): 13373–84.

38. Li, J., J.T. Wang and S.P. Whelan, A unique strategy for mRNA cap methylation used by vesicular stomatitis virus. *Proc Natl Acad Sci U S A*, 2006. **103**(22): 8493–8.

39. Rahmeh, A.A., *et al.*, Ribose 2'-O methylation of the vesicular stomatitis virus mRNA cap precedes and facilitates subsequent guanine-N-7 methylation by the large polymerase protein. *J Virol*, 2009. **83**(21): 11043–50.

40. Chandrika, R., *et al.*, Mutations in conserved domain I of the Sendai virus L polymerase protein uncouple transcription and replication. *Virology*, 1995. **213**(2): 352–63.

41. Smallwood, S., *et al.*, Mutations in conserved domain II of the large (L) subunit of the Sendai virus RNA polymerase abolish RNA synthesis. *Virology*, 1999. **262**(2): 375–83.

42. Duprex, W.P., F.M. Collins and B.K. Rima, Modulating the function of the measles virus RNA-dependent RNA polymerase by insertion of green fluorescent protein into the open reading frame. *J Virol*, 2002. **76**(14): 7322–8.

43. Ruedas, J.B. and J. Perrault, Insertion of enhanced green fluorescent protein in a hinge region of vesicular stomatitis virus L polymerase protein creates a temperature-sensitive virus that displays no virion-associated polymerase activity *in vitro*. *J Virol*, 2009. **83**(23): 12241–52.

44. Ogino, T., *et al.*, Sendai virus RNA-dependent RNA polymerase L protein catalyzes cap methylation of virus-specific mRNA. *J Biol Chem*, 2005. **280**(6): 4429–35.

45. Qanungo, K.R., *et al.*, Two RNA polymerase complexes from vesicular stomatitis virus-infected cells that carry out transcription and replication of genome RNA. *Proc Natl Acad Sci U S A*, 2004. **101**(16): 5952–7.

46. Emerson, S.U., Reconstitution studies detect a single polymerase entry site on the vesicular stomatitis virus genome. *Cell*, 1982. **31**(3 Pt 2): 635–42.

47. Whelan, S.P. and G.W. Wertz, Transcription and replication initiate at separate sites on the vesicular stomatitis virus genome. *Proc Natl Acad Sci U S A*, 2002. **99**(14): 9178–83.

48. Chuang, J.L. and J. Perrault, Initiation of vesicular stomatitis virus mutant polR1 transcription internally at the N gene *in vitro*. *J Virol*, 1997. **71**(2): 1466–75.

49. Stillman, E.A. and M.A. Whitt, The length and sequence composition of vesicular stomatitis virus intergenic regions affect mRNA levels and the site of transcript initiation. *J Virol*, 1998. **72**(7): 5565–72.

50. Fearns, R. and P.L. Collins, Role of the M2-1 transcription antitermination protein of respiratory syncytial virus in sequential transcription. *J Virol*, 1999. **73**(7): 5852–64.

51. Luk, D., *et al.*, Intergenic sequences of the vesicular stomatitis virus genome (New Jersey serotype): Evidence for two transcription initiation sites within the L gene. *Virology*, 1987. **160**(1): 88–94.

52. Stillman, E.A. and M.A. Whitt, Transcript initiation and 5′-end modifications are separable events during vesicular stomatitis virus transcription. *J Virol*, 1999. **73**(9): 7199–209.

53. Furuichi, Y. and A.J. Shatkin, Viral and cellular mRNA capping: Past and prospects. *Adv Virus Res*, 2000. **55**: 135–84.

54. Abraham, G., D.P. Rhodes and A.K. Banerjee, Novel initiation of RNA synthesis *in vitro* by vesicular stomatitis virus. *Nature*, 1975. **255**(5503): 37–40.

55. Testa, D. and A.K. Banerjee, Two methyltransferase activities in the purified virions of vesicular stomatitis virus. *J Virol*, 1977. **24**(3): 786–93.

56. Liuzzi, M., *et al.*, Inhibitors of respiratory syncytial virus replication target cotranscriptional mRNA guanylylation by viral RNA-dependent RNA polymerase. *J Virol*, 2005. **79**(20): 13105–15.

57. Saraste, M., P.R. Sibbald and A. Wittinghofer, The P-loop — A common motif in ATP- and GTP-binding proteins. *Trends Biochem Sci*, 1990. **15**(11): 430–4.

58. Shao, H., *et al.*, Characterization and mutational studies of equine infectious anemia virus dUTPase. *Biochim Biophys Acta*, 1997. **1339**(2): 181–91.

59. Hercyk, N., S.M. Horikami and S.A. Moyer, The vesicular stomatitis virus L protein possesses the mRNA methyltransferase activities. *Virology*, 1988. **163**(1): 222–5.

60. Ferron, F., *et al.*, Viral RNA-polymerases — A predicted 2′-O-ribose methyltransferase domain shared by all Mononegavirales. *Trends Biochem Sci*, 2002. **27**(5): 222–4.

61. Bujnicki, J.M. and L. Rychlewski, In silico identification, structure prediction and phylogenetic analysis of the 2′-O-ribose (cap 1) methyltransferase domain in the large structural protein of ssRNA negative-strand viruses. *Protein Eng*, 2002. **15**(2): 101–8.

62. Hager, J., *et al.*, Active site in RrmJ, a heat shock-induced methyltransferase. *J Biol Chem*, 2002. **277**(44): 41978–86.

63. Fabrega, C., *et al.*, Structure and mechanism of mRNA cap (guanine-N7) methyltransferase. *Mol Cell*, 2004. **13**(1): 77–89.

64. De la Pena, M., O.J. Kyrieleis and S. Cusack, Structural insights into the mechanism and evolution of the vaccinia virus mRNA cap N7 methyl-transferase. *EMBO J*, 2007. **26**(23): 4913–25.

65. Egloff, M.P., *et al.*, An RNA cap (nucleoside-2′-O-)-methyltransferase in the flavivirus RNA polymerase NS5: Crystal structure and functional characterization. *Embo J*, 2002. **21**(11): 2757–68.

66. Zhou, Y., *et al.*, Structure and function of flavivirus NS5 methyltransferase. *J Virol*, 2007. **81**(8): 3891–903.

67. Mastrangelo, E., *et al.*, Crystal structure and activity of Kunjin virus NS3 helicase; Protease and helicase domain assembly in the full length NS3 protein. *J Mol Biol*, 2007. **372**(2): 444–55.

68. Assenberg, R., *et al.*, Crystal structure of the Murray Valley encephalitis virus NS5 methyltransferase domain in complex with cap analogues. *J Gen Virol*, 2007. **88**(Pt 8): 2228–36.

69. Dong, H., *et al.*, West Nile virus methyltransferase catalyzes two methylations of the viral RNA cap through a substrate-repositioning mechanism. *J Virol*, 2008. **82**(9): 4295–307.

70. Kolakofsky, D., *et al.*, Viral RNA polymerase scanning and the gymnastics of Sendai virus RNA synthesis. *Virology*, 2004. **318**(2): 463–73.

71. Li, J., *et al.*, Opposing effects of inhibiting cap addition and cap methylation on polyadenylation during vesicular stomatitis virus mRNA synthesis. *J Virol*, 2009. **83**(4): 1930–40.

72. Sims 3rd, R.J., R. Belotserkovskaya and D. Reinberg, Elongation by RNA polymerase II: The short and long of it. *Genes Dev*, 2004. **18**(20): 2437–68.

73. Hammond, D.C. and J.A. Lesnaw, The fates of undermethylated mRNA cap structures of vesicular stomatitis virus (New Jersey) during *in vitro* transcription. *Virology*, 1987. **159**(2): 229–36.

74. Rose, J.K., H.F. Lodish and M.L. Brock, Giant heterogeneous polyadenylic acid on vesicular stomatitis virus mRNA synthesized *in vitro* in the presence of S-adenosylhomocysteine. *J Virol*, 1977. **21**(2): 683–93.

75. Galloway, S.E. and G.W. Wertz, S-adenosyl homocysteine-induced hyper-polyadenylation of VSV mRNA requires the methyltransferase activity of the L protein. *J Virol*, 2008. **82**(24): 12280–90.

76. Gershon, P.D. and B. Moss, Stimulation of poly(A) tail elongation by the VP39 subunit of the vaccinia virus-encoded poly(A) polymerase. *J Biol Chem*, 1993. **268**(3): 2203–10.

77. Li, J., J.S. Chorba and S.P. Whelan, Vesicular stomatitis viruses resistant to the methylase inhibitor sinefungin upregulate RNA synthesis and reveal mutations that affect mRNA cap methylation. *J Virol*, 2007. **81**(8): 4104–15.

78. Slomovic, S., *et al.*, Addition of poly(A) and poly(A)-rich tails during RNA degradation in the cytoplasm of human cells. *Proc Natl Acad Sci U S A*, 2010. **107**(16): 7407–12.

79. Rahmeh, A.A., A.D. Schenk, E.I. Danek, P.J. Kranzusch, B. Liang, T. Walz and S.P. Whelan, Molecular architecture of the vesicular stomatitis virus RNA polymerase. *Proc Natl Acad Sci U S A*, 2010. **107**: 20075–80.

80. Ferrer-Orta, C., *et al.*, A comparison of viral RNA-dependent RNA polymerases. *Curr Opin Struct Biol*, 2006. **16**(1): 27–34.

Chapter 8

Transcription of Vesicular Stomatitis Virus RNA Genome

Debasis Panda* and Asit K. Pattnaik*,†

1. Introduction

The observation that vesicular stomatitis virus (VSV) genome lacks infectivity[1] and the subsequent discovery of the presence of an RNA-dependent RNA polymerase (RdRp) within the purified virions of VSV[2] has led to the classification of single-stranded RNA virus world into positive- and negative-strand (NS) RNA viruses. The genomes of NS RNA viruses can exist in multiple RNA segments (segmented) or in a single RNA molecule (non-segmented). The four families of viruses with non-segmented genomes currently known to exist (*Rhabdoviridae*, *Paramyxoviridae*, *Filoviridae*, and *Bornaviridae*) have been grouped in the order *Mononegavirales*. Members of *Mononegavirales* are enveloped and have genomes that are non-segmented and negative-sense (or negative-polarity), being complementary to mRNA. These non-segmented negative-strand (NNS) RNA viruses are composed of a wide variety of human and animal pathogens and share many common features such as genome structure, arrangement of the genes, mode of genetic expression, and the presence of virion-associated RdRp. All members of the order *Mononegavirales*, with the exception of the family *Bornaviridae*, replicate in the cytoplasm. *Bornaviruses* replicate in the nucleus and differ from other members of the order in the manner of mRNA processing. This chapter provides an overview of the recent progresses made in our understanding of transcription of NNS RNA viruses with particular emphasis on the rhabdovirus, VSV.

---

*School of Veterinary Medicine and Biomedical Sciences, and the Nebraska Center for Virology, University of Nebraska-Lincoln, Lincoln, NE 68583, USA.

†E-mail: apattnaik2@unl.edu

Members of the family *Rhabdoviridae* can infect plants, vertebrates, as well as invertebrates. The distinct morphology of the viral particles (bullet-shaped) separates rhabdoviruses from other NNS RNA viruses in the order Mononegavirales. Animal rhabdoviruses constitute four genera: *Vesiculovirus, Lyssavirus, Ephemerovirus*, and *Novirhabdovirus*. The enveloped virions enter susceptible cells through receptor-mediated endocytosis and the viral nucleocapsid (NC) is released into the cytoplasm by low pH-dependent fusion of the envelope of the virion with the endosomal membrane. The viral NC, which contains the RNA genome tightly wrapped around by the viral NC (N) protein, serves as the template for primary transcription by the NC-associated RdRp. As a result of primary transcription, the viral mRNAs are synthesized, which are subsequently translated to produce the viral proteins. In the presence of the viral proteins, particularly, the N protein as well as the phosphoprotein (P) and the large (L) proteins, the latter two being the subunits of the viral RdRp, the input NS genomic NC is replicated to generate the full-length anti-genomic (positive-strand) NCs. The anti-genomic NCs serve as templates for further rounds of replication and amplification of genomic and anti-genomic NCs. Secondary transcription from the newly made genomic NCs results in synthesis of large amounts of mRNAs, which are translated to produce the viral proteins. Genomic NCs are subsequently assembled into progeny virions at the plasma membrane.

Establishment of reverse genetics system to recover infectious defective interfering (DI) particles of VSV entirely from cDNA clones[3] has led to subsequent recovery of infectious viruses from full-length cDNAs of members of NNS RNA viruses.[4–9] Armed with the ability to recover infectious full-length viral genomes as well as subgenomic replicons, it has been possible to engineer changes into genomes of these NNS RNA viruses to allow detailed investigations into the role(s) and requirements of genomic sequences in the process of mRNA transcription and RNA replication. Additionally, with the use of viral subgenomic replicons or the full-length viral genomes, it has been possible to examine the role(s) of individual viral proteins in replication and pathogenic mechanisms of the viruses. Majority of our current understanding of NNS RNA virus genome transcription and replication has been derived from the studies using the rhabdoviruses VSV and rabies virus and also several paramyxoviruses such as respiratory syncytial virus (RSV), simian virus 5, Sendai virus, and measles virus.

2.   Genome Organization and Gene Products

Members of *Mononegavirales* contain an NNS RNA genome. The genome size ranges from approximately 8900 nucleotide (nt) for bornaviruses to 19000 nt for filoviruses. They lack the 5′-cap and 3′-poly(A) tail. The rhabdoviruses with genome lengths of approximately 11 kb (VSV) or 15 kb (rabies virus) carry only five genes that produce the five viral proteins upon infection. These proteins are N, P, the matrix protein (M), the glycoprotein (G), and the L. The genes are organized in modular fashion in the order 3′-N-P-M-G-L-5′ (Fig. 1). Each protein-coding gene is flanked by highly conserved sequence elements that are required for synthesis of mature mRNAs with 5′-caps and 3′-polyadenylated tail. This arrangement of genes is highly conserved throughout Mononegavirales. Other members of Mononegavirales carry several additional genes that are required for virus attachment and entry, transcription and replication, virus assembly, or evasion of host immune response. The 3′- and 5′-termini of the viral genomes contain non-coding sequences, termed leader (Le), and trailer (Tr), respectively. The length of the Le region is approximately 40–50 nt, whereas Tr regions vary considerably, ranging in length from 20 nt to 600 nt. A high degree of sequence complementarity exists between Le and Tr. Both Le and Tr regions play multifunctional roles not only in genome transcription and replication but also in encapsidation and assembly of progeny virus.

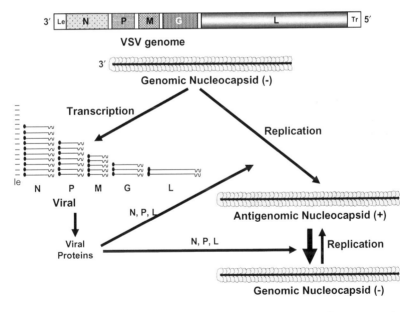

Fig. 1. (Upper) Organization of VSV genome. The protein-coding regions in VSV genome are shown in rectangular boxes. Le, leader region; Tr, trailer regions. (Lower) Transcription and replication events in infected cells. The genomic RNA in the NC is shown as a black line with the N protein (shaded bilobed structure) covering the RNA. Transcription results in synthesis of capped and polyadenylated mRNAs. Replication in the presence of N, P, and L proteins results in the synthesis of full-length products that are encapsidated with N protein.

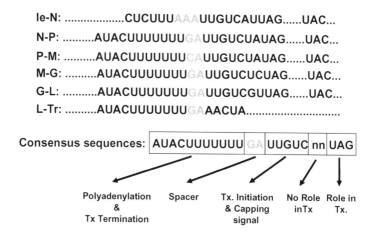

Fig. 2. Sequences at various gene junctions in the genome of VSV. The intergenic sequences that are not transcribed into mRNAs are shown in gray letters. Below are the consensus sequences and their role in transcription initiation, capping, polyadenylation, and transcription termination.

As mentioned earlier, highly conserved sequences are present at the beginning and end of each protein-coding gene that are required for initiation of transcription, capping, and polyadenylation of mRNAs, and termination of transcription. In VSV, the intergenic regions contain two nt that are not represented in the mRNAs. These sequences also play important roles during transcription of the viral genome. Although the gene junctions are highly conserved in sequence and length in VSV genome (Fig. 2), they are either conserved or highly variable in other members of NNS RNA viruses.

VSV encodes five proteins, namely, N, P, M, G, and L. All these proteins are required for the complete life cycle of the virus. Although another small protein encoded in the P mRNA has been detected, this protein is not involved in the viral replication cycle[10] and its role in pathogenesis remains unknown. The N protein encapsidates the full-length genomic and anti-genomic RNAs to form NCs or RNPs that serve as the functional templates for viral genome transcription and replication. The crystal structure of VSV and rabies virus RNPs reveals that each molecule of N protein sequesters nine nt of RNA in a cavity that is formed between two lobes of the protein.[11,12] The viral RNPs are also associated with the viral RdRp composed of the L and P proteins, which catalyzes transcription and replication of the viral genome.[13] The L protein has all the enzymatic activities that are required for synthesis of mature mRNAs with methylated 5′-cap and polyadenylated 3′-tail whereas the P protein acts as a cofactor for L protein functions. The M protein has been shown to condense the viral NCs into tightly coiled structure prior to virus assembly, imparting the virus a bullet-like shape.[14,15] It also plays a role in virus assembly by interacting with virus glycoprotein G at the plasma membrane. The G protein forms spikes on the viral envelope and functions in virus attachment and entry as well as virus assembly.

3. *cis*-Acting Signals for Transcription

All NNS RNA viruses contain Le and Tr regions at the 3′- and 5′-termini of the genomes, respectively, and a series of consensus sequences at the beginning and end of each protein-coding gene. These sequences are involved in transcription termination, polyadenylation, and transcription reinitiation of individual genes. With the advent of reverse genetics systems, much of these sequences have been well characterized for VSV and also for other NNS RNA viruses.[16–27] Among paramyxoviruses, Sendai virus, human parainfluenza virus type 1 (hPIV1), and hPIV3 show high degree of conservation in gene end and intergenic sequences, whereas the RSV, mumps virus, and SV5 genomes exhibit significant divergence in gene end and intergenic junction sequences. Numerous studies have been conducted to examine the role of these sequences in virus genome transcription and replication. However, due to space limitations, the primary focus will be on the involvement of various sequence elements in VSV genome transcription and replication.

3.1. *Role and requirement of leader and trailer regions*

The sequences at the termini of the genome (the Le and Tr regions) not only function in regulating transcription and replication but also play critical roles in encapsidation and packaging of the genome into virus particles. With the use of *in vitro* assembled synthetic NCs containing various lengths of Le sequences, it was shown that optimal transcription required the first 15–17 nt of the Le as the promoter.[28] Importantly, the first three nt (UGC) at the 3′-end were indispensable for transcription. However, the use of the *in vitro* assembled templates could not recapitulate the events of Le termination and N mRNA initiation. Subsequent studies using subgenomic replicons and DI particle genomes of VSV and engineering changes into the termini, it was possible to decipher the involvement of specific sequence elements in these regions in regulating VSV RNA transcription and replication.[23,24,29–31] It was demonstrated that the extent of terminal complementarity is a major determinant of replication.[30] In contrast, using the DI RNA template and deletion mutagenesis, it was shown that the loss of complementarity did not have any major effect on the level of replication; rather the presence of specific sequences at the termini of

DI template regulates replication efficiency of the templates.[24] It is possible that yet unidentified additional elements or factors may be involved in regulating the transcription and replication directed by the Le and Tr sequences and could potentially account for the disparate results obtained from these studies.[24,30] Deletion analyses of the Le sequences have also shown that the first 24 nt contain overlapping signals necessary for replication and transcription, whereas nt 25–47 appear to be necessary for optimal level of transcription.[23] It is possible that the downstream sequences in the leader region, particularly, nt 25–47, might be involved in interaction with the P protein of the polymerase complex for optimal level of transcription. A potential stem-loop structure has been predicted to exist within Le RNA (nt 16–33)[32] and P protein interacts with this region.[33] Using a subgenomic replicon and reciprocal swapping of leader and trailer region, also it was shown that nt 19–46 in the leader is necessary for optimal level of transcription and primary sequences rather than extent of terminal complementarity play a role in transcription.[31] Although a minigenome encoding Le and Le complement was highly active in replication, it was less active in transcription suggesting that sequence elements other than Le are also essential for optimal level of transcription.[30] It appears that sequences in the Le and Tr regions of VSV have evolved to generate a balance between replication and transcription.

3.2. *Role of leader-N gene junction sequences*

Unlike the protein-coding gene junction sequences, the Le-N gene junction sequences are somewhat different (Fig. 2). Following the leader, a trinucleotide sequence (AAA) is found immediately upstream of the N gene transcription start sequences. Additionally, the Le-N junction sequence also does not contain the stretch of U residues. Studies with deletion, substitution, and insertion of nt at the Le-N gene junction have identified essential roles of these nt in VSV transcription.[31] Deletion analyses using a minigenome template identified the role of AAA in transcription. Deletion of 5 nt (CUUUG) immediately upstream of AAA sequence or the dinucleotide (AU, Fig. 2) following UUGUC sequence did not adversely affect transcription or replication.[34] However, deletion of AAA, UUGUC, or UAG abrogated transcription without affecting replication. Interestingly, insertion of 5 heterologous nt (GCGCG) between AAA and UUGUC, UUGUC and AU, or AU and UAG did somewhat affect the levels of transcription, demonstrating that spacing of these various elements is critical for optimal transcription.[34] The Le-N junction does not contain the poly(A) signal sequence. Insertion of the poly (A) signal at the end of Le did not result in polyadenylation of the leader RNA,[34] indicating that the leader RNA may be too small to be polyadenylated by the viral

RdRp. This is consistent with the observation that the minimal size of the transcript that can be polyadenylated and terminated must be at least 51 nt in length.[35]

3.3. *Role of gene junction sequences*

3.3.1. *Signals for transcription initiation*

As mentioned earlier, the protein-coding gene junction sequences in VSV are highly conserved in sequence and length (Fig. 2). For VSV, the first 10 nt of each gene are highly conserved having sequences 3′UUGUCnnUAG5′ where n is any nt. It has been suggested that polymerase scans through the intergenic junction and initiate transcription at the first 'U' residue it encounters. Mutational analysis using a bicistronic minigenome has identified an absolute requirement for pyrimidine residue at position 1 of this conserved sequence.[26] The first three nt are critical for efficient transcription.[26,36] Using a bipartite VSV replication system, the roles of the gene start sequences in transcription initiation were examined.[36] It was demonstrated that in VSV infected cells as well as *in vitro*, the first three nt are involved not only in transcription initiation but also in mRNA capping and polyadenylation. Alterations in these conserved sequences affected the mRNA elongation and RdRp processivity and in some cases resulted in synthesis of prematurely terminated transcripts.[36] The exact mechanism by which the initiation signals affect transcription initiation and the polymerase processivity is unclear but it is likely that capping of the nascent transcript is critical. These studies suggested that during initiation, the nascent transcript maintains its contact with the polymerase until the 5′-end modification occurs by addition of cap while the polymerase continues to transcribe the first 200 nt or so. Failure of the polymerase to modify the 5′-end then could result in abortive transcription.[36] Recent studies demonstrating that the start sequences of mRNA are critical for capping[37] are consistent with the model that mRNA capping is intimately linked to polymerase processivity.

3.3.2. *Signals for mRNA capping*

All NS RNA virus mRNAs have cap structures at their 5′-termini and are translated in a cap-dependent manner. For the segmented genome NS RNA viruses (members of the family Orthomyxoviridae, Bunyaviridae, and Arenaviridae), short oligonucleotides containing the caps are derived from the host cell mRNAs in a process called "cap snatching" by the viral RdRp to initiate synthesis of their own mRNAs. However, for the NNS RNA viruses, the viral RdRp catalyzes *de novo* synthesis of capped mRNAs. In the conventional capping reaction employed for capping of cellular mRNAs as well as some viral (vaccinia virus and reovirus) mRNAs, the mRNA capping enzyme transfers the GMP moiety of GTP to the 5'-diphosphate end of the acceptor RNA to generate the cap structure. In contrast, the cap structure of the mRNAs of NNS RNA viruses are generated by transfer of the 5'-monophosphorylated nascent mRNA start sequence containing short RNAs onto the GDP moiety of GTP.[38–41] In a series of elegant experiments, Banerjee's group demonstrated that the VSV L protein catalyzes the transfer of the 5'-monophosphorylated RNA having start sequence 5'AACAG3' onto a GDP moiety by an RNA:GDP polyribonucleotidyltransferase (PRNTase) activity encoded within the protein.[37] Furthermore, the VSV L protein specifically capped the mRNA start sequence but not the leader start sequence 5'ACGAA3'.[37] These results have suggested that VSV polymerase recognizes the mRNA start sequence and the mRNA must contain triphosphorylated APuCNG sequence, where Pu is purine residue and N is any residue.[37] Further studies have shown that a histidine residue in the L protein of VSV, which is part of a highly conserved histidine-arginine (HR) motif found in L proteins of all NNS RNA viruses, is involved in transferring the 5'-monophosphorylated nascent mRNA sequences onto the GDP moiety to generate the cap structure.[42] These studies have provided a mechanistic understanding of the unique capping process in VSV mRNA and are consistent with previous studies, which showed that first three residues in mRNA are critical in transcription initiation and subsequent elongation following capping of the nascent chain.[26,36] The first four residues in the mRNAs of all members of Rhabdoviridae are conserved,[26] suggesting that similar capping reaction might be occurring in this family of viruses. Although the mRNA start sequences are not conserved among all NNS RNA viruses and each family contains their own unique sets of conserved mRNA start sequences,[26] the capping reaction appears to be similar. This observation coupled with the finding that all NNS RNA viruses contain the highly conserved HR motif[42] indicates that the L proteins of all NNS viruses may possess the RNA:GDP PRNTase activity and may employ this unique mechanism of capping of their mRNAs. Further details of this unique mechanism of capping can be found elsewhere in this book.

3.3.3. *Signals for polyadenylation and transcription termination*

Highly conserved sequences are present at the end of each protein-coding genes. In VSV, these sequences include a stretch of seven uridine residues (U tract), a conserved tetranucleotide sequence (3'AUAC5') upstream, and a conserved dinucleotide sequence (3'G/CA) downstream of the U tract. Mutational analysis of the conserved 3'AUACUUUUUUU5' region in VSV genome suggested that these 11 nt are required for polyadenylation and transcription termination of mRNA.[16,20]

Of the four nt ($3'$AUAC$5'$) upstream of the U tract, the C residue is absolutely critical and any change at this position abolishes the termination activity.[16] Changes at other three positions result in varying degree of termination of the upstream mRNA.[16] In all rhabdoviruses, the C residue is present immediately before the U tract, indicating the importance of this residue for efficient transcription termination. Increasing or decreasing the number of uridine residues in the U tract also results in inefficient termination and generation of read-through products.[16,20,34] Reducing the U tract by a single nt (from seven to six) results in loss of termination and concomitant increase in the amount of read-through transcripts, whereas increasing the U tract by one, three, or seven nt does not affect polyadenylation and termination significantly.[16,20,34] Similarly, disruption of the U tract also results in read-through products.[16,20,34] Overall, these studies indicate that the minimal length of U tract must be at least seven residues for the polymerase to polyadenylate and that polyadenylation of upstream mRNA is a prerequisite for transcription termination.[20]

The mechanism by which AUACUUUUUU sequence regulates polyadenylation is not fully understood. It has been proposed that viral transcriptase may polyadenylate the mRNA by reiterative transcription using the U tract as the template. The AU-rich nature of the tetranucleotide sequence upstream of U tract appears to be critical[43] suggesting the requirement for weak base-pairing between the template and the nascent RNA chain for reiterative transcription during polyadenylation. However, the importance of the critical requirement for the C residue immediately upstream of the U tract remains unknown. Although insertion of AUAC or U tract sequences in the middle of a gene does not result in polyadenylation or transcription termination, insertion of both of these elements in tandem results in efficient polyadenylation and transcription termination at the inserted site,[20] indicating that these two elements together are necessary and sufficient for polyadenylation and transcription termination. Insertion of the intergenic dinucleotide GA following AUAC and U tract did not significantly enhance termination suggesting that the dinucleotide is not required for termination *per se*.[20] However, a role for the dinucleotide in transcription termination was suggested from studies in which all possible combinations were examined for their ability to terminate transcription of the upstream gene and reinitiation of transcription of the downstream gene. Although termination of the upstream gene was not severely impaired with any combination of the dinucleotide sequence, initiation of transcription of the downstream gene was affected.[17,25,26] These results, taken together, indicate that the dinucleotide may in fact play a role in transcription initiation of the downstream gene rather than transcription termination *per se*. The role may be to function as a spacer sequence to separate the U tract from the initiation signal. This spacer could be just two residues long as present in the VSV genome or could be just one or six non-U nt without having any significant effect on transcription reinitiation of the downstream gene.[17] Although the exact role that the intergenic dinucleotide plays during VSV transcription remains controversial, it modulates transcription in such a way that downstream mRNA is expressed at a rate, which renders the virus most fit during evolution. From studies in which shortened U tract and longer intergenic junctions were engineered into the genome of VSV and the mutant viruses were examined,[44] it was shown that replication of VSV mutants with shorter U tracts was compromised regardless of the length of the IGR, which correlated directly with the levels of viral gene expression. When selection pressure was applied by passaging in culture, mutant viruses having better replication potential arose that possessed increased U tract,[44] indicating its importance in viral gene expression.

The relative distance between gene start and gene end also plays a role in transcription termination. Using a minireplicon system and inserting an extra transcription unit of various length, it was demonstrated that the levels of transcription termination decrease as the length of the transcription unit becomes less than 70 nt and transcription termination is completely abolished when the transcription unit is less than 51 nt.[35] Additionally, the size of the transcript rather than the sequence *per se* appears to be important for efficient transcription termination. VSV polymerase can initiate transcription at least 268 nt downstream or 200 nt upstream of the termination signal albeit with a reduced efficiency.[45] Moreover, transcription initiation for the second gene was progressively decreased as the length of the intervening region is increased. VSV RdRp could also initiate transcription from the upstream initiation sites following termination and this initiation is dependant on presence of the U tract immediately upstream of the initiation signal.[45] In VSV there is no gene overlap but the property of transcription initiation from an upstream start site could be due to retention of some of the ancestral property of the RdRp. It is noteworthy that another rhabdovirus, sigmavirus, and members of pneumovirus and filovirus possess transcriptionally active gene overlap and the viral RdRp accesses these initiation sites to transcribe the genes.

4.   *trans*-Acting Functions in Transcription

The viral N protein encapsidates the genomic and anti-genomic RNAs to form the NC complex, which serves as the template for transcription and replication by the viral RdRp, a multisubunit complex of the viral P and L proteins that also contains several host cell proteins. The N protein primarily plays a structural role in generating the NC template; however, studies also suggest that it is a component of the viral RdRp that replicates the viral genome. The L protein carries out the catalytic functions associated with the viral RdRp, whereas the P protein acts as an indispensable accessory subunit. All NNS RNA viruses encode proteins that are homologous to these three proteins. In addition, several NNS viruses encode additional proteins that modulate the functions of the viral RdRp. Although much is known about the structure and function of the P protein of VSV and other NNS RNA viruses, studies on the N and L protein structure and function are beginning to emerge.

4.1. *The P protein*

The phosphoprotein P is the non-catalytic cofactor of the viral RdRp complex. It interacts both with the L and N proteins[46,47] and with the termini of the viral genome.[33] As the name implies, it is phosphorylated and is found with different degrees of phosphorylation in infected cells and in the virions.[48–51] Based on early studies using deletion mutagenesis approach, three functional domains were mapped (reviewed in Ref. 52). The amino-terminal domain I spanning almost half of the molecule is phosphorylated by cellular casein kinase II at specific Ser and Thr residues.[48,53–55] Phosphorylation in this domain is necessary for transcription[56–58] and is indispensable for virus growth.[59] The carboxy-terminal domain II is also phosphorylated at specific residues[48,50,56] for optimal activity of the P protein in replication and virus growth.[59,60] At the extreme carboxy-terminus following the domain II is the domain III of about 25 residues, which is required for binding to L protein[61,62] and to the NC template.[47] Between domains I and II, a hypervariable central hinge region of 50–60 amino acid residues does not appear to play any major role in transcription but may play an essential role in packaging and/or assembly of infectious VSV.[63]

Although early deletion mapping studies suggested the domain structure of the P protein as described above, based on the recent structural and functional studies, we propose here a new domain organization of this protein (Fig. 3). The central region of the molecule (P central domain, P_{CD}; spanning residues

Fig. 3. Proposed domain structure of the VSV P protein. P_{NTD}, the amino-terminal domain; P_{CD}, the central domain; and P_{CTD}, the carboxy-terminal domain. The phosphorylation sites in P_{NTD} and P_{CTD} have been identified by filled circles. Although the structure of the residues from 178 to 182 were not part of the crystal structures of P_{CTD} or P_{CD}, for the sake of convenience and continuity of the residue numbers, they have been included as part of P_{CTD}.

107–177) containing parts of the hinge region folds independently and is involved in oligomerization of the protein.[64] The crystal structure P_{CD} predicts that an α-helix and two β-hairpins in this domain are critical for homodimer formation and that the N and L binding domains likely reside outside of this central domain. The C-terminal domain of P (P_{CTD}, spanning residues 183–265) also folds as an autonomously folding unit, whose structure has been recently solved alone[65] or in complex with the N protein.[66] Indeed, the structure of P_{CTD} in complex with N protein confirms that P_{CTD} interacts with N protein. For viral RNA synthesis to ensue, the P must interact with N and L proteins. The structure of the N-terminal domain (P_{NTD} spanning residues 1–106) of the protein is not known at this time; it is presumed to fold independently (Ming Luo, personal communication) and likely encode critical determinants of P protein functions, such as transcription activity and L protein binding activity. This domain also has been shown to interact with and maintain the N protein in an encapsidation-competent form.[67]

4.2. *The N protein*

Throughout replication cycle, encapsidation of RNA template by the N protein is required to form a biologically active NC template for transcription and replication of the viral genome. The N protein interaction with the RNA is not sequence specific and N-RNA complexes having cellular RNA can be readily detected.[12,68,69] The P protein confers specificity to the N protein to encapsidate the viral RNA.[69–71] Chemical probing studies suggested that the phosphodiester backbone of RNA in the viral NCs is protected by the N proteins whereas the bases are exposed, for the polymerase to recognize during transcription and replication.[72] The N-RNA complexes of VSV appear very similar in structure to those of the N-RNA complexes of rabies virus determined by cryoelectron microscopy[68,73] and X-ray crystallography.[11,12] These studies have suggested a bilobed structure for the N protein with each lobe representing the amino- and carboxy-terminal regions of the N protein. The RNA is sequestered in a central hydrophobic pocket formed between the two lobes and each N monomer sequesters nine nt of RNA. Additionally, a C-terminal loop and an N-terminal extended arm of the N protein were proposed to be important for interaction with the neighboring N molecules in the decameric N-RNA structure. The crystal structure of N-RNA complex[12] and N–P complex[66] has identified several N protein residues that are in contact with RNA and the P protein.

Structure- and homology-guided mutagenesis studies of a central conserved region of the N molecule and the C-terminal loop region have revealed that these regions play critical roles in the viral NC template functions in transcription and replication.[74,75] Mutagenesis studies in N protein of Indiana virus have identified

Y289 in a highly conserved central region of the molecule as a critical amino acid for encapsidation and replication of the RNA genome.[75] The hydrophobic and aromatic nature of this position for the functions of the N protein appears to be critical.[75] Mutations at several other positions in this conserved region showed that transcription activities of the resulting mutant N-RNA templates were impaired to varying degree without adversely affecting the replication functions of the templates.[75] The precise reasons for these differential replication and transcription activities are not known. It may be possible that mutation in these residues reduced their grip on the RNA in the RNP template such that these templates are more readily accessible to viral replication machinery. On the other hand, the association of the transcription machinery may be less efficient with the mutant N-RNA templates to drive transcription. These results suggest that the structure of the N protein and the resulting N-RNA template, in part, may play a role regulating the transcription and replication activities.[75] Similar conclusions have been recently obtained by examining the mutants that were predicted to disrupt the contacts between the C-terminal loop and the N-terminal extended arm of the N protein.[74] These studies suggested that disrupting the contacts between the C-terminal loop and the N-terminal arm may affect the N protein domain structure in such a manner that the resulting N-RNA template functions are disparately affected.

In early studies using a VSV polR1 mutant virus with a single amino acid change (Arg179His) in the N protein, it was demonstrated that in this virus, the N gene is transcribed in excess molar amounts over that of the leader RNA.[76,77] From these studies, the authors proposed that in VSV transcription initiation occurs internally from the N gene independent of the leader transcription and that the RNP template regulates the transcription and replication functions.[76,77] The recent structure-guided mutagenesis studies coupled with previous biochemical studies strengthen these observations and support the contention that the viral N-RNA complex plays a critical role in regulating the template functions.

4.3. *The L protein*

The L protein is the catalytic subunit of the viral RdRp and possesses all the enzymatic functions associated with the synthesis of mature mRNAs as well as replication of the viral genome. Early studies using temperature-sensitive mutants have suggested that the L protein not only carries the RdRp activity, but also is involved in capping and polyadenylation of the viral mRNAs. As VSV and several other NNS RNA viruses employ unique mechanisms of capping of their mRNA, recent studies have focused on understanding this capping mechanism. These studies have revealed that the L protein also possesses the PRNTase activity,[42] which is used to transfer the monophosphorylated nascent mRNA chains onto the GDP moiety during the capping reaction. The active site for the PRNTase has been mapped to a conserved HR motif in which the histidine residue is involved in a covalent interaction with the 5′-monophosphate end of the mRNA chain forming the intermediate L-pRNA complex that is used to transfer the 5′-pRNA to GDP to form the core cap structure.[42] In addition, the core cap structure is also methylated at guanine-N-7 (G-N-7) and ribose sugar 2′-O positions by the methyltransferase activity of the L protein.[78–81] Although the catalytic tetrad KDKE residues and the S-adenosyl methionine substrate binding residues in the L protein are responsible for both of these activities,[79,82] it appears that the 2′-O methylation facilitates efficient methylation at the G-N-7 site.[83] It should be noted that the 2′-O methylation is not a strict requirement for all other NNS RNA viruses, as Sendai virus and Newcastle disease virus mRNAs contain only G-N-7 methylated cap structures.[41,84]

4.4. *Other trans-acting viral proteins*

The matrix proteins of VSV and rabies virus have been shown to inhibit viral genome transcription.[85,86] Previous *in vitro* studies with VSV and other NNS RNA viruses have shown that the M protein inhibits viral transcription.[87–89] The M protein of rabies virus seems to play a differential role in regulation of gene expression and virus budding.[85] The mechanism of inhibition is not understood, but it is presumed to be mediated through interaction with the N-RNA template. In the case of RSV, the M2-1 protein has been shown to be involved in transcription processivity.[90] At low levels of expression, M2-1 protein confers processivity to the transcriptase to prevent intragenic termination but higher concentrations of M2-1 were required for read-through of diverse gene junction sequences of RSV.[91–93] These studies indicate that the M2-1 protein modulates the transcription and replication activities and may be a component of the viral RdRp. In contrast, the M2-2 protein appears to be involved in inhibiting mRNA synthesis and promoting genome replication.[94] Several other paramyxoviruses encode a number of small proteins that have also been shown to modulate transcription and replication of the viral genomes. Of significance are the V and C proteins of these viruses, which appear to inhibit both mRNA transcription and genome replication through interactions with the viral RNA, the NC protein, or the polymerase protein.[95–97]

4.5. *The transcriptase vs replicase*

The viral RdRp recognizes the gene junction sequences to initiate and terminate transcription resulting in generation of the subgenomic, capped, and polyadenylated mRNAs, whereas during replication, the RdRp ignores the gene junction sequences and generates the full-length and encapsidated RNAs. The L and P proteins are considered the core subunits of the viral RdRp but whether the same RdRp with the core subunits manifests itself as two separate polymerase complexes to carry out these disparate functions has been the subject of investigation in the past several years. Several recent studies have suggested, implicated, or demonstrated the existence of two different viral RNA polymerases in VSV. First, the P protein has been shown to exist in multiple phosphorylated forms and mutational studies have revealed that phosphorylation at different domains of the P proteins differentially regulates its functions in transcription and replication.[57,60] Phosphorylation within the P_{NTD} is necessary for transcription activity,[57] whereas phosphorylation within P_{CTD} is required for replication activity[60] of the P protein. The findings that transcriptionally inactive P protein can support efficient replication of a DI particle RNA[57] and P protein mutants with significantly reduced activity in RNA replication are as active as the wild-type P protein in transcription[60] suggest the presence of two polymerases. Second, studies with polR1 mutant virus[76] and recent studies with recombinant VSV having one small gene inserted at leader and N gene junction[98] have suggested two entry sites for the viral polymerase: one entering at the Le-N gene junction to initiate transcription and the other entering at the extreme 3'-end of the genome to initiate replication. These results implicate the existence of two different polymerases and support the two polymerase hypothesis. Perhaps, the strongest evidence yet in support of the two polymerase hypothesis has been obtained by the isolation and characterization of polymerase complexes from VSV-infected cells that support either transcription or replication.[99] In this study, by biochemical fractionation of VSV-infected cell extracts, Banerjee and co-workers purified and characterized two distinct polymerase complexes that carried out transcription and replication separately.[99] The transcriptase was shown to be a multiprotein complex consisting of viral proteins L and P along with cellular guanylyltransferase, translation elongation factor 1α and heat shock protein 60 (Hsp60) and directed *in vitro* synthesis of capped mRNA but not leader RNA.[99] In contrast, the replicase was found to be a complex of L, N, and P protein but lacked the cellular proteins that are present in the transcriptase complex and directed synthesis from the 3'-end of the genome RNA to generate products in the presence of N and P proteins, reflecting a bonafide replication reaction. These studies unequivocally establish that the transcriptase and replicase of VSV are distinct multiprotein complexes that differ from each other in protein composition and activity. Although the cellular proteins are part of the transcriptase complex, their role in the formation and/or activity in the transcription reaction remains speculative.

5. Model of Transcription

The common features in all NNS RNA viruses are that transcription is sequential[100,101] and that the genes located at the 3′-end of the genome are transcribed more frequently than the genes present at the 5′-end, resulting in a gradient in mRNA abundance.[102,103] Although several models, such as precursor-cleavage model, multiple entry site model, were proposed to account for the polar and sequential nature of transcription of VSV genomes, the most favored model[104] posits that the RdRp begins transcription from the extreme 3′-end of the genome to generate Le RNA and then stops and reinitiates transcription at the N gene start site. In this single-entry, stop–start model, as the RdRp moves along the template, it stops and starts at various gene junctions, generating the individual mRNAs. Transcriptional attenuation at the gene junctions due to the inability of a fraction of the RdRp to reinitiate transcription could result in the gradient in mRNA abundance. As the N protein becomes available, its association with the nascent Le RNA allows the RdRp to read-through the Le-N and other gene junction sequences, generating the N protein-encapsidated full-length replication products. How the N protein binding to Le sequences allows the polymerase to read-through the gene junction remains an open question. Although the model accommodates many of the available data, more recent data as well as some past data are incompatible with this single-entry model. If one considers single-entry for the RdRp at the 3′-end of the genome, Le RNA should be synthesized in excess over the N mRNA because transcription is sequential and polar. However, polR1 mutants of VSV synthesize excess N mRNA over Le RNA,[76] suggesting that the RdRp initiates transcription from the N mRNA start site independent of Le RNA synthesis. The results implicate the presence of two entry sites for the RdRp: one at the extreme 3′-end and the other at the Le-N junction, although they do not rule out the possibility of RdRp entering only at the 3′-end and reaching the N mRNA start site at the Le-N junction by a non-transcriptive scanning mode. The demonstration that transcription and replication initiate at separate sites in VSV[98] and the purification and characterization of two distinct RdRp complexes from VSV-infected cells that separately perform transcription and replication[99] also implicate the two-entry site model.

A consensus model for NNS RNA virus transcription and replication was proposed recently[105] to accommodate the currently available data. The basic tenets of the model are as follows: (i) the RdRp enters the template only at the 3′-end by displacing the terminal N molecule on the template, (ii) reaches the Le-N gene junction by scanning in the absence of *de novo* synthesized N protein and initiates transcription at the N mRNA start site, (iii) the newly synthesized N protein associates with newly made RdRp to initiate Le RNA synthesis from the 3′-end,

(iv) under conditions where Le RNA synthesis is not coupled to its encapsidation with the N protein, the Le RNA termination occurs at the Le-N junction and reinitiation of N mRNA synthesis occurs; however, if Le RNA encapsidation occurs due to the availability of N protein, Le RNA termination is suppressed and the encapsidated full-length anti-genomes are synthesized as products of replication. The 3'-entry model is proposed because it may be energetically more favorable for the RdRp to displace the terminal N protomers than the internal N protomers at the Le-N junction.

The existence of two forms (transcriptase and replicase) of RdRp[99] and the demonstration of separate initiation sites for transcription and replication[98] do not necessarily exclude the single 3'-entry model. As argued by Banerjee,[106] it is possible that both transcriptase and replicase enter at the 3'-end only but the transcriptase reaches the Le-N junction in a non-transcriptive scanning mode and initiates transcription without synthesizing Le RNA. The replicase having been assembled with *de novo* synthesized N protein could initiate Le RNA synthesis and encapsidation from the 3'-end and synthesize the full-length replication products. Data from our laboratory[34] have shown that under conditions when the N mRNA start site was made suboptimal, a fraction of the VSV RdRp initiated from this suboptimal site whereas another fraction of the RdRp initiated transcription at a cryptic UUGUC site that is located 220 nt downstream of the N start site, suggesting scanning by VSV RdRp to find the initiation signals. RdRp scanning for initiation signals has also been demonstrated for RSV.[107] Whether this occurs at the 3'-end genome in the absence of N protein remains to be seen. Although it appears that a single 3'-entry, two-polymerase model is more plausible, further studies are needed to develop an accurate and consensus model for transcription and replication in VSV and other NNS RNA viruses.

6. Concluding Remarks and Future Perspective

With the ability to introduce specific changes into the genomes of NNS RNA viruses, it has been possible to examine the roles and requirements of various sequence elements and viral proteins that regulate transcription and replication of these viral genomes. Identification and characterization of these *cis*-acting elements involved in genetic expression of the viral genomes have been mostly completed. However, the mechanistic understanding of how these sequence elements carry out the functions is poorly understood. The functions of various structural domains of the viral N, P, and L proteins are beginning to be addressed in greater detail. Now that the X-ray crystallographic structures of the N protein and the N-RNA complexes of VSV, rabies, and RSV[11,12,108] are available, further mutagenesis studies based on these structures will provide a better understanding of how the structure of the N protein regulates the template functions of the resulting N-RNA complex. Structure determination of the full-length L and P proteins will be important for further understanding of the viral RdRp functions.

The exact mechanism by which the VSV RdRp gains access to the NC template is still unknown and remains a contentious issue. Future studies to determine whether the viral RdRp enters only through the 3′-end or can enter directly at the Le-N junction will be important. It is also critical to more fully characterize the "transcriptase" and "replicase" complexes in terms of their subunit composition and properties. The roles that the identified host cell proteins play in the transcriptase complex remain largely speculative. With the availability of cell lines that constitutively express functional replication proteins of VSV,[109] it may be possible to purify the transcriptase and replicase complexes from these cells for further biochemical characterization.

One major avenue of research that has not been addressed yet is the identification and characterization of host cell functions that are required for transcription and replication of the VSV genome. So far, only a handful of host cell proteins involved in VSV gene expression have been identified through biochemical approaches. The use of genome-wide siRNA screens will be highly beneficial for identification of the cellular proteins that are involved viral RdRp functions. Such screens have been extremely useful in studies with several other viral pathogens including influenza virus.[110,111] Application of this technology to the studies with other NNS viruses will be critical for identification of unique targets for therapeutic intervention.

References

1. Huang, A.S. and R.R. Wagner. Comparative sedimentation coefficients of RNA extracted from plaque-forming and defective particles of vesicular stomatitis virus. *J Mol Biol*, 1966. **22**:381–4.
2. Baltimore, D., A.S. Huang and M. Stampfer. Ribonucleic acid synthesis of vesicular stomatitis virus, II. An RNA polymerase in the virion. *Proc Natl Acad Sci U S A*, 1970. **66**:572–6.
3. Pattnaik, A.K., L.A. Ball, A.W. LeGrone and G.W. Wertz. Infectious defective interfering particles of VSV from transcripts of a cDNA clone. *Cell*, 1992. **69**: 1011–20.
4. Garcin, D., T. Pelet, P. Calain, L. Roux, J. Curran and D. Kolakofsky. A highly recombinogenic system for the recovery of infectious Sendai paramyxovirus from cDNA: Generation of a novel copy-back nondefective interfering virus. *EMBO J*, 1995. **14**:6087–94.

5. Lawson, N.D., E.A. Stillman, M.A. Whitt and J.K. Rose. Recombinant vesicular stomatitis viruses from DNA. *Proc Natl Acad Sci U S A*, 1995. **92**:4477–81.

6. Radecke, F., P. Spielhofer, H. Schneider, K. Kaelin, M. Huber, C. Dotsch, G. Christiansen and M.A. Billeter. Rescue of measles viruses from cloned DNA. *EMBO J*, 1995. **14**:5773–84.

7. Schneider, U., M. Schwemmle and P. Staeheli. Genome trimming: A unique strategy for replication control employed by Borna disease virus. *Proc Natl Acad Sci U S A*, 2005. **102**:3441–6.

8. Schnell, M.J., T. Mebatsion and K.K. Conzelmann. Infectious rabies viruses from cloned cDNA. *EMBO J*, 1994. **13**:4195–203.

9. Volchkov, V.E., V.A. Volchkova, E. Muhlberger, L.V. Kolesnikova, M. Weik, O. Dolnik and H.D. Klenk. Recovery of infectious Ebola virus from complementary DNA: RNA editing of the GP gene and viral cytotoxicity. *Science*, 2001. **291**:1965–9.

10. Kretzschmar, E., R. Peluso, M.J. Schnell, M.A. Whitt and J.K. Rose. Normal replication of vesicular stomatitis virus without C proteins. *Virology*, 1996. **216**:309–16.

11. Albertini, A.A., A.K. Wernimont, T. Muziol, R.B. Ravelli, C.R. Clapier, G. Schoehn, W. Weissenhorn and R.W. Ruigrok. Crystal structure of the rabies virus nucleoprotein–RNA complex. *Science*, 2006. **313**:360–3.

12. Green, T.J., X. Zhang, G.W. Wertz and M. Luo. Structure of the vesicular stomatitis virus nucleoprotein–RNA complex. *Science*, 2006. **313**:357–60.

13. Emerson, S.U. and Y. Yu. Both NS and L proteins are required for *in vitro* RNA synthesis by vesicular stomatitis virus. *J Virol*, 1975. **15**:1348–56.

14. Lyles, D.S., M.O. McKenzie, P.E. Kaptur, K.W. Grant and W.G. Jerome. Complementation of M gene mutants of vesicular stomatitis virus by plasmid-derived M protein converts spherical extracellular particles into native bullet shapes. *Virology*, 1996. **217**:76–87.

15. Newcomb, W.W. and J.C. Brown. Role of the vesicular stomatitis virus matrix protein in maintaining the viral nucleocapsid in the condensed form found in native virions. *J Virol*, 1981. **39**:295–9.

16. Barr, J.N., S.P. Whelan and G.W. Wertz. *cis*-Acting signals involved in termination of vesicular stomatitis virus mRNA synthesis include the conserved AUAC and the U7 signal for polyadenylation. *J Virol*, 1997. **71**:8718–25.

17. Barr, J.N., S.P. Whelan and G.W. Wertz. Role of the intergenic dinucleotide in vesicular stomatitis virus RNA transcription. *J Virol*, 1997. **71**:1794–801.

18. Conzelmann, K.K. Nonsegmented negative-strand RNA viruses: Genetics and manipulation of viral genomes. *Annu Rev Genet*, 1998. **32**:123–62.

19. Cowton, V.M., D.R. McGivern and R. Fearns. Unravelling the complexities of respiratory syncytial virus RNA synthesis. *J Gen Virol*, 2006. **87**:1805–21.

20. Hwang, L.N., N. Englund and A.K. Pattnaik. Polyadenylation of vesicular stomatitis virus mRNA dictates efficient transcription termination at the intercistronic gene junctions. *J Virol*, 1998. **72**:1805–13.

21. Kuo, L., R. Fearns and P.L. Collins. Analysis of the gene start and gene end signals of human respiratory syncytial virus: Quasi-templated initiation at position 1 of the encoded mRNA. *J Virol*, 1997. **71**:4944–53.

22. Kuo, L., R. Fearns and P.L. Collins. The structurally diverse intergenic regions of respiratory syncytial virus do not modulate sequential transcription by a dicistronic minigenome. *J Virol*, 1996. **70**:6143–50.

23. Li, T. and A.K. Pattnaik. Overlapping signals for transcription and replication at the 3′ terminus of the vesicular stomatitis virus genome. *J Virol*, 1999. **73**:444–52.

24. Li, T. and A.K. Pattnaik. Replication signals in the genome of vesicular stomatitis virus and its defective interfering particles: Identification of a sequence element that enhances DI RNA replication. *Virology*, 1997. **232**:248–59.

25. Stillman, E.A. and M.A. Whitt. The length and sequence composition of vesicular stomatitis virus intergenic regions affect mRNA levels and the site of transcript initiation. *J Virol*, 1998. **72**:5565–72.

26. Stillman, E.A. and M.A. Whitt. Mutational analyses of the intergenic dinucleotide and the transcriptional start sequence of vesicular stomatitis virus (VSV) define sequences required for efficient termination and initiation of VSV transcripts. *J Virol*, 1997. **71**:2127–37.

27. Whelan, S.P., J.N. Barr and G.W. Wertz. Transcription and replication of nonsegmented negative-strand RNA viruses. *Curr Top Microbiol Immunol*, 2004. **283**:61–119.

28. Smallwood, S. and S.A. Moyer. Promoter analysis of the vesicular stomatitis virus RNA polymerase. *Virology*, 1993. **192**:254–63.

29. Pattnaik, A.K., L.A. Ball, A. LeGrone and G.W. Wertz. The termini of VSV DI particle RNAs are sufficient to signal RNA encapsidation, replication, and budding to generate infectious particles. *Virology*, 1995. **206**:760–4.

30. Wertz, G.W., S. Whelan, A. LeGrone and L. A. Ball. Extent of terminal complementarity modulates the balance between transcription and replication of vesicular stomatitis virus RNA. *Proc Natl Acad Sci U S A*, 1994. **91**:8587–91.

31. Whelan, S.P. and G.W. Wertz. Regulation of RNA synthesis by the genomic termini of vesicular stomatitis virus: Identification of distinct sequences essential for transcription but not replication. *J Virol*, 1999. **73**:297–306.

32. Grinnell, B.W. and R.R. Wagner. Nucleotide sequence and secondary structure of VSV leader RNA and homologous DNA involved in inhibition of DNA-dependent transcription. *Cell*, 1984. **36**:533–43.

33. Keene, J.D., B.J. Thornton and S.U. Emerson. Sequence-specific contacts between the RNA polymerase of vesicular stomatitis virus and the leader RNA gene. *Proc Natl Acad Sci U S A*, 1981. **78**:6191–5.

34. Hwang, L.N. Identification and characterization of *cis*-acting signals in the genome of vesicular stomatitis virus that control polyadenylation, transcription termination and transcription reinitiation. Ph.D. Thesis, University of Miami, Coral gables, 1999.

35. Whelan, S.P., J.N. Barr and G.W. Wertz. Identification of a minimal size requirement for termination of vesicular stomatitis virus mRNA: Implications for the mechanism of transcription. *J Virol*, 2000. **74**:8268–76.

36. Stillman, E.A. and M.A. Whitt. Transcript initiation and 5′-end modifications are separable events during vesicular stomatitis virus transcription. *J Virol*, 1999. **73**:7199–209.

37. Ogino, T. and A.K. Banerjee. Unconventional mechanism of mRNA capping by the RNA-dependent RNA polymerase of vesicular stomatitis virus. *Mol Cell*, 2007. **25**:85–97.

38. Abraham, G., D.P. Rhodes and A.K. Banerjee. Novel initiation of RNA synthesis *in vitro* by vesicular stomatitis virus. *Nature*, 1975. **255**:37–40.

39. Barik, S. The structure of the 5′ terminal cap of the respiratory syncytial virus mRNA. *J Gen Virol*, 1993. **74**(Pt 3):485–90.

40. Gupta, K.C. and P. Roy. Alternate capping mechanisms for transcription of spring viremia of carp virus: Evidence for independent mRNA initiation. *J Virol*, 1980. **33**:292–303.

41. Ogino, T., M. Kobayashi, M. Iwama and K. Mizumoto. Sendai virus RNA-dependent RNA polymerase L protein catalyzes cap methylation of virus-specific mRNA. *J Biol Chem*, 2005. **280**:4429–35.

42. Ogino, T., S.P. Yadav and A.K. Banerjee. Histidine-mediated RNA transfer to GDP for unique mRNA capping by vesicular stomatitis virus RNA polymerase. *Proc Natl Acad Sci U S A*, 2010. **107**:3463–8.

43. Barr, J.N. and G.W. Wertz. Polymerase slippage at vesicular stomatitis virus gene junctions to generate poly(A) is regulated by the upstream 3′-AUAC-5′ tetranucleotide: Implications for the mechanism of transcription termination. *J Virol*, 2001. **75**:6901–13.

44. Hinzman, E.E., J.N. Barr and G.W. Wertz. Selection for gene junction sequences important for VSV transcription. *Virology*, 2008. **380**:379–87.

45. Barr, J.N., X. Tang, E. Hinzman, R. Shen and G.W. Wertz. The VSV polymerase can initiate at mRNA start sites located either up or downstream of a transcription termination signal but size of the intervening intergenic region affects efficiency of initiation. *Virology*, 2008. **374**:361–70.

46. Canter, D.M. and J. Perrault. Stabilization of vesicular stomatitis virus L polymerase protein by P protein binding: A small deletion in the C-terminal domain of L abrogates binding. *Virology*, 1996. **219**:376–86.

47. Emerson, S.U. and M. Schubert. Location of the binding domains for the RNA polymerase L and the ribonucleocapsid template within different halves of the NS phosphoprotein of vesicular stomatitis virus. *Proc Natl Acad Sci U S A*, 1987. **84**:5655–9.

48. Barik, S. and A.K. Banerjee. Phosphorylation by cellular casein kinase II is essential for transcriptional activity of vesicular stomatitis virus phosphoprotein P. *Proc Natl Acad Sci U S A*, 1992. **89**:6570–4.

49. Bell, J.C. and L. Prevec. Phosphorylation sites on phosphoprotein NS of vesicular stomatitis virus. *J Virol*, 1985. **54**:697–702.

50. Chen, J.L., T. Das and A.K. Banerjee. Phosphorylated states of vesicular stomatitis virus P protein *in vitro* and *in vivo*. *Virology*, 1997. **228**:200–12.

51. Kingsford, L. and S.U. Emerson. Transcriptional activities of different phosphorylated species of NS protein purified from vesicular stomatitis virions and cytoplasm of infected cells. *J Virol*, 1980. **33**:1097–105.

52. Banerjee, A.K. and S. Barik. Gene expression of vesicular stomatitis virus genome RNA. *Virology*, 1992. **188**:417–28.

53. Barik, S. and A.K. Banerjee. Sequential phosphorylation of the phosphoprotein of vesicular stomatitis virus by cellular and viral protein kinases is essential for transcription activation. *J Virol*, 1992. **66**:1109–18.

54. Das, T., A.K. Gupta, P.W. Sims, C.A. Gelfand, J.E. Jentoft and A.K. Banerjee. Role of cellular casein kinase II in the function of the phosphoprotein (P) subunit of RNA polymerase of vesicular stomatitis virus. *J Biol Chem*, 1995. **270**:24100–7.

55. Gupta, A.K., T. Das and A.K. Banerjee. Casein kinase II is the P protein phosphorylating cellular kinase associated with the ribonucleoprotein complex of purified vesicular stomatitis virus. *J Gen Virol*, 1995. **76**(Pt 2):365–72.

56. Gao, Y. and J. Lenard. Multimerization and transcriptional activation of the phosphoprotein (P) of vesicular stomatitis virus by casein kinase-II. *EMBO J*, 1995. **14**: 1240–7.

57. Pattnaik, A.K., L. Hwang, T. Li, N. Englund, M. Mathur, T. Das and A.K. Banerjee. Phosphorylation within the amino-terminal acidic domain I of the phosphoprotein of vesicular stomatitis virus is required for transcription but not for replication. *J Virol*, 1997. **71**:8167–75.

58. Takacs, A.M., S. Barik, T. Das and A.K. Banerjee. Phosphorylation of specific serine residues within the acidic domain of the phosphoprotein of vesicular stomatitis virus regulates transcription *in vitro*. *J Virol*, 1992. **66**:5842–8.

59. Das, S.C. and A.K. Pattnaik. Phosphorylation of vesicular stomatitis virus phosphoprotein P is indispensable for virus growth. *J Virol*, 2004. **78**:6420–30.

60. Hwang, L.N., N. Englund, T. Das, A.K. Banerjee and A.K. Pattnaik. Optimal replication activity of vesicular stomatitis virus RNA polymerase requires phosphorylation of a residue(s) at carboxy-terminal domain II of its accessory subunit, phosphoprotein P. *J Virol*, 1999. **73**:5613–20.

61. Das, T., A.K. Pattnaik, A.M. Takacs, T. Li, L.N. Hwang and A.K. Banerjee. Basic amino acid residues at the carboxy-terminal eleven amino acid region of the phosphoprotein (P) are required for transcription but not for replication of vesicular stomatitis virus genome RNA. *Virology*, 1997. **238**:103–14.

62. Takacs, A.M., T. Das and A.K. Banerjee. Mapping of interacting domains between the nucleocapsid protein and the phosphoprotein of vesicular stomatitis virus by using a two-hybrid system. *Proc Natl Acad Sci U S A*, 1993. **90**:10375–9.

63. Das, S.C. and A.K. Pattnaik. Role of the hypervariable hinge region of phosphoprotein P of vesicular stomatitis virus in viral RNA synthesis and assembly of infectious virus particles. *J Virol*, 2005. **79**:8101–12.

64. Ding, H., T.J. Green, S. Lu and M. Luo. Crystal structure of the oligomerization domain of the phosphoprotein of vesicular stomatitis virus. *J Virol*, 2006. **80**: 2808–14.

65. Ribeiro Jr. E.A., A. Favier, F.C. Gerard, C. Leyrat, B. Brutscher, D. Blondel, R.W. Ruigrok, M. Blackledge and M. Jamin. Solution structure of the C-terminal nucleoprotein–RNA binding domain of the vesicular stomatitis virus phosphoprotein. *J Mol Biol*, 2008. **382**:525–38.

66. Green, T.J. and M. Luo. Structure of the vesicular stomatitis virus nucleocapsid in complex with the nucleocapsid-binding domain of the small polymerase cofactor, P. *Proc Natl Acad Sci U S A*, 2009. **106**:11713–8.

67. Chen, M., T. Ogino and A.K. Banerjee. Interaction of vesicular stomatitis virus P and N proteins: Identification of two overlapping domains at the N terminus of P that are involved in N0–P complex formation and encapsidation of viral genome RNA. *J Virol*, 2007. **81**:13478–85.

68. Chen, Z., T.J. Green, M. Luo and H. Li. Visualizing the RNA molecule in the bacterially expressed vesicular stomatitis virus nucleoprotein–RNA complex. *Structure*, 2004. **12**:227–35.

69. Masters, P.S. and A.K. Banerjee. Complex formation with vesicular stomatitis virus phosphoprotein NS prevents binding of nucleocapsid protein N to nonspecific RNA. *J Virol*, 1988. **62**:2658–64.

70. Howard, M. and G. Wertz. Vesicular stomatitis virus RNA replication: A role for the NS protein. *J Gen Virol*, 1989. **70**(Pt 10):2683–94.

71. Pattnaik, A.K. and G.W. Wertz. Replication and amplification of defective interfering particle RNAs of vesicular stomatitis virus in cells expressing viral proteins from vectors containing cloned cDNAs. *J Virol*, 1990. **64**:2948–57.

72. Iseni, F., F. Baudin, D. Blondel and R.W. Ruigrok. Structure of the RNA inside the vesicular stomatitis virus nucleocapsid. *RNA*, 2000. **6**:270–81.

73. Schoehn, G., F. Iseni, M. Mavrakis, D. Blondel and R.W. Ruigrok. Structure of recombinant rabies virus nucleoprotein–RNA complex and identification of the phosphoprotein binding site. *J Virol*, 2001. **75**:490–8.

74. Harouaka, D. and G.W. Wertz. Mutations in the C-terminal loop of the nucleocapsid protein affect vesicular stomatitis virus RNA replication and transcription differentially. *J Virol*, 2009. **83**:11429–39.

75. Nayak, D., D. Panda, S.C. Das, M. Luo and A.K. Pattnaik. Single-amino-acid alterations in a highly conserved central region of vesicular stomatitis virus N protein differentially affect the viral nucleocapsid template functions. *J Virol*, 2009. **83**: 5525–34.

76. Chuang, J.L. and J. Perrault. Initiation of vesicular stomatitis virus mutant polR1 transcription internally at the N gene *in vitro*. *J Virol*, 1997. **71**:1466–75.

77. Perrault, J., G.M. Clinton and M.A. McClure. RNP template of vesicular stomatitis virus regulates transcription and replication functions. *Cell*, 1983. **35**:175–85.

78. Grdzelishvili, V.Z., S. Smallwood, D. Tower, R.L. Hall, D.M. Hunt and S.A. Moyer. A single amino acid change in the L-polymerase protein of vesicular stomatitis virus completely abolishes viral mRNA cap methylation. *J Virol*, 2005. **79**:7327–37.

79. Li, J., E.C. Fontaine-Rodriguez and S.P. Whelan. Amino acid residues within conserved domain VI of the vesicular stomatitis virus large polymerase protein essential for mRNA cap methyltransferase activity. *J Virol*, 2005. **79**:13373–84.

80. Li, J., A. Rahmeh, M. Morelli and S.P. Whelan. A conserved motif in region V of the large polymerase proteins of nonsegmented negative-sense RNA viruses that is essential for mRNA capping. *J Virol*, 2008. **82**:775–84.

81. Li, J., J.T. Wang and S.P. Whelan. A unique strategy for mRNA cap methylation used by vesicular stomatitis virus. *Proc Natl Acad Sci U S A*, 2006. **103**:8493–8.

82. Galloway, S.E. and G.W. Wertz. *S*-adenosyl homocysteine-induced hyperpolyadenylation of vesicular stomatitis virus mRNA requires the methyltransferase activity of L protein. *J Virol*, 2008. **82**:12280–90.

83. Rahmeh, A.A., J. Li, P.J. Kranzusch and S.P. Whelan. Ribose $2'$-O methylation of the vesicular stomatitis virus mRNA cap precedes and facilitates subsequent guanine-N-7 methylation by the large polymerase protein. *J Virol*, 2009. **83**:11043–50.

84. Colonno, R.J. and H.O. Stone. Newcastle disease virus mRNA lacks $2'$-O-methylated nucleotides. *Nature*, 1976. **261**:611–4.

85. Finke, S., R. Mueller-Waldeck and K.K. Conzelmann. Rabies virus matrix protein regulates the balance of virus transcription and replication. *J Gen Virol*, 2003. **84**: 1613–21.

86. Li, Y., L.Z. Luo and R.R. Wagner. Transcription inhibition site on the M protein of vesicular stomatitis virus located by marker rescue of mutant tsO23(III) with M-gene expression vectors. *J Virol*, 1989. **63**:2841–3.

87. Carroll, A.R. and R.R. Wagner. Role of the membrane (M) protein in endogenous inhibition of *in vitro* transcription by vesicular stomatitis virus. *J Virol*, 1979. **29**:134–42.

88. Pal, R., B.W. Grinnell, R.M. Snyder and R.R. Wagner. Regulation of viral transcription by the matrix protein of vesicular stomatitis virus probed by monoclonal antibodies and temperature-sensitive mutants. *J Virol*, 1985. **56**:386–94.

89. Suryanarayana, K., K. Baczko, V. ter Meulen and R.R. Wagner. Transcription inhibition and other properties of matrix proteins expressed by M genes cloned from measles viruses and diseased human brain tissue. *J Virol*, 1994. **68**: 1532–43.

90. Collins, P.L., M.G. Hill, J. Cristina and H. Grosfeld. Transcription elongation factor of respiratory syncytial virus, a nonsegmented negative-strand RNA virus. *Proc Natl Acad Sci U S A*, 1996. **93**:81–5.

91. Fearns, R. and P.L. Collins. Role of the M2-1 transcription antitermination protein of respiratory syncytial virus in sequential transcription. *J Virol*, 1999. **73**:5852–64.

92. Hardy, R.W., S.B. Harmon and G.W. Wertz. Diverse gene junctions of respiratory syncytial virus modulate the efficiency of transcription termination and respond differently to M2-mediated antitermination. *J Virol*, 1999. **73**:170–6.

93. Hardy, R.W. and G.W. Wertz. The product of the respiratory syncytial virus M2 gene ORF1 enhances readthrough of intergenic junctions during viral transcription. *J Virol*, 1998. **72**:520–6.

94. Bermingham, A. and P.L. Collins. The M2-2 protein of human respiratory syncytial virus is a regulatory factor involved in the balance between RNA replication and transcription. *Proc Natl Acad Sci U S A*, 1999. **96**:11259–64.

95. Horikami, S.M., R.E. Hector, S. Smallwood and S.A. Moyer. The Sendai virus C protein binds the L polymerase protein to inhibit viral RNA synthesis. *Virology*, 1997. **235**:261–70.

96. Horikami, S.M., S. Smallwood and S.A. Moyer. The Sendai virus V protein interacts with the NP protein to regulate viral genome RNA replication. *Virology*, 1996. **222**:383–90.

97. Parks, C.L., S.E. Witko, C. Kotash, S.L. Lin, M.S. Sidhu and S.A. Udem. Role of V protein RNA binding in inhibition of measles virus minigenome replication. *Virology*, 2006. **348**:96–106.

98. Whelan, S.P. and G.W. Wertz. Transcription and replication initiate at separate sites on the vesicular stomatitis virus genome. *Proc Natl Acad Sci U S A*, 2002. **99**:9178–83.

99. Qanungo, K.R., D. Shaji, M. Mathur and A.K. Banerjee. Two RNA polymerase complexes from vesicular stomatitis virus-infected cells that carry out transcription and replication of genome RNA. *Proc Natl Acad Sci U S A*, 2004. **101**:5952–7.

100. Abraham, G. and A.K. Banerjee. Sequential transcription of the genes of vesicular stomatitis virus. *Proc Natl Acad Sci U S A*, 1976. **73**:1504–8.

101. Ball, L.A. and C.N. White. Order of transcription of genes of vesicular stomatitis virus. *Proc Natl Acad Sci U S A* 1976. **73**:442–6.

102. Iverson, L.E. and J.K. Rose. Localized attenuation and discontinuous synthesis during vesicular stomatitis virus transcription. *Cell*, 1981. **23**:477–84.

103. Villarreal, L.P., M. Breindl and J.J. Holland. Determination of molar ratios of vesicular stomatitis virus induced RNA species in BHK21 cells. *Biochemistry*, 1976. **15**:1663–7.

104. Emerson, S.U. Reconstitution studies detect a single polymerase entry site on the vesicular stomatitis virus genome. *Cell*, 1982. **31**:635–42.

105. Curran, J. and D. Kolakofsky. Nonsegmented negative-strand RNA virus RNA synthesis *in vivo*. *Virology*, 2008. **371**:227–30.

106. Banerjee, A.K. Response to Non-segmented negative-strand RNA virus RNA synthesis *in vivo*. *Virology*, 2008. **371**:231–3.

107. Bukreyev, A., B.R. Murphy and P.L. Collins. Respiratory syncytial virus can tolerate an intergenic sequence of at least 160 nucleotides with little effect on transcription or replication *in vitro* and *in vivo*. *J Virol*, 2000. **74**:11017–26.

108. Tawar, R.G., S. Duquerroy, C. Vonrhein, P.F. Varela, L. Damier-Piolle, N. Castagne, K. MacLellan, H. Bedouelle, G. Bricogne, D. Bhella, J.F. Eleouet and F.A. Rey. Crystal structure of a nucleocapsid-like nucleoprotein–RNA complex of respiratory syncytial virus. *Science*, 2009. **326**:1279–83.

109. Panda, D., P.X. Dinh, L.K. Beura and A.K. Pattnaik. Induction of interferon and interferon signaling pathways by replication of defective interfering particle RNA in cells constitutively expressing VSV replication proteins. *J Virol*, 2010. **84**:4826–31.

110. Brass, A.L., I.C. Huang, Y. Benita, S.P. John, M.N. Krishnan, E.M. Feeley, B.J. Ryan, J.L. Weyer, L. van der Weyden, E. Fikrig, D.J. Adams, R.J. Xavier, M. Farzan and S.J. Elledge. The IFITM proteins mediate cellular resistance to influenza A H1N1 virus, West Nile virus, and dengue virus. *Cell*, 2009. **139**:1243–54.

111. Konig, R., S. Stertz, Y. Zhou, A. Inoue, H.H. Hoffmann, S. Bhattacharyya, J.G. Alamares, D.M. Tscherne, M.B. Ortigoza, Y. Liang, Q. Gao, S.E. Andrews, S. Bandyopadhyay, P. De Jesus, B.P. Tu, L. Pache, C. Shih, A. Orth, G. Bonamy, L. Miraglia, T. Ideker, A. Garcia-Sastre, J.A. Young, P. Palese, M.L. Shaw and S.K. Chanda. Human host factors required for influenza virus replication. *Nature*, 2010. **463**:813–7.

Chapter 9

Assembly of Vesicular Stomatitis Virus

Ming Luo[*], Todd J. Green[*] and Z. Hong Zhou[†]

1. Introduction

The assembly of vesicular stomatitis virus (VSV) begins with the assembly of the nucleocapsid concomitant with genomic replication. The notion that VSV replication could not take place if the nucleocapsid protein, N (422 amino acids in length), was not simultaneously produced was first established with experiments where translation was inhibited by cycloheximide.[1,2] The model for replication provides that the assembly of the N protein on the nascent genomic RNA would promote continuous RNA synthesis by the viral RNA-dependent RNA polymerase to complete the genome replication. This process occurs for both the plus and minus strands of the genome.[3,4] Without enwrapping of the nascent genomic RNA by N, viral RNA synthesis would continue. In addition, efficient replication of the genomic RNA by the VSV polymerase requires the formation of a proper complex between the N protein and the phosphoprotein, P (265 amino acids in length). This complex is also a subunit of the VSV polymerase.[5,6] It was suggested that the formation of an N–P complex prevents the nonspecific aggregation of N so a sufficient soluble pool of the N protein is available for encapsidation of the genomic RNA.[7,8] More recent data suggest, as discussed later in this chapter, that the P protein plays a more active role in the assembly of the nucleocapsid, as well as the assembly of the virion.

[*]Department of Microbiology, The University of Alabama at Birmingham, Birmingham, AL 35294, USA.

[†]Department of Microbiology, Immunology & Molecular Genetics, University of California at Los Angeles, Los Angeles, CA 90095, USA.

2.   Structure of the N Protein

The crystal structure of a nucleocapsid-like particle that consists of 10 subunits of N and a single strand of RNA containing 90 bases revealed the detailed architecture of the nucleocapsid.[9] This nucleocapsid-like particle is formed when the N protein is expressed together with the P protein in *Escherichia coli*.[10] In the presence of the P protein, a single thread of RNA with random sequence is encapsidated by the N protein to form a ring. The ring structure must be considered an artifact of the heteroexpression in *E. coli* because the viral nucleocapsid is a linear structure. However, the intermolecular interactions of the N protein and its encapsidation of RNA is reminiscent of the viral nucleocapsid. A single N protein subunit contains two structural lobes: the N-terminal lobe and the C-terminal lobe (Fig. 1). Each lobe is composed primarily of α-helices. The N-terminal lobe contains seven helices, named α-1 through α-7. There is a long N-terminal arm (the N-arm) of 28 residues preceding α-1, containing a motif of β-strand, turn, β-strand near the N-terminus. One additional β-strand is located between α-1 and α-2, and three consecutive β-strands between α-3 and α-4. A large flexible loop connects the last β-strand and α-4. Secondary structure elements from α-1 to α-7 form a compact structural lobe that is linked to the C-terminal lobe by a single polypeptide. The C-terminal lobe contains only α helices, α-8 through α-15. A compact domain is formed by α-8 through α-12, with a large loop (the C-loop) composed of 36 residues connecting

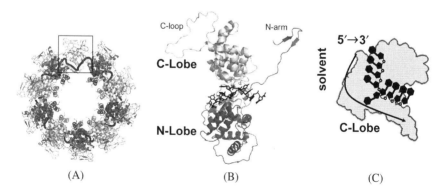

(A)   (B)   (C)

Fig. 1.   (A) A decamer of N in complex with random RNA in the crystal structure. Each N subunit is shown as a ribbon drawing and the RNA is represented by a tube in the center. The C-lobe of one N subunit (see boxed subunit) was removed to show the location of encapsidated RNA. (B) A ribbon drawing viewed from inside of the ring shows the N-lobe (bottom half) and the C-lobe (top half) of the N protein, with nine nucleotides in the cavity between the two lobes. (C) A cartoon is drawn to demonstrate how the RNA is sequestered in the nucleocapsid. Orientation of the bases is shown according to its sequence from the 5′ end to the 3′ end.

this domain to the last three helices, which seems to form a small structural domain. The single polypeptide between the two lobes may be considered as a hinge that allows the N-terminal lobe to swing open and closed relative to the C-terminal lobe. However, a tyrosine residue, Tyr289, on the back of the hinge region is central to a cluster of hydrophobic residues that may restrict the relative motion between the two lobes. The orientation of the two lobes renders the N protein a unique V-shaped structural motif with a capsid at the center to accommodate its captive target, the viral genomic RNA.

This V-shaped structural motif is present in all the nucleocapsid proteins of negative strand RNA viruses when their available crystal structures are superimposed.[11] The structural conservation is also consistent with the homology found in their amino acid sequence alignment.[12,13] The basic motif conserved in the nucleocapsid protein has been named as (5H + 3H) (Fig. 2), which suggests that there are five α-helices in the N-terminal lobe and three α-helices in the C-terminal lobe conserved in all nucleocapsid proteins of negative strand RNA virus. The two clusters of α-helices located on two distinct lobes are linked by the single polypeptide. The (5H + 3H) motif forms the cavity in which the viral genomic RNA may be encapsided. The amino acid sequences within the cavity appear to be relatively more conserved in comparison to amino acids on the rest of the surface of the nucleocapsid protein,[14] which is consistent with the essential function of the nucleocapsid protein to encapsidate the viral genomic RNA. This topological homology in the N protein structure is similar among other viral structural proteins, such as the β-barrel motif observed in the capsid proteins of spherical viruses.[14]

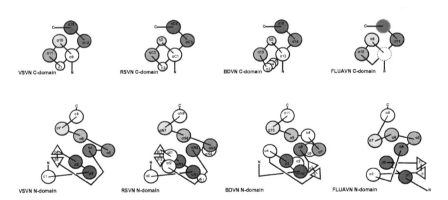

Fig. 2. Topology representation of the core secondary structure elements in the N protein from VSV, RSV, BDV, and FLUAV. Helices are shown as circles, whereas β-strands are shown as triangles.

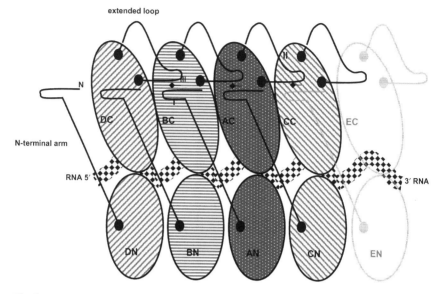

Fig. 3. A drawing viewed from the outside of the N-RNA ring to show the cross-molecular interactions among the N subunits in the nucleocapsid. Each N subunit is represented as an N-lobe and a C-lobe with an N arm and a C-loop. The N subunits are labeled A, B, C, D, and E, respectively. Each of the three interactions (I, II, and III) formed by the N-arm and C-loop is illustrated across A, B, C, and D subunits. Subunit E is not involved in any additional unique contact. The encapsidated RNA is represented by a ribbon.

3. Assembly of the Nucleocapsid

To extrapolate from the ring structure of the nucleocapsid-like particle, the N protein may be viewed as lining up side-by-side throughout the length of the VSV nucleocapsid (Fig. 3). In this linear structure, the C-terminal lobe makes contact with its neighbor burying a surface area of 1469 Å^2. The N-terminal lobe also makes contact with its neighbor burying a lesser surface area of 507 Å^2. It appears that the C-terminal lobe is the major component that constructs the nucleocapsid. This side-by-side arrangement of the N protein subunits builds a tunnel throughout the capsid, which is derived from the space of the V-shaped N protein subunit. The viral genomic RNA threads through the tunnel so that it is completely sequestered in the nucleocapsid. In the nucleocapsid, each N protein subunit covers nine nucleotides. Counting from the $5'$ end, the bases of nucleotides 1–4 are stacked to form a motif similar to half of a type A RNA duplex. The bases of these nucleotides face the opening of the cavity. The bases of nucleotides 5, 7, and 8 are also stacked, but the bases face the interior of the cavity. Nucleotide 6 swings out of this stack and is the nucleotide with its base pointing outwards the most. Nucleotide 9 is located between two neighboring subunits of N. It is likely that the orientation of the base in nucleotide 9 may change when the nucleocapsid assumes a more extended conformation, either in the virion or in the cytoplasm after being released from the virion at the end of the entry process. This notion is supported by comparison with a homologous structure of the nucleocapsid-like particle of rabies virus (RABV), which has 11 subunits in the ring.[15] When the curvature of the nucleocapsid increases, the small C-terminal domain may move inwards to cover the space.[14] At this time, it could not be clearly resolved which residues of the N protein interact with the bases of the encapsidated RNA because the RNA in the nucleocapsid-like particle has random sequences. However, basic residues that interact with the backbone phosphate groups are identified in the cavity. In the VSV N protein, six positively charged residues are found to bind phosphate groups. These residues include Arg143, Arg146, Lys155, Lys286, Arg317, and Arg408.

For a long period of time, the nucleocapsid protein of negative strand RNA viruses was considered an RNA binding protein.[16,17] The VSV nucleocapsid was described as a ribonucleoprotein complex in which the RNA is resistant to digestion by RNA nucleases, such as micrococcal nuclease.[18] However, the RNA in the VSV nucleocapsid could be in fact digested with a sufficient amount of RNase A (unpublished data from the author's lab). The cavity formed by the N protein where the RNA is accommodated does not have features consistent with an RNA binding protein because no specific patterns of conserved positively charged residues are found.[14] In the VSV N protein, sidechains of the six positively charged residues make salt bridges with the phosphate groups of nucleotides 8, (4 + 5), 9, 3, 4, and 4, respectively. There are additional positively charged residues in the N protein cavity, but not all phosphate groups in the viral RNA interact with a positively charged residue. The structural positions of the positively charged residues are not conserved among the N proteins of rhabdoviruses. Similarly, there are no specific interactions with the bases for conserved hydrogen bonds or hydrophobic interactions with a certain base type or a specific nucleotide sequence. Alternatively, the crystal structure of the nucleocapsid-like particle suggests that the RNA is encapsidated in the capsid analogous to any other RNA virus nucleocapsid, such as a poliovirus particle.[19] The VSV N protein appears to form a stable capsid of multiple subunits to encapsidate the genomic RNA in its center similar to poliovirus. The only difference between the polymeric capsid of VSV and that of poliovirus is that the VSV capsid has a linear rather than icosahedral symmetry. The VSV capsid may be assembled in an analogous way to that of poliovirus with cross-molecular interactions involving polypeptide termini and extended loops in the capsid protein subunit. Results of a series of experiments indeed support such reasoning.[20] In the nucleocapsid, there are three unique types of cross-molecular interactions spanning four N protein subunits in addition to the side-by-side interactions.[9] First, the N-arm of an N protein subunit interacts with the left neighboring C-terminal lobe on the proximal surface, if the nucleocapsid is viewed with the opening of the RNA cavity facing away from the reader and the C-terminal lobe is on top of the N-terminal lobe (Fig. 3). Second, the C-loop of this subunit interacts with the right neighboring C-terminal lobe. Lastly, the N-arm of this subunit also interacts with the C-loop of the N protein subunit two units away on the left. This pattern of cross-molecular interactions is present across the length of the nucleocapsid. The types of interactions include salt bridges, hydrogen bonds, and hydrophobic interactions. However, there are no conserved residues involved in these interactions among the N proteins of rhabdoviruses even though the extent of the cross-molecular interactions is similar.[14] When deletions were made to the N-arm or the C-loop, which destroyed two of the three unique cross-molecular interactions each time, the nucleocapsid-like particle could no longer be assembled, nor could RNA be encapsidated by the mutant N protein.[20] The same result was obtained when a stretch of amino acids involved in the side-by-side interactions of the C-terminal lobe was changed to polyalanine (N 320–324, Ala$_5$). This mutant N protein can, however, form a stable complex with a dimer of the P protein, which mimicks the N°P$_2$ complex as the precursor for assembly of the nucleocapsid.[21] This further confirmed that the N protein is not an RNA binding protein because the precursor does not bind RNA by itself. The only structure in which the viral RNA is associated with regard to the N protein is that of the polymeric nucleocapsid. These results indicated that the formation of a stable VSV nucleocapsid requires both the side-by-side interactions of the N protein subunits and their cross-molecular interactions involving extended N-termini and the large loop in the C-terminal lobe. This is analogous to the requirements of the assembly of an icosahedral capsid.[22] On the other hand, empty capsids with very much a similar structure as the nucleocapsid were formed when residue 290 was mutated from serine to tryptophan.[20] The enlarged sidechain of tryptophan imposes steric hindrance to the RNA so it can no longer be encapsidated. This observation further establishes that the N protein is not an RNA binding protein, but a subunit of a polymeric capsid.

If the N protein is not a RNA binding protein, a critical question arises regarding how the viral genomic RNA is specifically encapsidated in the nucleocapsid. It has been speculated that the 3′ or 5′ sequence of the viral genome may contain a packaging signal that may be recognized by the N protein. Previous studies suggested that the association of the N protein with the P protein prevents it from association with nonspecific RNAs.[7,23,24] However, such a complex may not render any recognition of the specific RNA sequence by the N protein even though some earlier investigations suggested that the N protein could recognize the leader sequence of the VSV genome.[3,25] It was also speculated that a host factor could be involved in the recognition of the packaging signal.[26] As the crystal structure does not support that the N protein could recognize any specific RNA sequence, we postulate that the encapsidation of the viral genomic RNA takes place at the site of replication by the viral RNA polymerase, with the P protein playing a critical role (Fig. 4). The P protein is associated with the L protein to form the active viral RNA polymerase.[27,28] In this polymerase complex, the P protein delivers the L protein to the nucleocapsid, which serves as the template for replication.[29,30] The C-terminal domain of the P protein forms a small compact domain containing α-helices and β-strands.[31] When the P protein binds the nucleocapsid, this domain sits between two adjacent C-terminal lobes, interacting with the two C-loops from opposite sides

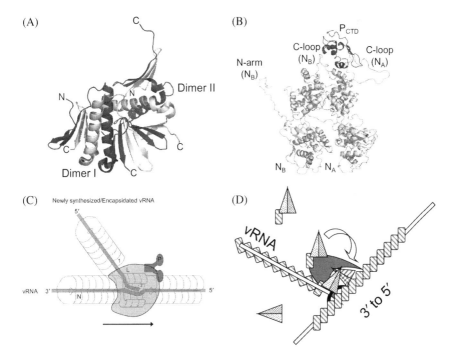

Fig. 4. (A) A ribbon drawing showing the structure of the middle domain of the P protein as a tetramer. (B) The binding of the C-terminal domain of the P protein with the nucleocapsid. (C) A model for the replication complex that contains the P-anchored L protein that induces opening of the N protein to unveil the genomic RNA sequence as the template for RNA synthesis. (D) A schematic drawing to show the mechanism of the N protein-driven replication. The P dimer (triangle) shuttles the N protein (rectangle) to the replication site to encapsidate the newly copied viral RNA.

as well as helix α13 of the N protein. Such a binding site could only be present in the nucleocapsid because only in the nucleocapsid are the N protein subunits lined up side-by-side. Upon recognition of this unique binding site, the P protein therein renders the specific recognition of the nucleocapsid to the viral polymerase. At the same time, the P protein binds the free N protein as an $N°P_2$ complex to keep it unassociated before the N protein is committed to the assembly of the nucleocapsid. In addition, we suggest that the P protein actually places the N protein at the replication site by the association of the P dimer with the P protein that is already part of the viral polymerase (Fig. 4). The region that is responsible for the dimer association of the P protein is located in the middle of the P protein (the middle domain).[32] This region contains a central α-helix with one β-hairpin preceding the helix, and one after. The α-helix forms a super twist with another in the P dimer. The β-hairpins, on the other hand, form a sheet with four anti-parallel β-strands by domain swapping between the two P protein subunits in the dimer. The biological unit of the P protein is likely to be the P dimer. The anti-parallel β-sheet may also be involved in the formation of an extended β-sheet with eight β-strands for multimerization of the P protein. The P dimer in the $N°P_2$ complex is therefore capable of shuffling the newly synthesized N protein to the replication site by forming a sheet of the eight β-strands between the P dimer in the viral polymerase and that in the $N°P_2$ complex. At this location, the N protein may be used to assemble the nucleocapsid by encapsidating the nascent viral genomic RNA. The viral genome is specifically encapsidated because the assembly site is located where it is being synthesized. The assembly of the nucleocapsid supports genomic length synthesis of RNA by the viral polymerase. The P protein not only maintains a pool of the N protein available for nucleocapsid assembly but also transports it to the correct location for nucleocapsid assembly.

4. Assembly of the Virion

When the replication of the viral genome and the assembly of the nucleocapsid are simultaneously completed, the nucleocapsid has to be transported from the site of replication to the site of virion assembly.[33] Nucleocapsid assembly occurs at the site of viral replication near the nucleus. Here, P and L are attached to the assembled nucleocapsid with each subsequently packaged inside the virion.[34] This is a common feature of negative strand RNA viruses that carry their viral polymerase within the assembled virion. The nucleocapsid–polymerase complex is then transported to the plasma membrane via a microtubule-mediated route. This process is believed to be facilitated by mitochondria, as experiments to visualize intracellular transport of VSV nucleocapsids within cells have shown that nucleocapsids are closely associated with mitochondria.[33] Once the nucleocapsid is delivered to the plasma membrane, it must be transformed from the coiled, highly extended state that is observed in the cell to the condensed state that is observed in the virion. The key viral protein that directs this condensation is the matrix protein, M (229 amino acids in length).[35,36] The M protein may be divided into two domains: a small domain (encompassing the first 43 amino acids of the N-terminus) and a large domain composed of the remaining residues of M protein.[37] The small domain of the M protein is responsible for association with the nucleocapsid, whereas the large domain appears to be key for interaction with the plasma membrane alone.[37–40] The crystal structure of the large domain has been determined, which shows a compact structure containing three α-helices sandwiched between two sets of β-sheets.[41] A portion of the surface of M has a hydrophobic face that is complemented with conserved positively charged residues.[42] The M protein condenses the nucleocapsid yielding a cylinder with a diameter of 51–55 nm.[35,43] The separation of each turn of the supercoiled nucleocapsid is about 5 nm. The M protein appears to connect the adjacent turns of the coiled nucleocapsid.[36] The association of the M protein with the nucleocapsid not only condenses the nucleocapsid into the coiled conformation, but also allows the M–nucleocapsid complex to bind with the plasma tail of the surface glycoprotein, G, which is embedded in the cytoplasmic membrane at the virion assembly site.[44] Free M proteins in the cytoplasm do not associate with plasma membrane[45,46] and the aggregation of the M proteins involves the small domain of the M protein.[42,47] The polymerization of the M subunits, a process that is likely to be initiated by association with the nucleocapsid at budding, promotes membrane association by the M proteins. The nucleocapsid must adopt a conformation that can recruit the M protein at the site of budding.[48,49] An assembly model was proposed as follows,[35] which is supported by other observations: The 5′ end of the nucleocapsid initiates the formation of a conformation that allows the M protein to cross-connect adjacent turns of the supercoiled nucleocapsid. This early complex anchors the nucleocapsid at the site in the plasma membrane where the glycoprotein G has been located as trimeric clusters.[50] Such a step ensures that all three major structural proteins are properly incorporated into the mature virion.[51] The next step is the assembly of the cone nose portion of the supercoiled nucleocapsid in the bullet-shaped virion, which is controlled by the M protein.[52] Budding continues with condensation of the nucleocapsid supercoil by the M protein and recruitment of the G protein that is embedded in the plasma membrane as a network of trimers.[53] The cytoplasmic domain (the C-terminus) of the G protein interacts with the polymerized M protein that is connected to the supercoiled nucleocapsid. When the entire nucleocapsid is packaged into the

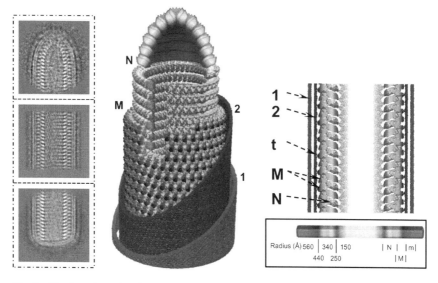

Fig. 5.   The three-dimensional structure of VSV virion. Representative class averaged images from the 2D classfications of VSV are shown on the left. The three-dimensional model is presented by exposing each layer of VSV virion: 1, the outer leaflet of the envelope; 2, the inner leaflet of the envelope; M, the matrix protein; N, the nucleocapsid protein. A cross-section of the virion trunk is also presented in the right panel with the radii of each layer shown.

virion, the virus particle pinches off the plasma membrane with the help of host factors.[54]

The recent structure of the VSV virion determined by cryoEM revealed details of the assembled virion, and provides more insight on the assembly process[55] (Fig. 5). Based on the superposition of the N-RNA crystal structure with the cryoEM density of the nucleocapsid in the virion, the nucleocapsid forms a left-handed superhelix with the 3′ end of the viral RNA genome at the tip of the cone-shaped nose and the 5′ end at the blunt end of the virion. At the cone tip, the first turn of the nucleocapsid superhelix contains 10 subunits of the N protein. This implies that the nucleocapsid tends to form a ring-like structure if there are no other stabilizing factors. The ring-like structure of 10 N subunits is similar to that of the recombinant N protein ring in which random RNA is encapsidated in *E. coli.*[10] The ring-like structure formed by 10 N subunits may represent the most stable structure when the interactions between the N subunits provide the stabilizing forces. Between the first and the second turns of the nucleocapsid superhelix, M proteins are found to be associated with the N-terminal lobe of the N proteins in the adjacent turns. The M proteins are exterior to the nucleocapsid superhelix and beneath the membrane envelope. The association with the M protein induces a tilt in the N protein from a more vertical orientation in the first turn to a more lateral orientation in the second turn. The lateral orientation of the N protein expands the radius of the nucleocapsid superhelix so that the N-terminal lobes are more separated and more N subunits are included in the second turn. The radial expansion presumably reduces the interactions between the N subunits so the circular structure becomes less stable. However, the interactions with the M proteins may compensate for the loss of stabilizing energy due to the reduction of N protein interactions. The radial expansion continues through the eighth turn of the nucleocapsid superhelix with addition of the M proteins in every groove between adjacent turns. The number of N subunits per turn reaches a maximum of 37.5, or 75 per two turns, at the eighth turn. The radius and number of N subunits per turn remain constant for all subsequent turns in the trunk of the virion. The interactions between the M and N proteins in the virion support the notion that the M protein is responsible for maintaining the bullet-shape and the condensed nucleocapsid in the VSV virion.[35,36,52]

The structure of the nucleocapsid in the trunk region was determined to 10.6 Å resolution by cryoEM, which allows for a fairly unambiguous fit of the crystal structure of the N-RNA complex into the cryoEM density. The N subunit in each turn is stacked atop yet between two adjacent N subunits in the subsequent turn, giving rise to 75 subunits in every two turns. Each turn of the nucleocapsid does not actually make any close contact in the vertical direction parallel to the axis of the superhelix, although the N subunit is vertically positioned exactly between the two subunits in the next turn. The connection between turns of the nucleocapsid superhelix is mainly made by the M proteins, as discussed later. The outer radius of the nucleocapsid superhelix is 225 Å, the inner radius 154 Å, and the rise of each turn is 50.8 Å. The N subunit tilts upwards by 27° to the cross-section plane, compared with 55° in the more vertical orientation in the ring-like crystal structure with 10 subunits. The more flat orientation requires that the N-terminal lobes fit in a larger radius compared with the C-terminal lobes, thus the N-lobes make less contact. Fitting the crystal structure of N into the cryoEM density suggests that the contacts between the C-terminal lobes are mostly maintained as observed in the crystal structure. The N subunit is slightly rotated outwards while keeping the RNA in the same position in the N cavity. It is speculated that the base in the ninth nucleotide in each N subunit may assume a more relaxed conformation and the small C-terminal domain may also rotate to cover any gap because of the rotation of the N subunit, based on the comparison of the 10 subunit ring of the VSV N-RNA complex with the 11 subunit ring of the RABV N-RNA complex.[14] The C-terminal lobe is where the cross-molecular interactions are maintained along the length of the nucleocapsid, including the side-by-side interactions, interactions between the N-arm and the C-loop, as well as interactions of the C-terminal lobe with the

N-arm and the C-loop. The type of interactions between the N-terminal lobes in the nucleocapsid superhelix is unclear.

As observed in the cryoEM structure of the VSV virion, M proteins form a two-dimensional mesh surrounding the nucleocapsid superhelix. There is one M subunit for each N subunit with the exception of the first and last turns. The M protein mesh folds into a cylinder that is coaxial with the nucleocapsid superhelix. As the nucleocapsid does not make close vertical contacts between turns, it is the M protein that renders the rigid architecture of the bullet-shaped virion. In the cylindrical mesh of M proteins, each M subunit interacts with four neighboring subunits. The large domain of the M protein faces the membrane envelope, whereas the small domain whose atomic structure is unknown faces the nucleocapsid superhelix. The large domains of adjacent M proteins do not make direct contacts between one another in the cylinder. The small domain, on the other hand, makes two unique contacts with its neighboring M subunits. Viewing with the conical tip of the virion pointing up, the small domain of the M protein makes a contact with the large domain on the left. The surface of the large domain on the left contains a depression that can accommodate the protruding small domain. This contact maintains the lateral contact between the M subunits along each turn of the nucleocapsid superhelix. At the same time, the small domain also makes a vertical contact with the M subunit below, which sets the rise between the M subunits. This contact involves residues 41–52 and a large depression, mainly hydrophobic, in the large domain of the M subunit below as observed in the crystal structure of full-length recombinant M protein.[56] Residues 41–52 form a small fold that is stabilized by interactions with the large depression that includes residues 120–124. Residues 120–124 have been shown to play a role in M self-association and virion assembly.[42,57] In addition to building the architecture of the M protein cylinder, the interactions of the small domain with the N-terminal lobes of the N protein also dictate the architecture of the nucleocapsid superhelix. The extreme N-terminal end of the small domain makes a close contact with the N-terminal lobe of an N subunit in the upper turn, whereas the region between the residues in the small domain that contact the neighboring large domains of the M subunits makes a close contact with the N-terminal lobe on the left in the lower turn. Thus, the M protein also keeps the N subunits on register from turn to turn and dictates that each turn is separated by a fixed rise. The virion is therefore rigidly organized by the M protein with interdigitated interactions between M subunits, and between M and N subunits (Fig. 6).

In the final virion, the M protein interacts with both the G protein and the superhelix of the nucleocapsid. Before budding, trimers of the G protein are localized in microdomains (or rafts) on the plasma membrane.[51,58,59] In order for the M protein to interact with the cytoplasmic tail of the G protein, the helical

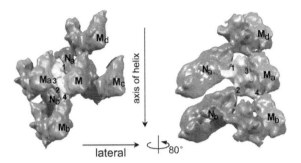

Fig. 6. Unique contacts between M and N within the virion trunk. A central copy of the M protein is shown with four adjacent copies of M (labeled Ma–d) and two copies N (labeled Na,b) subunits in the neighborhood of one M are shown. The solid surface of the cryo-EM map is rendered at 1.0 σ. The contact points on the M-hub that mediate interactions with N and M are labeled 1 to 4.

symmetry of the M mesh must match the hexagonal symmetry of the G protein in the membrane envelope of the final virion. Indeed, the size of the M subunits and their radial orientation constitute a perfect cylindrical lattice that allows the C-domain of M to match the hexagonal symmetry of the G trimers in the membrane envelope, and the superhelical symmetry of the nucleocapsid inside the M cylinder. The ratio among the three proteins, G:M:N, in the trunk of the virion is 1:1:1. The tight association of these three proteins requires a highly coordinated budding process. Recently published data suggest that the nucleocapsid is initially transported to the location on the plasma membrane where the G protein is localized.[50] The M protein, on the other hand, is transported to microdomains different from those localized with the G protein. The association of the nucleocapsid with the G protein occupied microdomains promotes the recruitment of the M protein clustered in the nearby microdomains. One of the possible budding initiation structures may be the formation of a two-turn superhelix of the nucleocapsid (Fig. 6). When the first-turn of the M subunits is recruited, the M protein may tighten the association by interacting with the G protein and the N protein simultaneously to initiate budding. The continuation of the budding process may follow a fast, cooperative mode involving association of the G:M:N with matching symmetries. If the M protein was not recruited as in the case of a temperature-sensitive mutant of M, spherical virions would be formed containing almost no M protein.[52] These particles released without complementation by the wild-type M protein had more than two-fold reduction in infectivity. In addition, the amino acid sequence to the transmembrane domain of the G protein plays a critical role in virus budding.[60]

References

1. Rubio, C., *et al.*, Replication and assembly of VSV nucleocapsids: Protein association with RNPs and the effects of cycloheximide on replication. *Virology*, 1980. **105**(1): 123–35.

2. Davis, N.L. and G.W. Wertz, Synthesis of vesicular stomatitis virus negative-strand RNA *in vitro*: Dependence on viral protein synthesis. *J Virol*, 1982. **41**(3): 821–32.

3. Blumberg, B.M., M. Leppert and D. Kolakofsky, Interaction of VSV leader RNA and nucleocapsid protein may control VSV genome replication. *Cell*, 1981. **23**(3): 837–45.

4. Hill, V.M., L. Marnell and D.F. Summers, *In vitro* replication and assembly of vesicular stomatitis virus nucleocapsids. *Virology*, 1981. **113**(1): 109–18.

5. Peluso, R.W. and S.A. Moyer, Viral proteins required for the *in vitro* replication of vesicular stomatitis virus defective interfering particle genome RNA. *Virology*, 1988. **162**(2): 369–76.

6. La Ferla, F.M. and R.W. Peluso, The 1:1 N–NS protein complex of vesicular stomatitis virus is essential for efficient genome replication. *J Virol*, 1989. **63**(9): 3852–7.

7. Masters, P.S. and A.K. Banerjee, Complex formation with vesicular stomatitis virus phosphoprotein NS prevents binding of nucleocapsid protein N to nonspecific RNA. *J Virol*, 1988. **62**(8): 2658–64.

8. Howard, M. and G. Wertz, Vesicular stomatitis virus RNA replication: A role for the NS protein. *J Gen Virol*, 1989. **70**(Pt 10): 2683–94.

9. Green, T.J., *et al.*, Structure of the vesicular stomatitis virus nucleoprotein-RNA complex. *Science*, 2006. **313**(5785): 357–60.

10. Green, T.J., *et al.*, Study of the assembly of vesicular stomatitis virus N protein: Role of the P protein. *J Virol*, 2000. **74**(20): 9515–24.

11. Luo, M., *et al.*, Structural comparisons of the nucleoprotein from three negative strand RNA virus families. *Virol J*, 2007. **4**: 72.

12. Jambou, R.C., *et al.*, Complete sequence of the major nucleocapsid protein gene of human parainfluenza type 3 virus: Comparison with other negative strand viruses. *J Gen Virol*, 1986. **67**(Pt 11): 2543–8.

13. Walker, P.J., *et al.*, Structural and antigenic analysis of the nucleoprotein of bovine ephemeral fever rhabdovirus. *J Gen Virol*, 1994. **75**(Pt 8): 1889–99.

14. Luo, M., *et al.*, Conserved characteristics of the rhabdovirus nucleoprotein. *Virus Res*, 2007. **129**(2): 246–51.

15. Albertini, A.A., *et al.*, Crystal structure of the rabies virus nucleoprotein–RNA complex. *Science*, 2006. **313**(5785): 360–3.

16. Pons, M.W., Isolation of influenza virus ribonucleoprotein from infected cells. Demonstration of the presence of negative-stranded RNA in viral RNP. *Virology*, 1971. **46**(1): 149–60.

17. Compans, R.W., J. Content and P.H. Duesberg, Structure of the ribonucleoprotein of influenza virus. *J Virol*, 1972. **10**(4): 795–800.

18. Chanda, P.K. and A.K. Banerjee, Two distinct populations of vesicular stomatitis virus ribonucleoprotein cores with differential sensitivities to micrococcal nuclease. *Biochem Biophys Res Commun*, 1979. **91**(4): 1337–45.

19. Hogle, J.M., M. Chow and D.J. Filman, Three-dimensional structure of poliovirus at 2.9 A resolution. *Science*, 1985. **229**(4720): 1358–65.

20. Zhang, X., *et al.*, Role of intermolecular interactions of vesicular stomatitis virus nucleoprotein in RNA encapsidation. *J Virol*, 2008. **82**(2): 674–82.

21. Mavrakis, M., *et al.*, Isolation and characterisation of the rabies virus N degrees–P complex produced in insect cells. *Virology*, 2003. **305**(2): 406–14.

22. Bertolotti-Ciarlet, A., *et al.*, Structural requirements for the assembly of Norwalk virus-like particles. *J Virol*, 2002. **76**(8): 4044–55.

23. Chen, M., T. Ogino and A.K. Banerjee, Interaction of vesicular stomatitis virus P and N proteins: Identification of two overlapping domains at the N terminus of P that are involved in N0–P complex formation and encapsidation of viral genome RNA. *J Virol*, 2007. **81**(24): 13478–85.

24. Mavrakis, M., *et al.*, Rabies virus chaperone: Identification of the phosphoprotein peptide that keeps nucleoprotein soluble and free from non-specific RNA. *Virology*, 2006. **349**(2): 422–9.

25. Moyer, S.A., *et al.*, Assembly and transcription of synthetic vesicular stomatitis virus nucleocapsids. *J Virol*, 1991. **65**(5): 2170–8.

26. Gupta, A.K., J.A. Drazba and A.K. Banerjee, Specific interaction of heterogeneous nuclear ribonucleoprotein particle U with the leader RNA sequence of vesicular stomatitis virus. *J Virol*, 1998. **72**(11): 8532–40.

27. Emerson, S.U. and Y. Yu, Both NS and L proteins are required for *in vitro* RNA synthesis by vesicular stomatitis virus. *J Virol*, 1975. **15**(6): 1348–56.

28. Naito, S. and A. Ishihama, Function and structure of RNA polymerase from vesicular stomatitis virus. *J Biol Chem*, 1976. **251**(14): 4307–14.

29. Gill, D.S., D. Chattopadhyay and A.K. Banerjee, Identification of a domain within the phosphoprotein of vesicular stomatitis virus that is essential for transcription *in vitro*. *Proc Natl Acad Sci U S A*, 1986. **83**(23): 8873–7.

30. Green, T.J. and M. Luo, Structure of the vesicular stomatitis virus nucleocapsid in complex with the nucleocapsid-binding domain of the small polymerase cofactor, P. *Proc Natl Acad Sci U S A*, 2009. **106**(28): 11713–8.

31. Ribeiro Jr., E.A., *et al.*, Solution structure of the C-terminal nucleoprotein–RNA binding domain of the vesicular stomatitis virus phosphoprotein. *J Mol Biol*, 2008. **382**(2): 525–38.

32. Ding, H., *et al.*, Crystal structure of the oligomerization domain of the phosphoprotein of vesicular stomatitis virus. *J Virol*, 2006. **80**(6): 2808–14.

33. Das, S.C., *et al.*, Visualization of intracellular transport of vesicular stomatitis virus nucleocapsids in living cells. *J Virol*, 2006. **80**(13): 6368–77.

34. Hsu, C.H., D.W. Kingsbury and K.G. Murti, Assembly of vesicular stomatitis virus nucleocapsids *in vivo*: A kinetic analysis. *J Virol*, 1979. **32**(1): 304–13.

35. Newcomb, W.W., *et al.*, *In vitro* reassembly of vesicular stomatitis virus skeletons. *J Virol*, 1982. **41**(3): 1055–62.

36. Odenwald, W.F., *et al.*, Stereo images of vesicular stomatitis virus assembly. *J Virol*, 1986. **57**(3): 922–32.

37. Ogden, J.R., R. Pal and R.R. Wagner, Mapping regions of the matrix protein of vesicular stomatitis virus which bind to ribonucleocapsids, liposomes, and monoclonal antibodies. *J Virol*, 1986. **58**(3): 860–8.

38. Chong, L.D. and J.K. Rose, Interactions of normal and mutant vesicular stomatitis virus matrix proteins with the plasma membrane and nucleocapsids. *J Virol*, 1994. **68**(1): 441–7.

39. Ye, Z., *et al.*, Membrane-binding domains and cytopathogenesis of the matrix protein of vesicular stomatitis virus. *J Virol*, 1994. **68**(11): 7386–96.

40. Justice, P.A., *et al.*, Membrane vesiculation function and exocytosis of wild-type and mutant matrix proteins of vesicular stomatitis virus. *J Virol*, 1995. **69**(5): 3156–60.

41. Gaudier, M., Y. Gaudin and M. Knossow, Crystal structure of vesicular stomatitis virus matrix protein. *EMBO J*, 2002. **21**(12): 2886–92.

42. Gaudier, M., Y. Gaudin and M. Knossow, Cleavage of vesicular stomatitis virus matrix protein prevents self-association and leads to crystallization. *Virology*, 2001. **288**(2): 308–14.

43. Newcomb, W.W. and J.C. Brown, Role of the vesicular stomatitis virus matrix protein in maintaining the viral nucleocapsid in the condensed form found in native virions. *J Virol*, 1981. **39**(1): 295–9.

44. Lyles, D.S., M. McKenzie and J.W. Parce, Subunit interactions of vesicular stomatitis virus envelope glycoprotein stabilized by binding to viral matrix protein. *J Virol*, 1992. **66**(1): 349–58.

45. Knipe, D.M., D. Baltimore and H.F. Lodish, Maturation of viral proteins in cells infected with temperature-sensitive mutants of vesicular stomatitis virus. *J Virol*, 1977. **21**(3): 1149–58.

46. Knipe, D., H.F. Lodish and D. Baltimore, Analysis of the defects of temperature-sensitive mutants of vesicular stomatitis virus: Intracellular degradation of specific viral proteins. *J Virol*, 1977. **21**(3): 1140–8.

47. Dancho, B., *et al.*, Vesicular stomatitis virus matrix protein mutations that affect association with host membranes and viral nucleocapsids. *J Biol Chem*, 2009. **284**(7): 4500–9.

48. Lyles, D.S. and M.O. McKenzie, Reversible and irreversible steps in assembly and disassembly of vesicular stomatitis virus: Equilibria and kinetics of dissociation of nucleocapsid–M protein complexes assembled *in vivo*. *Biochemistry*, 1998. **37**(2): 439–50.

49. Flood, E.A. and D.S. Lyles, Assembly of nucleocapsids with cytosolic and membrane-derived matrix proteins of vesicular stomatitis virus. *Virology*, 1999. **261**(2): 295–308.

50. Swinteck, B.D. and D.S. Lyles, Plasma membrane microdomains containing vesicular stomatitis virus M protein are separate from microdomains containing G protein and nucleocapsids. *J Virol*, 2008. **82**(11): 5536–47.

51. Knipe, D.M., D. Baltimore and H.F. Lodish, Separate pathways of maturation of the major structural proteins of vesicular stomatitis virus. *J Virol*, 1977. **21**(3): 1128–39.

52. Lyles, D.S., *et al.*, Complementation of M gene mutants of vesicular stomatitis virus by plasmid-derived M protein converts spherical extracellular particles into native bullet shapes. *Virology*, 1996. **217**(1): 76–87.

53. Doms, R.W., *et al.*, Differential effects of mutations in three domains on folding, quaternary structure, and intracellular transport of vesicular stomatitis virus G protein. *J Cell Biol*, 1988. **107**(1): 89–99.

54. Jayakar, H.R., K.G. Murti and M.A. Whitt, Mutations in the PPPY motif of vesicular stomatitis virus matrix protein reduce virus budding by inhibiting a late step in virion release. *J Virol*, 2000. **74**(21): 9818–27.

55. Ge, P., J. Tsao, S. Schein, T.J. Green, M. Luo and Z.H. Zhou, CryoEM model of the bullet-shaped vesicular stomatitis virus. *Science*, 2009. Revision.

56. Graham, S.C., *et al.*, Rhabdovirus matrix protein structures reveal a novel mode of self-association. *PLoS Pathog*, 2008. **4**(12): e1000251.

57. Connor, J.H., M.O. McKenzie and D.S. Lyles, Role of residues 121 to 124 of vesicular stomatitis virus matrix protein in virus assembly and virus–host interaction. *J Virol*, 2006. **80**(8): 3701–11.

58. Luan, P., L. Yang and M. Glaser, Formation of membrane domains created during the budding of vesicular stomatitis virus. A model for selective lipid and protein sorting in biological membranes. *Biochemistry*, 1995. **34**(31): 9874–83.

59. Kreis, T.E. and H.F. Lodish, Oligomerization is essential for transport of vesicular stomatitis viral glycoprotein to the cell surface. *Cell*, 1986. **46**(6): 929–37.

60. Robison, C.S. and M.A. Whitt, The membrane-proximal stem region of vesicular stomatitis virus G protein confers efficient virus assembly. *J Virol*, 2000. **74**(5): 2239–46.

Chapter 10

Paramyxovirus Budding Mechanisms

Megan S. Harrison*, Takemasa Sakaguchi[†]
and Anthony P. Schmitt[*,‡]

1. Introduction

The paramyxoviruses are composed of a diverse group of enveloped viruses having negative-sense, nonsegmented RNA genomes that are relatively large (15–19 kb in length). These viruses are the causative agents for a number of significant human and animal diseases. For example, substantial fractions of respiratory illnesses in young children are caused by infections with human respiratory syncytial virus (HRSV) and human parainfluenza viruses types 1–4 (hPIV 1–4).[1,2] Resurgences of disease caused by measles and mumps viruses have occurred recently in the United States, Canada, and the United Kingdom, with the numerous outbreaks being attributed to declining vaccination rates in these countries.[3−6] Newly-discovered zoonotic paramyxoviruses, Nipah virus and Hendra virus, have caused outbreaks of fatal disease in Asia and Australia as humans and their domesticated animals have increasingly come into contact with the fruit bat species that naturally harbor these pathogens.[7] Paramyxoviruses of agricultural significance include Newcastle disease virus (NDV) that infects chickens and other avian species and rinderpest virus that infects cattle.[8,9] Other paramyxoviruses have been widely studied as models of infection. These include Sendai virus that infects mice and parainfluenza virus 5 (PIV5, formerly SV5).

*Department of Veterinary and Biomedical Sciences, and Center for Molecular Immunology and Infectious Disease, The Pennsylvania State University, University Park, PA 16802, USA.

†Department of Virology, Graduate School of Biomedical Sciences, Hiroshima University, 1-2-3 Kasumi, Minami-ku, Hiroshima 734-8551, Japan.

‡E-mail: aps13@psu.edu

Paramyxovirus infections are transmitted by virus particles, which function to enclose viral nucleocapsids, also known as ribonucleoproteins (RNPs), and mediate their delivery into target cells. The particles are enveloped by membrane bilayers, which are derived from infected host cells during budding. Particle shape is usually roughly spherical, as illustrated in the electron micrographs of Fig. 1, but can also be filamentous. Particle size varies considerably, but typically ranges from 150 to 300 nm in diameter. Viral fusion (F) and attachment (HN, H, or G) proteins penetrate the virion envelopes, forming dense spike layers. Contained within the particles are the viral RNA genomes, bound and encapsidated with viral nucleocapsid proteins (N or NP), to form helical RNPs in association with viral RNA-dependent RNA polymerase complexes. Encapsidation of the viral RNA is necessary for viral transcription and genome replication to occur. Bound to the interior sides of virion envelopes are the viral matrix (M) proteins, which link together the RNP cargoes and the glycoprotein spike layers. M proteins coordinate virus assembly, directing RNPs and viral glycoproteins to concentrate at specific locations on plasma membranes from which virus particles will bud. Virus particles are formed as cellular membranes bulge outwards at these locations, eventually pinching off virions as a result of membrane fission (see Fig. 2). Here, we review events that occur during the assembly and release of paramyxovirus particles. Roles of viral matrix proteins, nucleocapsid proteins, glycoproteins, and accessory proteins in virus particle formation are discussed, as well as the recruitment of host proteins for particle budding. In addition, advances in our understanding of mechanisms that allow trafficking of viral components to selected sites on cellular membranes for efficient production of infectious virions are described.

2. Paramyxovirus M Proteins: The Organizers of Virus Assembly

Viral M proteins orchestrate the formation of infectious paramyxovirus particles by binding to cellular membranes and recruiting viral and cellular components to sites of budding. A threshold quantity of functional M protein must accumulate within infected cells for paramyxoviral particles to be produced efficiently. Recombinant Sendai and recombinant measles viruses completely lacking M genes both exhibit near-complete defects in particle formation.[10,11] These findings reinforce earlier studies using temperature-sensitive Sendai viruses, in which a correlation between poor M protein stability and reduced particle production was noted.[12,13] Similarly, recombinant measles virus encoding an unstable M protein in which a well-conserved valine residue at position 101 is mutated to alanine exhibited a severe defect in particle production.[14] Depletion of Sendai virus M protein using siRNA prevented M protein accumulation and resulted in a pronounced defect in

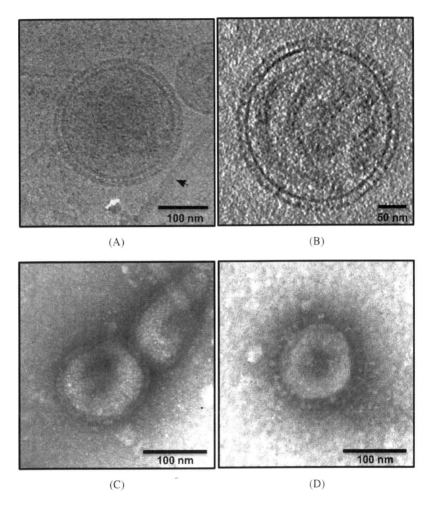

Fig. 1. Morphology of paramyxovirus virions and VLPs. (A) Purified PIV5 particle produced from LLCMK2 cells and observed by cryoelectron microscopy. The black arrow indicates a viral nucleocapsid outside of the virion. (B) Sendai virion purified from an embryonated chicken egg and imaged by cryoelectron tomography. (C) Mumps virions purified from 293T cells and visualized by electron microscopy after negative staining. (D) Mumps VLP purified from 293T cells transfected to produce the mumps virus M, NP, and F proteins, and visualized by electron microscopy after negative staining. Image for panel A was adapted from Terrier *et al.*[158]; image for panel B was adapted from Loney *et al.*[159]; images for panels C and D were adapted from Li *et al.*[27]

particle production.[15] Additional support for the notion that robust accumulation of functional M protein is critical for particle production has been obtained through the study of measles virus isolates derived from patients with subacute sclerosing panencephalitis (SSPE), a rare condition resulting from persistent measles

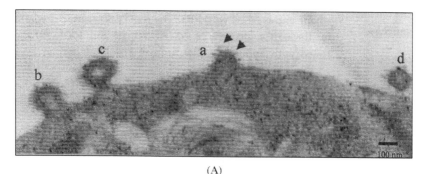

(A)

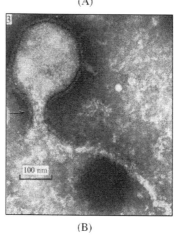

(B)

Fig. 2. Paramyxovirus particles budding from cell plasma membranes. (A) Thin section of an HRSV-infected BHK cell visualized by electron microscopy. (a)–(d) depict HRSV budding structures at the plasma membrane in various stages of formation, ranging from the initiation of membrane curvature (a) to the ultimate release of an HRSV particle (d). Black arrows indicate glycoproteins extending from the exterior of an HRSV budding structure, which is at an early stage of formation. (B) Thin section of a bovine PIV3-infected calf kidney cell, visualized by electron microscopy. A bPIV3 particle with glycoprotein spikes clearly visible on its surface is connected to the plasma membrane through a membrane stalk (black arrow). Image for panel A was adapted from Brown *et al.*[148]; image for panel B was adapted from Höglund and Morein.[160]

virus infection. These viruses exhibit severely defective particle production, which in many cases is caused by hypermutated and unstable M proteins (reviewed in Ref. 16).

The central role of paramyxovirus M proteins in virus assembly has been confirmed by studies in which virus-like particles (VLPs) have been produced in transfected cells. VLPs are particles resembling authentic virions (see Fig. 1D) and are produced from cells that have been manipulated to express the viral proteins that participate in particle formation. Paramyxovirus-like particles can be produced in many cases upon expression of M proteins alone. When expressed alone by trans-fection, the M proteins of hPIV1,[17] Sendai virus,[18,19] NDV,[20] Nipah virus,[21,22] and measles virus[14,23] each elicits efficient production of particles. Intriguingly, NDV M protein independently adsorbs onto phospholipid liposomes *in vitro*, lead-ing to deformation of the lipid bilayer and eventual release of lipid vesicles into the liposomes, indicating that the NDV M protein alone is sufficient to induce mem-brane curvature and fission.[24] For some paramyxoviruses, M protein expression alone in transfected cells is insufficient for VLP production; PIV5 and rinderpest virus M proteins are unable to induce detectable VLP production, and mumps virus M protein induces a barely detectable level of VLP production.[25-27] For PIV5 and mumps virus, VLP production becomes efficient when additional viral proteins are expressed together with M protein, although M protein is essential.[25,27] Thus, the requirements for efficient VLP production differ among paramyxoviruses, but the M proteins seem to play a fundamental role in all cases (see Table 1). This is in contrast to the production of influenza A VLPs, which is driven mainly by the influenza virus hemagglutinin glycoprotein.[28]

Structural studies of paramyxoviral M proteins, which are relatively hydropho-bic and prone to self-aggregation, have been challenging in the past. The recently-solved HRSV M protein atomic structure is a triumph towards understand-ing how the M protein structure may dictate its function in virus assembly and budding.[29] The HRSV M protein structure consists of two beta sheet-rich domains connected by a short, unstructured linker. Despite the lack of sequence homology, the overall structure of HRSV M protein is strikingly similar to that of the Ebola virus matrix protein, VP40.[29,30] A large positively-charged sur-face extends across both domains and the linker region, likely driving M pro-tein interactions with negatively-charged membrane surfaces or the negatively-charged nucleocapsid proteins. Comparing pneumovirus M protein sequences reveals similar sequence patterns in the C-termini, but more disparate sequences in the N-terminal and linker regions. More sequence diversity in the N-terminus may favor species-specific interactions, such as for M protein oligomerization or interactions with species-matched nucleocapsids and glycoproteins.[29] Previ-ous biochemical studies have implicated both electrostatic and hydrophobic inter-actions in mediating membrane binding of paramyxovirus M proteins,[26,31-33] Ebola virus VP40,[34,35] and vesicular stomatitis virus (VSV) M protein.[36,37] The C-terminal domain of HRSV M protein is proposed to interact predominantly with membranes,[38] as was posited for the Ebola virus VP40 C-terminus,[39] leaving the N-terminal region free to mediate interactions with other viral or cellular components.

Table 1.

Virus	Viral proteins important for VLP production	M protein sequences important for particle production	Notes	References
PIV5	M, NP, and F or M, NP, and HN	20-FPIV-24	No VLP production is observed when M protein is expressed alone. F and HN proteins are required and interchangeable for VLP production. NP protein enhances VLP production.	25, 91
Mumps virus	M, NP, and F	24-FPVI-27	Very little VLP production is observed when M protein is expressed alone. F protein strongly enhances VLP production, whereas HN protein does not. NP protein enhances VLP production.	27
Sendai virus	M	49-YLDL-52, 343-KIRKI-348	C protein enhances VLP production and binds to host protein Aip1/Alix. F protein enhances VLP production, but HN protein does not.	18–120
Nipah virus	M	62-YMYL-65, 92-YPLGVG-97	M-driven VLP production is not enhanced by either glycoprotein or N protein expression, but glycoproteins and N protein are incorporated into VLPs when they are co-expressed with M protein.	21, 22, 70
hPIV1	M	None reported	VLP production is driven by M protein. Nucleocapsid-like structures are incorporated into VLPs when NP protein is expressed.	17
NDV	M	None reported	M-driven VLP production is not enhanced by glycoprotein or NP protein expression. Glycoproteins and NP protein, when expressed in combination, are efficiently incorporated into VLPs.	20
Measles virus	M	None reported	M-driven VLP production is not enhanced by F protein expression.	14, 23

3. Paramyxovirus Glycoproteins and Nucleocapsid Proteins: Roles in Virus Assembly

3.1. *Viral glycoproteins*

Paramyxoviral glycoproteins (attachment and fusion proteins) decorate virion surfaces, forming spike layers visible by electron microscopy (see Fig. 1). Viral glycoproteins cluster together with M proteins on plasma membranes of infected cells in discrete patches from which budding is presumed to occur. Paramyxoviral M proteins are thought to assemble together with glycoproteins at sites of budding by interacting with the glycoprotein cytoplasmic tails. Evidence for interactions between paramyxoviral M proteins and glycoproteins has been obtained. Wild-type HRSV M and G proteins expressed together by transfection exhibited co-localization, which was lost upon removal of the first six amino acid residues of the G protein cytoplasmic tail.[40] In addition, specific interactions between the NDV HN and M proteins have been detected by co-immunoprecipitation from purified VLPs.[20]

Removal of paramyxovirus glycoprotein cytoplasmic tails disrupts M protein-glycoprotein interactions, and this has been correlated with reduced viral particle production. In the case of PIV5, assembly of viral M protein together with glycoproteins has been visualized by fluorescence microscopy. Clusters of M protein and HN glycoprotein are readily detected at the plasma membranes of cells infected with wt PIV5, but are absent from cells infected with recombinant viruses in which the HN protein cytoplasmic tail has been removed.[25,41] In cells infected with these altered viruses, HN protein is distributed randomly on the cell surface, M protein is dispersed throughout the cell, and virus particle production is greatly reduced (see Fig. 3).[25,41] The reduction in particle production is even more severe in the case of double-mutant virus in which both the HN and F protein cytoplasmic tails have been truncated.[42] This suggests that paramyxovirus glycoproteins can function redundantly during virus assembly, paralleling earlier results obtained with influenza A virus, in which simultaneous truncation of the HA and NA protein cytoplasmic tails resulted in severe defects in particle assembly and virion morphology, whereas individual truncations caused only mild or moderate defects.[43] The measles virus M protein has been shown to co-localize with the F and H glycoproteins in cells infected with the Edmonston vaccine strain. Co-localization was lost in cells infected with recombinant measles virus encoding glycoproteins with truncated cytoplasmic tails.[44] SSPE strains of measles virus often carry F proteins with mutations in their cytoplasmic tails, in correlation with the defective assembly and extensive syncytia formation characteristic of these strains.[45,46] Some recombinant measles viruses

200   M. S. Harrison, T. Sakaguchi and A. P. Schmitt

rPIV5   rPIV5 HNΔ2-9

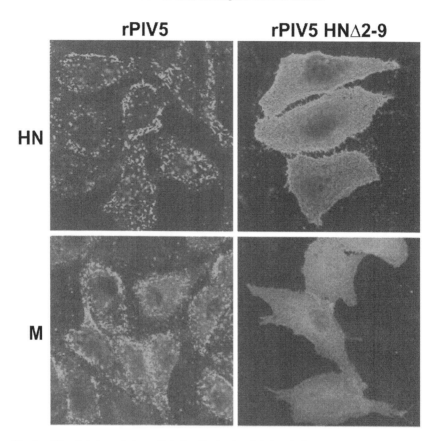

Fig. 3. Altered viral protein intracellular localization in cells infected with recombinant PIV5 encoding cytoplasmic tail-truncated HN protein. CV-1 cells were infected with the indicated viruses, fixed, and the HN and M proteins were visualized by immunofluorescence microscopy. Adapted from Schmitt et al.[41]

with mutations to the F and H glycoprotein cytoplasmic tails exhibit reduced F and H glycoprotein incorporation into virions as well as nonspecific incorporation of cellular proteins.[47] Cells infected with these altered viruses also undergo rapid and extensive syncytia formation.[47] The two Sendai virus glycoproteins appear to play different roles in particle production. Truncations to the Sendai virus F protein cytoplasmic tail reduced viral particle release,[48] suggesting that F protein–M protein interactions are important for Sendai virus particle production. However, it has long been recognized that Sendai virus HN protein is dispensable for particle production,[49–51] though more recent studies indicate that virions produced without HN protein may be morphologically altered.[52] Certain mutations to the Sendai virus HN protein reduced its incorporation into virions, though the efficiency of particle production was not affected.[48] Severe truncations to the HN protein cytoplasmic tail did reduce particle release, potentially by localizing to assembly complexes and interfering with particle production.[48] Overall, these findings indicate that paramyxoviral M proteins associate with viral glycoproteins via their cytoplasmic tails, resulting in efficient assembly and release of viral particles from specific sites on cellular membranes.

Roles for paramyxoviral glycoproteins in particle assembly have also been explored in the context of VLPs produced from transfected cells (see Table 1). Even in cases where M protein expression is sufficient for VLP production, glycoproteins expressed with M protein are often incorporated into the VLPs, resulting in particles that more closely resemble authentic paramyxovirus virions.[20,22] In some instances, viral glycoproteins enhance the efficiency of VLP production, reinforcing the idea that M proteins and glycoproteins can cooperate during particle production. For example, M protein-driven production of Sendai VLPs was made more efficient when Sendai virus F protein was co-expressed together with M protein.[18,19] Sendai virus HN protein did not enhance the efficiency of VLP production, however.[18,19] Similarly, the most optimal production of mumps VLPs was achieved when the mumps virus M, NP, and F proteins were co-expressed, whereas co-expression of mumps virus M, NP, and HN proteins yielded VLP production that was markedly less efficient.[27] Hence, in some cases paramyxovirus fusion and attachment proteins play unequal roles in particle assembly. For PIV5, on the other hand, the F and HN glycoproteins appear to play redundant and interchangeable roles in VLP production, as optimal particle release could be achieved either through co-expression of the M, NP, and F proteins or through co-expression of the M, NP, and HN proteins.[25] Neither glycoprotein could function for PIV5-like particle production if its cytoplasmic tail was truncated, reinforcing earlier results obtained with recombinant viruses.[41,42] In summary, co-expression of paramyxovirus glycoproteins together with M proteins generally leads to incorporation of the glycoproteins into budding VLPs, and in some cases leads to enhanced VLP production.

The same interactions between paramyxoviral M proteins and glycoproteins that drive virus assembly have been found to negatively regulate cell–cell fusion in many cases. This was illustrated in experiments using siRNA to knockdown assorted measles virus proteins in the context of a virus infection, and assessing the effects on cell–cell fusion. As expected, knockdown of the measles virus glycoproteins, F and H, decreased cell–cell fusion.[53] Knockdown of NP, P, and L proteins also decreased cell–cell fusion.[53] However, knockdown of M protein increased cell–cell fusion and enhanced viral transcription, consistent with the possibility that M protein is multifunctional and that assembly of M protein together with viral glycoproteins restrains F protein activity.[53] Mutations to M protein found in several measles vaccine strains, including the Edmonston strain, confer enhanced virus budding and decreased cell–cell fusion. This phenomenon has been attributed to enhanced interactions between M protein and the cytoplasmic tail of H protein derived from these strains as judged by co-localization studies.[54] Measles virus may acquire mutations to modulate the strength of M protein–H protein interactions to direct its mode of spread (cell–cell fusion *versus* budding).[54] Similarly, canine distemper virus (CDV) cell–cell fusion is inhibited when CDV M protein is co-expressed together with the F and H proteins. CDV syncytia formation is further decreased by co-expression of the N protein with F, H, and M, suggesting not only that M protein can restrain fusion function when assembled together with viral glycoproteins, but also that M-mediated recruitment of nucleocapsid-like structures to these assemblies restrains membrane fusion even further.[55] These findings illuminate the dynamic modulation of interactions between paramyxovirus M proteins, glycoproteins, and nucleocapsid proteins in controlling the mode of virus spread by virus budding or direct cell-to-cell transfer.

3.2.   *Viral nucleocapsid proteins*

The paramyxovirus nucleocapsid proteins (N or NP) form RNase resistant RNP complexes with viral RNA. The recent determination of the crystal structure of HRSV N protein bound to RNA by X-ray diffraction and cryoelectron microscopy revealed that RNA belts around a basic groove that runs along the N protein decameric ring, leaving bases exposed and potentially allowing RNA recognition by RNA-dependent RNA polymerase.[56] Direct interactions between viral M proteins and nucleocapsid proteins, shown biochemically,[57−59] likely drive the incorporation of viral genomes into virions. Targeting of measles virus RNPs to potential sites of budding at the plasma membrane is observed in cells infected with measles virus, and recombinant measles virus encoding an unstable M protein loses this phenotype.[14] In transfected cells, measles virus M protein targets N protein to the plasma membrane, and the C-terminus of N protein is important for this targeting to occur.[57] These studies indicate that paramyxovirus M proteins target RNPs to sites of virus assembly on plasma membranes.

Paramyxoviruses are able to selectively incorporate genomes into budding particles. This selection can occur on the basis of species homology; for example, cells transfected to produce hPIV1 NP protein and also infected with Sendai virus produced hPIV1 nucleocapsid-like structures, but these were not efficiently incorporated into the Sendai virions.[60] Selective incorporation of genomes can also occur on the basis of genome length. Short, defective-interfering genomes are incorporated into Sendai virions with poor efficiency compared with full-length genomes.[61] Selectivity of incorporation based on genome polarity is stringent for some (−)-sense RNA viruses such as rhabdoviruses, which incorporate (−)-sense RNA genomes into particles more efficiently than (+)-sense antigenomes.[62] Paramyxoviruses are also considered to favor packaging of (−)-sense genomes, although the packaging bias of Sendai virus is relatively weak.[62] The sense of viral RNAs incorporated into paramyxovirus virions corresponds with the ratios of (+)-sense to (−)-sense viral RNAs found in the infected cells, which for Sendai virus appears to be biased in favor of (−)-sense genomes at least in part through regulation of viral RNA synthesis by C protein.[63] These findings demonstrate the ability of paramyxoviruses to modulate the efficiency of genome incorporation into particles to favor the production of infectious virions.

Paramyxovirus nucleocapsid-like structures produced in transfected cells are often found to incorporate efficiently into VLPs (see Table 1). For example, VLPs of hPIV1, Sendai virus, and Nipah virus all are found to enclose nucleocapsid-like structures when the viral nucleocapsid proteins are co-expressed together with M proteins.[17,18,22] Similarly, NDV-like particles were found to incorporate NDV N protein, but in this case the N protein was incorporated efficiently only when NDV glycoproteins were expressed in addition to the M and N proteins.[20] PIV5 and mumps virus NP proteins were incorporated into VLPs when co-expressed with the viral M proteins and glycoproteins in transfected cells, and in these cases NP protein expression enhanced the efficiency of particle release.[25,27] The mechanism by which these NP proteins enhance particle release is unknown, though it is postulated that direct participation of nucleocapsid proteins in virus particle release may serve to minimize the release of noninfectious particles lacking RNPs.

4. Role of Host Proteins in Paramyxovirus Budding

4.1. *ESCRT machinery*

Topological similarities between budding of intraluminal vesicles (ILVs) into late endosomes, forming multivesicular bodies (MVBs), and virus budding have led to investigations of MVB biogenesis machinery in virus budding. MVB biogenesis occurs as a part of the vacuolar protein sorting (Vps) pathway of the cell and is carried out in large part by Class E proteins that form endosomal sorting complexes required for transport (ESCRTs) (reviewed in Ref. 64). In the final step of ILV release into late endosomes, the ESCRT machinery is dismantled by the AAA-ATPase Vps4. The involvement of the Vps pathway in the budding of many retroviruses (reviewed in Ref. 65) and several negative-strand RNA viruses, including

Ebola virus[66] and Lassa fever virus,[67] was substantiated from experiments in which disruption of this pathway through expression of dominant-negative (DN) Vps4 proteins blocked budding. The involvement of ESCRT machinery in the budding of paramyxoviruses has been suggested by findings that PIV5 budding[25] and mumps VLP budding[27] were inhibited upon expression of DN Vps4 proteins.

Although host ESCRT machinery is utilized by a range of distantly-related enveloped viruses during budding, some enveloped viruses can bud through ESCRT-independent mechanisms. For example, budding of VSV[68] and influenza virus[28] occurs normally in the presence of DN VPS4 proteins. Among paramyxoviruses, HRSV budding[69] and Nipah VLP budding[70] occur efficiently when DN Vps4 protein is expressed. These observations raise the possibility that some viruses may carry out budding using host machinery that is separate from ESCRTs, or alternatively may bud even without the assistance of host machinery (reviewed in Ref. 71).

The budding of Sendai VLPs is linked to ESCRT machinery through the viral C proteins, which bind to the Class E protein Aip1/Alix.[72] C proteins are accessory proteins expressed by a subset of paramyxoviruses and are produced from the viral P genes via overlapping reading frames. Sendai virus encodes a set of four C proteins from its P gene: C, C', Y1, and Y2.[73] C and C' proteins harbor N-terminal membrane-targeting sequences, which the shorter Y1 and Y2 proteins lack.[74] Expression of Sendai virus C protein enhanced the production of M-driven VLPs from transfected cells.[18] Both C and C' proteins were capable of enhancing Sendai VLP production, whereas Y1 and Y2 proteins were not, suggesting that membrane-targeting is critical to the VLP production function of C proteins.[75] DN Vps4 protein expression did not inhibit Sendai virus M-alone VLP production, but blocked the enhanced VLP production observed upon C protein co-expression.[75] This finding is consistent with the possibility that C protein may enhance VLP production by directly recruiting Aip1/Alix, and thereby indirectly recruiting additional ESCRT machinery, to VLP assembly sites on membranes.

Various studies investigating the importance of ESCRT machinery in the context of Sendai virus infection have produced results that are seemingly in conflict with one another. In Sakaguchi *et al.*,[72] DN Vps4 protein expression was shown to inhibit the budding of Sendai virus from cells, which had been transfected with infectious Sendai virus nucleocapsids. However, in Gosselin-Grenet *et al.*,[76] no inhibition in virion release was observed from cells infected with Sendai virus and transfected to express DN Vps4, even though PIV5 virion release was inhibited in a parallel experiment. Similar discrepancies exist in the results obtained from experiments in which Sendai virus budding was measured after siRNA-mediated depletion of

Aip1/Alix from cells,[72,76] and from experiments using C protein knockout and knockdown viruses. 4C(−) recombinant Sendai virus that does not produce any of the four C proteins was shown to bud particles inefficiently.[77] However, no effect of C protein depletion on Sendai virus particle release could be demonstrated in experiments by Gosselin-Grenet *et al.*[76] using a 4C knockout virus encoding GFP-fused C protein as a separate transcriptional unit in the virus genome, with siRNA-mediated knockdown of GFP/C protein.[76] 4C(−) recombinant virus is severely debilitated and passaging of this virus may have led to unanticipated changes in its genome. While this concern is alleviated in experiments using the GFP/C protein-expressing virus, incomplete siRNA-mediated knockdown could yield misleading results if C protein is normally present in functional excess. Taken together, these findings indicate that Sendai virus C proteins may contribute directly to Sendai virus assembly, though their role in the context of virus infection warrants further investigation.

4.2. *Ubiquitin*

Ubiquitin acts as a sorting signal for recognition of target membrane proteins by ESCRT proteins (reviewed in Refs. 78 and 79), many of which possess ubiquitin-binding domains.[80,81] Consequently, ubiquitin may play an important role in virus budding analogous to its role in the sorting of cargoes that are to be incorporated into MVBs. Additional observations linking ubiquitin to virus budding include (1) incorporation of high levels of ubiquitin inside retrovirus particles[82,83]; (2) multiple monoubiquitination of retroviral Gag proteins[84]; and (3) inhibition of Gag monoubiquitination[85] and retrovirus budding[86,87] upon treatment of cells with proteasome inhibitors, which deplete cellular pools of free ubiquitin. Ubiquitin has been linked specifically to the budding of negative-strand RNA viruses. VLP production driven by VSV M protein is proteasome inhibitor sensitive[88] and VSV M and Ebola VP40 proteins act as substrates for ubiquitin attachment *in vitro*.[88,89] Ebola VP40 ubiquitination can be detected in cells transfected to express VP40 together with hemagglutinin-tagged ubiquitin (HA-Ub).[90]

The ubiquitin-proteasome pathway may also be involved in paramyxovirus budding. PIV5 virion and VLP release from 293T cells is significantly reduced upon proteasome inhibitor treatments[91] and potential measles virus M protein ubiquitination was observed in cells transfected to express measles virus M protein together with HA-Ub.[23] Interestingly, differences may exist among paramyxoviruses in the degree of dependence on ubiquitin for budding. Treatments with proteasome inhibitors failed to inhibit the budding of HRSV or the budding of Nipah VLPs.[69,70] Sendai virus budding was found to be proteasome inhibitor sensitive in some cell types and proteasome inhibitor resistant in other cell types.[92] These findings suggest involvement of ubiquitin in the assembly and budding of some paramyxoviruses, though an understanding of the details of this involvement will require further investigation.

4.3. *Angiomotin-like 1*

An interaction between PIV5 M protein and cellular protein angiomotin-like 1 (AmotL1) was identified by yeast two-hybrid experiments.[93] Subsequently, PIV5 M protein and, to a lesser extent, mumps virus M protein were found to associate with AmotL1 in co-immunoprecipitation experiments.[93] AmotL1 is functionally involved in PIV5 budding, as siRNA-mediated depletion of AmotL1 reduced PIV5 budding. Overexpression of AmotL-1-derived polypeptides that bind to PIV5 M protein inhibited VLP production.[93] AmotL1 localizes to tight junctions and apical membranes in epithelial cells[94] and could be involved in the apical assembly of PIV5 and mumps virus. Interestingly, AmotL1 harbors two PPxY motifs that are similar to viral late domains (discussed below), raising the possibility that additional host factors for budding could be recruited to PIV5 assembly sites via AmotL1.[93]

4.4. *The cytoskeleton*

The actin cytoskeleton was first implicated in paramyxovirus budding based on the detection of large amounts of actin in purified Sendai virions and measles virions.[95,96] Incorporation of actin into NDV and HRSV virions has also been observed, as well as incorporation into Sendai VLPs.[19,97,98] The M proteins of Sendai virus and NDV both interact with actin[99] and a tight association between actin filaments and budding virions/viral filaments has been observed by an EM-analysis of both measles virus-infected and HRSV-infected cells.[100,101] The major components of the cytoskeleton, actin microfilaments and the microtubule network, control cellular processes including cell motility, membrane dynamics, protein trafficking, and signaling.[102] Disrupting the cytoskeleton using drug treatments can impair paramyxovirus replication. In HRSV-infected cells, disruption of the actin or microtubule networks severely impairs virus assembly and release.[103,104] Visual tracking of the movement of RNPs in living HRSV-infected cells using molecular beacons revealed filament migration and rotational motion consistent with myosin motor-driven transport on the actin cytoskeleton.[105] For hPIV3, release of infectious virions is impaired by disruption of microtubules, though depolymerization of actin microfilaments did not affect hPIV3 release.[106]

For Sendai virus, disruption of microtubules induced asymmetric budding, similar to that observed with the pantropic F1-R mutant.[107] These findings suggest that intact microtubules are required for efficient and/or polarized budding of some paramyxoviruses.[106,107]

4.5. *Viral sequence motifs important for virion release*

Late domains are protein–protein interaction sequences within viral proteins that mediate recruitment of host factors to virus assembly sites (reviewed in Refs. 65 and 108). Mutations to late domains can induce defects at the late stages of viral budding, preventing the membrane fission event that allows separation of newly-formed virions from infected cells (reviewed in Ref. 109). The well-characterized retroviral late domain sequences, P(T/S)AP, PPxY, and YP(x)$_n$L, bind to cellular components linked to ESCRT machinery. P(T/S)AP late domains bind to the ESCRT-I component Tsg101. PPxY late domains bind to Nedd4-like E3 ubiquitin ligases. YP(x)$_n$L late domains bind to Aip1/Alix, which is associated with ESCRT-III.[79,109,110] Similar late domain sequences are found in the matrix proteins of some negative-strand RNA viruses, including VSV,[111–113] rabies virus,[114] Ebola virus,[66,89,115] LCMV,[116] and Lassa fever virus,[116] suggesting that the strategy of recruiting host machinery via late domains to assist in virus budding is conserved even among distantly-related viruses. These late domain sequences are generally absent from paramyxovirus proteins, implying that paramyxoviruses that utilize ESCRT machinery for budding must access this machinery in alternative ways.

Several short sequence motifs within paramyxovirus M proteins have been shown to be important for budding activity and may function to recruit host factors, analogous to well-characterized late domains (see Table 1). PIV5 M protein contains a sequence 20-FPIV-24 that is critical to its budding function. This sequence was able to restore VLP production activity to PTAP-mutated HIV-1 Gag protein.[91] Budding function was lost upon mutation of either the phenylalanine or proline residues within the FPIV motif.[91] FPIV-like sequences are found in the M proteins of viruses closely related to PIV5, including mumps virus M protein, which contains the sequence 21-FPVI-24. Mumps VLP production was inhibited upon mutation of FPVI.[27] FPIV and FPIV-like sequences are presumed to function by mediating host factor recruitment to virus assembly sites, but the binding partner(s) for these sequences has not yet been identified.

Nipah virus M protein possesses two recently-identified sequences that are important for VLP production: 62-YMYL-65 and 92-YPLGVG-97.[21,70] Appending the YMYL sequence to PTAPPEY-mutated Ebola virus VP40 restored VP40 VLP production, suggesting that YMYL can function as a late domain.[21]

Mutating YMYL inhibited Nipah VLP production and also induced re-localization of the protein to the nuclei of transfected cells.[21] The YPLGVG sequence within Nipah virus M protein contains several residues that are well-conserved among paramyxoviral M proteins. Mutation of this sequence both disrupted Nipah VLP production and also induced re-localization of the M protein to the nucleus.[70] The YPLGVG sequence did not rescue Ebola VLP production when appended to late domain-defective VP40, but did induce the formation of filamentous projections similar to those observed upon wt VP40 or wt Nipah virus M protein expression.[70] Interestingly, sequences resembling YPLGVG have recently been identified in the Ebola virus and Marburg virus VP40 proteins, and mutation of these sequences disrupted VP40 stability, intracellular localization, and VLP production,[117] suggesting that YPLGVG-like sequences may carry out similar functions in both paramyxovirus M proteins and filovirus VP40 proteins. YMYL and YPLGVG sequences are both proposed to bind host factors, and it will be important to identify the corresponding partner proteins in the future.

Sendai virus M protein contains multiple sequence motifs important for its budding function. One of these is a C-terminal sequence, 343-KIRKI-348, necessary for efficient M protein-induced VLP release.[19] Interestingly, Sendai virus F protein contains a similar sequence in the cytoplasmic region, 524-RLKR-527, which is also important for the release of F protein-containing vesicles from transfected cells.[19] These sequences resemble the KLKK actin-binding domain found in some F-actin-binding proteins.[118,119] The actin cytoskeleton has been implicated in paramyxovirus assembly, as discussed earlier, and both M protein and F protein-containing VLPs/vesicles contain actin, supporting a model in which M protein and F protein both interact with actin during virus assembly, leading to the incorporation of actin into virus particles.[19] An additional sequence motif within the Sendai virus M protein, 49-YLDL-52, is also important for M-driven VLP production.[120] This sequence was shown to direct binding to Aip1/Alix, the same host factor that is bound by the Sendai virus C protein. Aip1/Alix interaction mediated by YLDL appears to be distinct from that mediated by $YP(x)_nL$-type late domains, and also distinct from the interaction mediated by C protein, which involves neither YLDL nor $YP(x)_nL$.[72,120] Poor budding of Sendai VLPs caused by mutation of YLDL was not restored through the addition of P(T/S)AP, PPxY, or $YP(x)_nL$ sequences to M protein.[120] As discussed earlier, some unresolved questions exist regarding the importance of Aip1/Alix to virus particle production in the context of Sendai virus infection.[76] Further investigation is warranted to clarify the role of this host factor and to determine whether recruitment mediated by M and C proteins is simply additive, or whether each of these proteins has a specialized role relating to Aip1/Alix recruitment to virus assembly sites.

5. Trafficking of Viral Components to Virus Assembly Sites

5.1. *Paramyxovirus M protein trafficking through the nucleus*

Although paramyxovirus genome replication and transcription take place in the cytoplasm, the M proteins of HRSV, NDV, and Sendai virus have all been observed in the nucleus early during infection. At later points during infections, the M proteins were localized primarily in the cytoplasm and associated with cell membranes.[59,121−123] Mutations to the 62-YMYL-65 or 92-YPLGVG-97 sequences in Nipah virus M protein led to poor VLP production and localization of the mutant proteins in the nucleus, consistent with nuclear-cytoplasmic shuttling of Nipah virus M protein.[21,70] Additional study is needed to clarify the roles of nuclear-localized M proteins. Transient nuclear localization of M proteins could effectively delay the budding of virus particles until late in the infectious cycle, when all viral components have accumulated to high levels in the cell. While inside the nucleus, M proteins could exert secondary functions, such as inhibition of host nuclear processes, similar to the multifunctional VSV M protein.[124,125]

Functional nuclear localization signals (NLSs) within the NDV and HRSV M proteins have been characterized. The NDV M protein possesses a bipartite NLS composed of basic amino acid clusters.[126] The HRSV M protein NLS is directly recognized by a nuclear import receptor, importin beta 1, for recruitment into the nucleus.[127] Late in infection, large amounts of paramyxovirus M protein must be present outside of the nucleus at sites of virus assembly. HRSV M protein nuclear export is directed by a leucine-rich, Crm-1 (exportin-1)-dependent nuclear export signal (NES). Recombinant HRSV encoding M protein with mutations in the NES could not be recovered, suggesting the importance of the NES for virus replication. Treatment of HRSV-infected cells with leptomycin B, a Crm-1 inhibitor, prevented nuclear export of HRSV M protein and impaired virus production.[128] These findings support a model in which paramyxoviral M proteins transiently accumulate in the nucleus, but must eventually exit and accumulate at virus assembly sites to coordinate virus budding.

5.2. *Budding of virus particles from apical membranes of polarized cells*

Polarized epithelial cells lining the body's surfaces possess two sides. The apical sides face outwards whereas the basolateral sides face inward towards the underlying tissues. The polarized epithelia of the respiratory tract are targeted for infection by many paramyxoviruses, and particle release often occurs from the apical surfaces of these cells. For example, preferential budding from apical surfaces of polarized cells has been observed during infections with HRSV, Sendai virus,

PIV5, hPIV3, and measles virus.[106,129–131] Influenza A virus, also a respiratory virus, buds apically,[131] whereas VSV and Marburg virus bud basolaterally.[132] The polarized budding of viruses from cells can influence the outcome of infection, with apical budding favoring localized infection of the respiratory tract and facilitating transmission between hosts, and basolateral budding favoring systemic dissemination of infection and the establishment of persistent infections (reviewed in Refs. 16, 133, and 134).

The polarity of paramyxovirus budding appears to be largely determined by viral M proteins, which intrinsically localize to apical membranes and in some cases recruit viral glycoproteins to these membranes. Measles virus glycoproteins F and H, for example, are targeted to basolateral surfaces when expressed alone, but are observed at apical cell surfaces in virus-infected cells.[135] M protein appears to re-route F and H proteins to apical surfaces, as F and H proteins are observed on basolateral surfaces in cells infected with recombinant measles virus lacking the M gene.[136] Targeting of measles virus glycoproteins to basolateral surfaces in the absence of functional M proteins may facilitate cell-to-cell transmission of infection in the event that virus particle formation cannot occur, contributing to systemic spread *in vivo*.[44] Consistent with this idea, the bipolar budding of a pantropic mutant Sendai virus, which causes a systemic infection, has been attributed to mutations in the M protein. The mutant F1-R M protein was unable to re-route F1-R F protein to apical cell surfaces, and expression of the F1-R M protein disrupted the microtubule network, altering the maintenance of cell polarity.[137] HRSV F glycoprotein is intrinsically targeted to apical cell surfaces, but the apical release of HRSV is unlikely to be directed by F protein, and instead is likely directed by internal viral proteins such as the M protein. Recombinant HRSVs lacking F protein or lacking all three glycoproteins (F, G, and SH) still bud apically.[138] These findings all support a key role for paramyxovirus M proteins in the polarized assembly and release of virus particles.

HRSV utilizes the apical recycling endosome (ARE) for efficient polarized budding.[69,139] The ARE is required for apical transport and recycling of host proteins in polarized cells and is enriched in Rab11a. Two ARE-associated components, myosin Vb and Rab 11 family interacting protein 2 (FIP2), have been implicated in apical budding of HRSV. Expression of altered myosin Vb protein prevented HRSV replication and assembly at the apical cell surface.[139] Interestingly, expression of DN Rab11-FIP2 protein induced a late domain-like phenotype of HRSV, characterized by retention of elongated viral filaments on the surfaces of infected cells (see Fig. 4).[69] Hence, HRSV budding, known to occur independently of ESCRT machinery, instead appears to rely on ARE-associated machinery during the final stage of membrane fission.

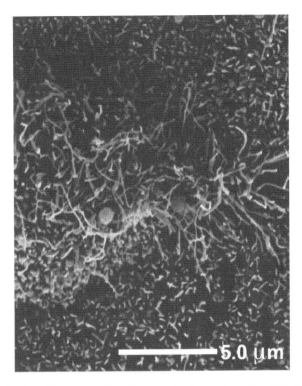

Fig. 4. Late budding defect observed in cells expressing dominant-negative FIP2-ΔC2 protein and infected with HRSV. Elongated HRSV filaments, visualized by scanning electron microscopy, are observed budding from the apical surface of an HRSV-infected polarized MDCK cell expressing FIP2 protein, which lacks its C2 domain. Adapted from Utley *et al.*[69]

5.3. *Budding of virus particles from lipid raft membranes*

Lipid raft microdomains are patches of cellular membrane enriched in cholesterol and sphingolipids that allow concentration of selected proteins together on membranes (reviewed in Refs. 140–143). Lipid rafts are more generally referred to as detergent resistant microdomains (DRMs), defined as membrane that fails to become solubilized in the presence of cold nonionic detergents such as Triton-X 100. Many viral proteins associate with DRMs, suggesting that lipid rafts may serve as sites for viral assembly. For example, in cells infected with Sendai virus, the F, HN, M, N, and P proteins were all found at least partially associated with DRMs.[144] Similar observations have been made with a variety of paramyxovirus proteins in infected cells, including the H and F proteins of CDV,[153] the HN, F, and NP proteins of NDV,[97] the F, G, and N proteins of HRSV,[98, 145] and the H, F, M, and N proteins of measles virus.[146,147] Additionally, several lipid raft markers were found associated with NDV virions,[97] HRSV virions,[145,148] and budding HRSV filaments.[145,148,149] When expressed alone in transfected cells, NDV F protein was associated with DRMs, and deletions of the NDV F protein cytoplasmic tail inhibited F protein DRM association.[150] HRSV M protein expressed in transfected cells associated with raft membrane only when it was co-expressed with the viral F protein,[151] and Sendai virus M protein associated with raft membrane only when it was co-expressed with the Sendai virus HN or F proteins.[152] In the case of measles virus, both F and M proteins independently associated with DRMs[146,147] and M protein DRM association was increased upon co-expression of F protein.[23] These findings support a model in which paramyxoviruses use lipid rafts as platforms for concentration of viral proteins on cellular membranes during virus assembly.

Functional involvement of lipid rafts during paramyxovirus infections has been investigated in experiments using methyl-β-cyclodextin (MβCD) to deplete cholesterol from cells during infection. Sendai virus particle production is not detectably affected by MβCD treatment, though the infectivity of Sendai virus particles released from MβCD treated cells is reduced.[144] MβCD treatment had no effect on CDV particle release or infectivity, but treatment of CDV virions with MβCD after they had been released resulted in a loss of infectivity that was restored upon the addition of cholesterol.[153] NDV particle release was actually enhanced in cells treated with MβCD, though the quality of these particles was impaired judged by abnormal morphology and reduced infectivity.[97] Similarly, NDV particles released from Niemann-Pick syndrome-type C cells, a cell-type lacking lipid rafts, were irregular in polypeptide composition, exhibit altered density, and were deficient in infectivity.[154] In sum, these findings indicate the importance of lipid rafts for assembly of correctly-formed and infectious paramyxovirus particles.

Recently, Yeo *et al.*[155] and Robinzon *et al.*[156] reported microarray analyses of the host cell transcriptome in paramyxovirus-infected cells *versus* uninfected cells, revealing upregulation of enzymes involved in cholesterol biosynthesis during infection. An enzyme involved in the formation of endogenous cholesterol, 3-hydroxy-3-methylglutaryl-coenzyme A reductase (HMGCR), and low-density lipoprotein receptor, involved in cholesterol homeostasis, were both upregulated during HRSV infection.[155,157] Similar findings were made in experiments with measles virus infection, with alterations in HMGCR levels, as well as the levels of squalene monooxygenase, which functions in sterol biosynthesis.[156] During acute measles virus infection, simvastatin-mediated inhibition of HMGCR reduced the amount of particle release.[156] Several genes involved in cholesterol biosynthesis were downregulated in persistently-infected cells, compared with cells undergoing acute measles virus infection. This has led to the interesting suggestion that innate antiviral mechanisms may exist in cells directed at inhibiting cholesterol biosynthesis, with the aim of restricting the budding of infectious viral particles.

6. Conclusions and Outlook

Much has been learned in recent years about events that occur during the dynamic processes of paramyxovirus assembly and budding. The role of paramyxovirus M protein as the key organizer of virus assembly has been reaffirmed, and additional roles of M protein in directing the apical budding of virus, and modulating cell–cell spread, have been characterized. Roles for viral glycoproteins, nucleocapsid proteins, and accessory proteins during the virus assembly process have been defined. Many details of lipid raft involvement in paramyxovirus assembly have been clarified. However, additional aspects of the paramyxovirus assembly process remain poorly understood. Structural information, necessary for a more complete understanding of M protein function, is lacking for most paramyxovirus M proteins. Although a variety of cellular components have been implicated in paramyxovirus budding, and several short sequence motifs within paramyxoviral M proteins that are crucial for budding function have been defined, the mechanisms of host factor recruitment and the precise roles of these factors in the budding process remain unclear. We look forward to future advances that enhance our understanding of these and related topics, and facilitate development of new antiviral strategies aimed at restricting paramyxovirus budding.

Acknowledgments

The authors are supported in part by the Middle Atlantic Regional Center of Excellence (MARCE) for Biodefense and Emerging Infectious Disease Research NIH grant AI057168, and research grant AI070925 from the National Institute of Allergy and Infectious Disease to A.P.S. The authors were also supported, in part, under a grant with the Pennsylvania Department of Health using Tobacco Settlement funds to A.P.S. The Department specifically disclaims responsibility for any analyses, interpretations, or conclusions.

References

1. Counihan, M.E., D.K. Shay, R.C. Holman, S.A. Lowther and L.J. Anderson, *Pediatr Infect Dis J*, 2001. 646.
2. Welliver, R.C., *J Pediatr*, 2003. S112.

3. CDC, *MMWR Morb Mortal Wkly Rep*, 2006. 173.

4. CDC, *MMWR Morb Mortal Wkly Rep*, 2006. 559.

5. CDC, *MMWR Morb Mortal Wkly Rep*, 2008. 893.

6. Hviid, A., S. Rubin and K. Mühlemann, *Lancet*, 2008. 932.

7. Eaton, B.T., C.C. Broder, D. Middleton and L.F. Wang, *Nat Rev Microbiol*, 2006. 23.

8. Alexander, D.J., in *OIE Manual of Diagnostic Tests and Vaccines for Terrestrial Animals*, Ed. Office of International Des Epizooties: Paris, 2009.

9. Roeder, P.L. and W.P. Taylor, *Vet Clin North Am Food Anim Pract*, 2002. 515.

10. Cathomen, T., B. Mrkic, D. Spehner, R. Drillien, R. Naef, J. Pavlovic, A. Aguzzi, M.A. Billeter and R. Cattaneo, *EMBO J*, 1998. 3899.

11. Inoue, M., Y. Tokusumi, H. Ban, T. Kanaya, M. Shirakura, T. Tokusumi, T. Hirata, Y. Nagai, A. Iida and M. Hasegawa, *J Virol*, 2003. 6419.

12. Kondo, T., T. Yoshida, N. Miura and M. Nakanishi, *J Biol Chem*, 1993. 21924.

13. Yoshida, T., Y. Nagai, K. Maeno, M. Iinuma, M. Hamaguchi, T. Matsumoto, S. Nagayoshi and M. Hoshino, *Virology*, 1979. 139.

14. Runkler, N., C. Pohl, S. Schneider-Schaulies, H.-D. Klenk and A. Maisner, *Cell Microbiol*, 2007. 1203.

15. Mottet-Osman, G., F. Iseni, T. Pelet, M. Wiznerowics, D. Garcin and L. Roux, *J Virol*, 2007. 2861.

16. Rima, B.K. and W.P. Duprex, *Virus Res*, 2005. 132.

17. Coronel, E.C., K.G. Murti, T. Takimoto and A. Portner, *J Virol*, 1999. 7035.

18. Sugahara, F., T. Uchiyama, H. Watanabe, Y. Shimazu, M. Kuwayama, Y. Fujii, K. Kiyotani, A. Adachi, N. Kohno, T. Yoshida and T. Sakaguchi, *Virology*, 2004. 1.

19. Takimoto, T., K.G. Murti, T. Bousse, R.A. Scroggs and A. Portner, *J Virol*, 2001. 11384.

20. Pantua, H.D., L.W. McGinnes, M.E. Peeples and T.G. Morrison, *J Virol*, 2006. 11062.

21. Ciancanelli, M.J. and C.F. Basler, *J Virol*, 2006. 12070.

22. Patch, J.R., G. Crameri, L.F. Wang, B.T. Eaton and C.C. Broder, *Virol J*, 2007. 1.

23. Pohl, C., W.P. Duprex, G. Krohne, B.K. Rima and S. Schneider-Schaulies, *J Gen Virol*, 2007. 1243.

24. Shnyrova, A.V., J. Ayllon, I.I. Mikhalyov, E. Villar, J. Zimmerberg and V.A. Frolov, *Cell Biol*, 2007. 627.

25. Schmitt, A.P., G.P. Leser, D.L. Waning and R.A. Lamb, *J Virol*, 2002. 3952.

26. Subhashri, R. and M.S. Shaila, *Biochem Biophys Res Commun*, 2007. 1096.

27. Li, M., P.T. Schmitt, Z. Li, T.S. McCrory, B. He and A.P. Schmitt, *J Virol*, 2009. 7261.

28. Chen, B.J., G.P. Leser, E. Morita and R.A. Lamb, *J Virol*, 2007. 7111.

29. Money, V.A., H.K. McPhee, J.A. Mosley, J.M. Sanderson and R.P. Yeo, *Proc Natl Acad Sci U S A*, 2009. 4441.

30. Dessen, A., V. Volchkov, O. Dolnik, H.-D. Klenk and W. Weissenhorn, *EMBO J*, 2000. 4228.

31. Caldwell, S.E. and D.S. Lyles, *J Virol*, 1986. 678.

32. Riedl, P., M. Moll, H.-D. Klenk and A. Maisner, *Virus Res*, 2002. 1.

33. Stricker, R., G. Mottet and L. Roux, *J Gen Virol*, 1994. 1031.

34. Jasenosky, L.D., G. Neumann, I. Lukashevich and Y. Kawaoka, *J Virol*, 2001. 5205.

35. Ruigrok, R.W., G. Schoehn, A. Dessen, E. Forest, V. Volchkov, O. Dolnik, H.-D. Klenk and W. Weissenhorn, *J Mol Biol*, 2000. 103.

36. Chong, L.D. and J.K. Rose, *J Virol*, 1993. 407.

37. Ye, Z., W. Sun, K. Suryanarayana, P. Justice, D. Robinson and R.R. Wagner, *J Virol*, 1994. 7386.
38. Greenfield, N.J., *Nat Protoc*, 2006. 2876.
39. Hoenen, T., V. Volchkov, L. Kolesnikova, E. Mittler, J. Timmins, M. Ottmann, O. Reynard, S. Becker and W. Weissenhorn, *J Virol*, 2005. 1898.
40. Ghildyal, R., D. Li, I. Peroulis, B. Shields, P.G. Bardin, M.N. Teng, P.L. Collins, J. Meanger and J. Mills, *J Gen Virol*, 2005. 1879.
41. Schmitt, A.P., B. He and R.A. Lamb, *J Virol*, 1999. 8703.
42. Waning, D.L., A.P. Schmitt, G.P. Leser and R.A. Lamb, *J Virol*, 2002. 9284.
43. Jin, H., G.P. Leser, J. Zhang and R.A. Lamb, *EMBO J*, 1997. 1236.
44. Moll, M., H.-D. Klenk and A. Maisner, *J Virol*, 2002. 7174.
45. Cattaneo, R., A. Schmid, D. Eschle, K. Baczko, V. ter Meulen and M.A. Billeter, *Cell*, 1988. 255.
46. Schmid, A., P. Spielhofer, R. Cattaneo, K. Baczko, V. ter Meulen and M.A. Billeter, *Virology*, 1992. 910.
47. Cathomen, T., H.Y. Naim and R. Cattaneo, *J Virol*, 1998. 1224.
48. Fouillot-Coriou, N. and L. Roux, *Virology*, 2000. 464.
49. Markwell, M.A., A. Portner and A.L. Schwartz, *Proc Natl Acad Sci U S A*, 1985. 978.
50. Portner, A., P.A. Marx and D.W. Kingsbury, *J Virol*, 1974. 298.
51. Stricker, R. and L. Roux, *J Gen Virol*, 1991. 1703.
52. Hirayama, E., M. Hattori and J. Kim, *Virus Res*, 2006. 199.
53. Reuter, T., B. Weissbrich, S. Schneider-Schaulies and J. Schneider-Schaulies, *J Virol*, 2006. 5951.
54. Tahara, M., M. Takeda and Y. Yanagi, *J Virol*, 2007. 6827.
55. Wiener, D., P. Plattet, P. Cherpillod, L. Zipperle, M. Doherr, M. Vandevelde and A. Zurbriggen, *Virus Res*, 2007. 145.
56. Tawar, R.G., S. Duquerroy, C. Vonrhein, P.F. Varela, L. Damier-Piolle, N. Castagné, K. MacLellan, H. Bedouelle, G. Bricogne, D. Bhella, J.F. Eléouët and F.A. Rey, *Science*, 2009. 1279.
57. Iwasaki, M., M. Takeda, Y. Shirogane, Y. Nakatsu, T. Nakamura and Y. Yanagi, *J Virol*, 2009. 10374.
58. Markwell, M.A. and C.F. Fox, *J Virol*, 1980. 152.
59. Yoshida, T., Y. Nagai, S. Yoshii, K. Maeno and T. Matsumoto, *Virology*, 1976. 143.
60. Coronel, E.C., T. Takimoto, K.G. Murti, N. Varich and A. Portner, *J Virol*, 2001. 1117.
61. Mottet, G. and L. Roux, *Virus Res*, 1989. 175.
62. Kolakofsky, D. and A. Bruschi, *Virology*, 1975. 185.
63. Irie, T., N. Nagata, T. Yoshida and T. Sakaguchi, *Virology*, 2008. 495.
64. Stuffers, S., A. Brech and H. Stenmark, *Exp Cell Res*, 2009. 1619.
65. Bieniasz, P.D., *Virology*, 2006. 55.
66. Licata, J.M., M. Simpson-Holley, N.T. Wright, Z. Han, J. Paragas and R.N. Harty, *J Virol*, 2003. 1812.
67. Urata, S., T. Noda, Y. Kawaoka, H. Yokosawa and J. Yasuda, *J Virol*, 2006. 4191.
68. Irie, T., J.M. Licata, J.P. McGettigan, M.J. Schnell and R.N. Harty, *J Virol*, 2004. 2657.
69. Utley, T.J., N.A. Ducharme, V. Varthakavi, B.E. Shepherd, P.J. Santangelo, M.E. Lindquist, J.R. Goldenring and J.E. Crowe, Jr., *Proc Natl Acad Sci U S A*, 2008. 10209.

70. Patch, J.R., Z. Han, S.E. McCarthy, L. Yan, L.F. Wang, R.N. Harty and C.C. Broder, *Virol J*, 2008. 1.

71. Chen, B.J. and R.A. Lamb, *Virology*, 2008. 221.

72. Sakaguchi, T., A. Kato, F. Sugahara, Y. Shimazu, M. Inoue, K. Kiyotani, Y. Nagai and T. Yoshida, *J Virol*, 2005. 8933.

73. Lamb, R.A. and G.D. Parks, in *Fields Virology*, D.M. Knipe and P.M. Howley, Editors. Lippincott, Williams and Wilkins: Philadelphia, 2006. 1449.

74. Marq, J.-B., A. Brini, D. Kolakofsky and D. Garcin, *J Virol*, 2007. 3187.

75. Irie, T., T. Nagata, T. Yoshida and T. Sakaguchi, *Virology*, 2008. 108.

76. Gosselin-Grenet, A.S., J.-B. Marq, L. Abrami, D. Garcin and L. Roux, *Virology*, 2007. 101.

77. Kurotani, A., K. Kiyotani, A. Kato, T. Shioda, Y. Sakai, K. Mizomoto, T. Yoshida and Y. Nagai, *Genes Cells*, 1998. 111.

78. Martin-Serrano, J., *Traffic*, 2007. 1297.

79. Morita, E. and W.I. Sundquist, *Annu Rev Cell Dev Biol*, 2004. 395.

80. Hicke, L. and R. Dunn, *Annu Rev Cell Dev Biol*, 2003. 141.

81. Shields, S.B., A.J. Oestreich, S. Winistorfer, D. Nguyen, J.A. Payne, D.J. Katzmann and R. Piper, *J Cell Biol*, 2009. 213.

82. Ott, D.E., L.V. Coren, E.N. Chertova, T.D. Gagliardi and U. Schubert, *Virology*, 2000. 111.

83. Putterman, D., R.B. Pepinsky and V.M. Vogt, *Virology*, 1990. 633.

84. Gottwein, E. and H.G. Krausslich, *J Virol*, 2005. 9134.

85. Strack, B., A. Calistri, M.A. Accola, G. Palu and H.G. Göttlinger, *Proc Natl Acad Sci U S A*, 2000. 13063.

86. Patnaik, A., V. Chau and J.W. Wills, *Proc Natl Acad Sci U S A*, 2000. 13069.

87. Schubert, U., D.E. Ott, E.N. Chertova, R. Welker, U. Tessmer, M.F. Princiotta, J.R. Bennink, H.G. Krausslich and J.W. Yewdell, *Proc Natl Acad Sci U S A*, 2000. 13057.

88. Harty, R.N., M.E. Brown, J.P. McGettigan, G. Wang, H.R. Jayakar, J.M. Huibregtse, M.A. Whitt and M.J. Schnell, *J Virol*, 2001. 10623.

89. Harty, R.N., M.E. Brown, G. Wang, J. Huibregtse and F.P. Hayes, *Proc Natl Acad Sci U S A*, 2000. 13871.

90. Okumura, A., P.M. Pitha and R.N. Harty, *Proc Natl Acad Sci U S A*, 2008. 3974.

91. Schmitt, A.P., G.P. Leser, E. Morita, W.I. Sundquist and R.A. Lamb, *J Virol*, 2005. 2988.

92. Watanabe, H., Y. Tanaka, Y. Shimazu, F. Sugahara, M. Kuwayama, A. Hiramatsu, K. Kiyotani, T. Yoshida and T. Sakaguchi, *Microbiol Immunol*, 2005. 835.

93. Pei, Z., Y. Bai and A.P. Schmitt, *Virology*, 2010. 155.

94. Sugihara-Mizuno, Y., M. Adachi, Y. Kobayashi, Y. Hamazaki, M. Nishimura, T. Imai, M. Furuse and S. Tsukita, *Genes Cells*, 2007. 473.

95. Lamb, R.A., B.W. Mahy and P.W. Choppin, *Virology*, 1976. 116.

96. Tyrrell, D.L. and E.J. Norrby, *J Gen Virol*, 1978. 219.

97. Laliberte, J.P., L.W. McGinnes, M.E. Peeples and T.G. Morrison, *J Virol*, 2006. 10652.

98. Marty, A., J. Meanger, J. Mills, B. Shields and R. Ghildyal, *Arch Virol*, 2004. 199.

99. Giuffre, R.M., D.R. Tovell, C.M. Kay and D.L. Tyrrell, *J Virol*, 1982. 963.

100. Bohn, W., G. Rutter, H. Hohenber, K. Mannweiler and P. Nobis, *Virology*, 1986. 91.

101. Jeffree, C.E., G. Brown, J.D. Aitken, D.Y. Su-Yin, B.-H. Tan and R.J. Sugrue, *Virology*, 2007. 309.

102. Gouin, E., M.D. Welch and P. Cossart, *Curr Opin Microbiol*, 2005. 35.

103. Burke, E., L. Dupuy, C. Wall and S. Barik, *Virology*, 1998. 137.

104. Kallewaard, N.L., A.L. Bowen and J.E. Crowe, Jr., *Virology*, 2005. 73.

105. Santangelo, P.J. and G. Bao, *Nucleic Acids Res*, 2007. 3602.

106. Bose, S., A. Malur and A.K. Banerjee, *J Virol*, 2001. 1984.

107. Tashiro, M., J.T. Seto, H.-D. Klenk and R. Rott, *J Virol*, 1993. 5902.

108. Calistri, A., C. Salata, C. Parolin and G. Palù, *Rev Med Virol*, 2009. 21.

109. Demirov, D.G. and E.O. Freed, *Virus Res*, 2004. 87.

110. Freed, E.O., *J Virol*, 2002. 4679.

111. Craven, R.C., R.N. Harty, J. Paragas, P. Palese and J.W. Wills, *J Virol*, 1999. 3359.

112. Harty, R.N., J. Paragas, M. Sudol and P. Palese, *J Virol*, 1999. 2921.

113. Jayakar, H.R., K.G. Murti and M.A. Whitt, *J Virol*, 2000. 9818.

114. Wirblich, C., G.S. Tan, A. Papaneri, P.J. Godlewski, J.M. Orenstein, R.N. Harty and M.J. Schnell, *J Virol*, 2008. 9730.

115. Martin-Serrano, J., T. Zang and P.D. Bieniasz, *Nat Med*, 2001. 1313.

116. Perez, M., R.C. Craven and J.C. de la Torre, *Proc Natl Acad Sci U S A*, 2003. 12978.

117. Liu, Y., L. Cocka, A. Okumura, Y.A. Zhang, J.O. Sunyer and R.N. Harty, *J Virol*, 2010. 2294.

118. Taylor, J.M., A. Richardson and J.T. Parsons, *Curr Top Microbiol Immunol*, 1998. 135.

119. Van Troys, M., D. Dewitte, M. Goethals, M.-F. Carlier, J. Vandekerckhove and C. Ampe, *EMBO J*, 1996. 201.

120. Irie, T., Y. Shimazu, T. Yoshida and T. Sakaguchi, *J Virol*, 2007. 2263.

121. Ghildyal, R., C. Baulch-Brown, J. Mills and J. Meanger, *Arch Virol*, 2003. 1419.

122. Peeples, M.E., C. Wang, K.C. Gupta and N. Coleman, *J Virol*, 1992. 3263.

123. Peeples, M.E., *Virology*, 1988. 255.

124. Ghildyal, R., A. Ho and D.A. Jans, *FEMS Microbiol Rev*, 2006. 692.

125. Jayakar, H.R., E. Jeetendra and M.A. Whitt, *Virus Res*, 2004. 117.

126. Coleman, N.A. and M.E. Peeples, *Virology*, 1993. 596.

127. Ghildyal, R., A. Ho, K.M. Wagstaff, M.M. Dias, C.L. Barton, P. Jans, P. Bardin and D.A. Jans, *Biochemistry (Mosc)*, 2005. 12887.

128. Ghildyal, R., A. Ho, M. Dias, L. Soegiyono, P.G. Bardin, K.C. Tran, M.N. Teng and D.A. Jans, *J Virol*, 2009. 5353.

129. Blau, D.M. and R.W. Compans, *Virology*, 1995. 91.

130. Roberts, S.R., R.W. Compans and G.W. Wertz, *J Virol*, 1995. 2667.

131. Rodriguez-Boulan, E. and D.D. Sabatini, *Proc Natl Acad Sci U S A*, 1978. 5071.

132. Sänger, C., E. Mühlberger, E. Ryabchikova, L. Kolesnikova, H.-D. Klenk and S. Becker, *J Virol*, 2001. 1274.

133. Schmitt, A.P. and R.A. Lamb, *Curr Top Microbiol Immunol*, 2004. 145.

134. Takimoto, T. and Portner, A. *Virus Res*, 2004. 133.

135. Maisner, A., H.-D. Klenk and G. Herrler, *J Virol*, 1998. 5276.

136. Naim, H.Y., E. Ehler and M.A. Billeter, *EMBO J*, 2000. 3576.

137. Tashiro, M., N.L. McQueen, J.T. Seto, H.-D. Klenk and R. Rott, *J Virol*, 1996. 5990.

138. Batonick, M., A.G.P. Oomens and G.W. Wertz, *J Virol*, 2008. 8664.

139. Brock, S.C., J.M. Heck, P.A. McGraw and J.E. Crowe, Jr., *J Virol*, 2005. 12528.

140. Brown, D.A. and E. London, *J Biol Chem*, 2000. 17221.

141. Chazal, N. and D. Gerlier, *Microbiol Mol Biol Rev*, 2003. 226.

142. Simons, K. and E. Ikonen, *Nature*, 1997. 569.

143. Simons, K. and D. Toomre, *Nat Rev Mol Cell Biol*, 2000. 31.

144. Gosselin-Grenet, A.S., G. Mottet-Osman and L. Roux, *Virology*, 2006. 296.

145. Brown, G., C.E. Jeffree, T. McDonald, H.W.M. Rixon, J.D. Aitken and R.J. Sugrue, *Virology*, 2004. 175.

146. Manié, S.N., S. de Breyne, S. Vincent and D. Gerlier, *J Virol*, 2000. 305.

147. Vincent, S., D. Gerlier and S.N. Manié, *J Virol*, 2000. 9911.

148. Brown, G., J.D. Aitken, H.W.M. Rixon and R.J. Sugrue, *J Gen Virol*, 2002. 611.

149. Jeffree, C.E., H.W.M. Rixon, G. Brown, J.D. Aitken and R.J. Sugrue, *Virology*, 2003. 254.

150. Dolganiuc, V., L. McGinnes, E.J. Luna and T.G. Morrison, *J Virol*, 2003. 12968.

151. Henderson, G., J. Murray and R.P. Yeo, *Virology*, 2002. 244.

152. Ali, A. and D.P. Nayak, *Virology*, 2000. 289.

153. Imhoff, H., V. von Messling, G. Herrler and L. Haas, *J Virol*, 2007. 4158.

154. Laliberte, J.P., L.W. McGinnes and T.G. Morrison, *J Virol*, 2007. 10636.

155. Yeo, D.S.-Y., R. Chan, G. Brown, L. Ying, R. Sutejo, J.D. Aitken, B.-H. Tan, M.R. Wenk and R.J. Sugrue, *Virology*, 2009. 168.

156. Robinzon, S., A. Dafa-Berger, M.D. Dyer, B. Paeper, S.C. Proll, T.H. Teal, S. Rom, D. Fishman, B. Rager-Zisman and M.G. Katze, *J Virol*, 2009. 5495.

157. Martinez, I., L. Lombardia, B. Garcia-Barreno, O. Dominguez and J.A. Melero, *J Gen Virol*, 2007. 570.

158. Terrier, O., J.-P. Rolland, M. Rosa-Calatrava, B. Lina, D. Thomas and V. Moules, *Virus Res*, 2009. 200.

159. Loney, C., G. Mottet-Osman, L. Roux and D. Bhella, *J Virol*, 2009. 8191.

160. Höglund, S. and B. Morein, *J Gen Virol*, 1973. 359.

Chapter 11

Virus–Host Interaction by Members of the Family *Rhabdoviridae* and *Filoviridae*

Douglas S. Lyles*

1. Introduction

In recent years, virus–host interactions have been one of the most actively investigated aspects of the biology of negative-strand RNA viruses. This is particularly true of the induction and suppression of host innate antiviral responses. Many host signaling mechanisms that detect the presence of viruses have been defined, as have many viral mechanisms to evade or suppress these host responses. The molecular mechanisms of the induction and suppression of host antiviral responses have been the subjects of many recent reviews, examples of which are given here.[1–3] The goal of this review is to begin to place these molecular mechanisms in the context of viral tissue tropism and pathogenesis in intact animal hosts. The focus of this review will be on two families of nonsegmented negative-strand RNA viruses, the *Rhabdoviridae* and the *Filoviridae*, which share several common principles that may be broadly applicable to the pathogenesis of other virus groups. The *Rhabdoviridae* family contains more than 175 different viruses, which are able to infect vertebrates, invertebrates, and plants.[4] This review will cover research with vesicular stomatitis virus (VSV), the prototype of the *Vesiculovirus* genus, and rabies virus (RV), the prototype of the *Lyssavirus* genus. The *Filoviridae* family currently contains six different viruses, five species of Ebola virus (EBOV), and one species of Marburg virus (MARV), which are classified into two separate genera.[5,6]

*Department of Biochemistry, Wake Forest University School of Medicine, Medical Center Boulevard, Winston-Salem, NC 27157, USA. E-mail: dlyles@wfubmc.edu

One of the striking features of RV and VSV is their pronounced neurotropism. Rabies is one of the oldest recognized infectious diseases, and even the earliest descriptions point to the behavioral symptoms in humans and dogs resulting from RV infection of the central nervous system (CNS).[4] Rabies is almost always fatal after the onset of disease. Despite the progress that has been made in control of rabies through RV vaccines, the annual number of deaths caused by RV worldwide is estimated to be between 40,000 and 70,000. However, the number of deaths from rabies may be considerably higher due to substantial under-reporting.

The neurotropism of VSV is primarily evident in the laboratory setting, in which VSV-induced neurological disease in rodents is widely studied by virologists and immunologists as models for viral pathogenesis and immunity.[4] In the natural setting, VSV is transmitted from insect vectors to domestic livestock, such as cattle, horses, and pigs, where it causes vesicular lesions in the oral mucosa and other exterior sites that resemble foot-and-mouth disease. Although VSV is not known to be neurotropic in these natural hosts, the occasional transmission of field strains of VSV to humans causes an acute febrile disease that may have some of the manifestations of viral neuroinvasion.

In contrast to rabies, viral hemorrhagic fevers caused by Marburg and EBOVs are relatively recently identified infectious diseases, having been first described in 1967 and 1976, respectively.[5] The natural hosts have not been identified with certainty, although the available evidence points to several bat species as potential reservoirs.[7,8] While EBOV and MARV do not appear to be virulent in bats, the occasional transmission of these viruses to humans and nonhuman primates causes a severe acute febrile disease characterized by multiple organ involvement and a high case mortality rate, approaching 90% for some strains of EBOV. Concern over the severity of the disease in these outbreaks and the potential for the use of these viruses as agents of bioterrorism has led to a substantial amount of recent research on these viruses, which has provided new information on their virus–host interactions.

This review will focus primarily on the interaction of the members of the *Rhabdoviridae* and *Filoviridae* with the innate immune system, as most of the recent research has been in this area. These viruses also have mechanisms to suppress elements of the adaptive immune system, though many of these are related to the interaction with the innate immune system. For example, inhibition of activation of STAT1 can suppress the response to type I interferons (IFNs) produced by innate immune cells, but also the response to IFNγ produced by activated T cells. Similarly, inhibition of IFN signaling can prevent up-regulation of class I and class II major histocompatibility proteins, thus inhibiting presentation of viral antigens to T cells.

This review will highlight the following themes that can be developed from recent research on VSV, RV, EBOV, and MARV that may apply to the pathogenesis of other viruses as well:

- The role of the innate antiviral response in determining viral tissue tropism and species specificity.
- The function of viral suppressors of innate antiviral signaling in viral pathogenesis.
- The role of specialized cell types in the innate immune system in determining the outcome of virus infection.

2.   The Role of the Innate Antiviral Response in Determining Viral Tissue Tropism and Species Specificity

There are many viruses for which the presence or the absence of host cell receptors for virus attachment and penetration (or other host factors necessary for virus replication) plays a major role in determining which cells support virus replication in the host. An excellent example is the use of CD4 as a receptor by human immunodeficiency viruses and the division into T cell-tropic and macrophage-tropic viruses according to usage of different chemokine receptors as co-receptors for virus penetration. However, there are many viruses that use as receptors either ubiquitously expressed host molecules or a wide variety of different host molecules. Such viruses typically establish productive infections in a wide variety of cell types in cell culture. However, in intact animal hosts, these viruses typically replicate in a much smaller number of different cell types. VSV and RV are excellent examples of such viruses.

It has been difficult to identify a cellular receptor for VSV, which binds to many different cell types in culture by interactions that appear to be of low affinity and often are not easily saturable. The viral envelope glycoprotein (G protein) appears to bind to many different cell surface components through relatively non-specific electrostatic and hydrophobic interactions.[9–11] Consistent with this lack of specificity in virus attachment, VSV establishes productive infections in many different cell types in culture. This includes cells from many different species as well as cells from many different tissue sources within a single species. While different cell types may vary in the amount of infectious VSV they can produce, there are very few cell types that would be considered nonpermissive for VSV replication.

In contrast to results in cell culture, there are very few tissues in intact animal hosts that produce detectable levels of infectious virus. In mice, susceptibility to

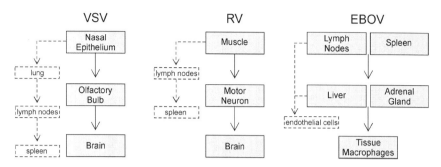

Fig. 1. Pathogenesis of VSV, RV, and EBOV. Major sites of virus replication and routes of spread in the host are indicated by solid lines and shaded symbols. Minor routes are indicated by dashed lines, open symbols, and smaller font.

VSV depends on the age and strain of mice and on the route of inoculation.[12–14] Infection by the intranasal route results in virus replication in the nasal mucosa, followed by neuronal spread of virus through the olfactory tract to the CNS (Fig. 1), resulting in viral encephalitis.[13,15] The brain is the primary site of virus replication, with lesser amounts of virus produced in the lungs and lymphoid organs, such as the spleen. In contrast to intranasal infection, systemic inoculation of mice, such as intravenous or intraperitoneal routes, rarely results in disease, and it is difficult to detect infectious virus in tissues of these mice.[16] The narrow tissue distribution of VSV replication in wild-type strains of mice contrasts sharply with the results of infection of transgenic mice that lack important elements of the innate antiviral response, such as the receptor for type I IFNs (IFN α/β) or the STAT1 transcription factor. VSV infection of these mice results in high virus titers in every organ system examined regardless of the route of inoculation, with the lowest virus titers in the brain.[16,17] These results illustrate the principle that the tissue tropism of VSV is not determined by which tissue can replicate virus best, but by which tissue can defend itself the least.

Similar to VSV, the RV G protein can mediate virus infection by low-affinity interactions with a wide variety of cell surface components.[18–20] RV is thus able to infect many different cell types in culture, with no particular preference for cells of neuronal origin. However, in intact animals, RV replication is primarily restricted to neurons of the peripheral and central nervous systems. Unlike VSV, the neurovirulence of RV is largely independent of the route of inoculation. For example, intramuscular inoculation is a model for RV transmission by animal bites (Fig. 1). While the potential of RV to replicate in other tissues in mice that are deficient in innate antiviral responses has not been explored, it is likely that the innate response serves to restrict virus replication in non-neuronal tissues, similar to VSV. That being said, there may be a role for host neuron-specific receptors in promoting virus spread in the CNS.[3] Three different host proteins have been identified whose expression can promote virus replication in otherwise resistant cell types, although their contribution to viral pathogenesis is controversial — the nicotinic acetylcholine receptor (nAchR),[21,22] neural cell adhesion molecule 1 (NCAM1 or CD56),[23] and low-affinity nerve growth factor receptor (p75NTR).[24] The nAchR is located at the postsynaptic muscle membrane and not in the motor neuron. Thus, it may play a role in initial infection of muscle cells, but is not likely to mediate neuronal spread. The role of CD56 and p75NTR in viral pathogenesis has also been questioned. Transgenic mice that lack p75NTR are similar to wild-type mice in their susceptibility to RV.[25,26] Mice that lack CD56 do show a delay in spread of RV through the CNS, but the mice still succumb to virus infection, suggesting that other receptors are also involved in viral neuroinvasion.[23]

Filoviruses also illustrate the principle that a wide variety of host receptors can be used to mediate infection. The filovirus glycoprotein spikes contain two disulfide-linked polypeptides, the extracellular GP_1, and membrane-spanning GP_2, which are derived from a common precursor.[5] GP_1 contains a mucin-like domain near its C-terminus that is highly glycosylated with N- and O-linked glycans. As a result, cellular lectins that bind these carbohydrates can serve as receptors to enhance virus uptake.[2] Some of these lectins are specific to particular cell types that are relevant to virus infection *in vivo*. These include DC-SIGN[27–29] and hMGL[30] that are expressed on dendritic cells and macrophages, L-SIGN and L-SECtin that are expressed on different classes of endothelial cells,[27,31] and the liver asialoglycoprotein receptor.[32] In addition to lectins, a variety of receptors for filoviruses have been identified, which mediate binding through protein–protein interactions. Expression of folate receptor α,[33] integrins containing a β1 subunit,[34] and the members of the Tyro3 family of receptor tyrosine kinases[35] have all been shown to enhance the uptake into otherwise resistant cells of EBOV and MARV, or VSV and retrovirus pseudotypes containing filovirus glycoproteins.

As expected of viruses that can use a wide variety of receptors, filoviruses can replicate in a wide variety of cell types in culture, with lymphocytes being a prominent exception.[36] However, as with VSV and RV, there are very few cell types that are major sites of filovirus replication in intact animal hosts. In nonhuman primate models, macrophages and dendritic cells in lymphoid tissues and parenchymal cells and Kuppfer cells in the liver appear to be the major sites of virus replication particularly early in infection (Fig. 1), as shown by immunohistochemistry, *in situ* hybridization, and viral infectivity assays.[37] The coagulopathy and multiple organ failure that is characteristic of hemorrhagic fevers in the case of filoviruses appear to be due to the processes that resemble disseminated intravascular coagulation, which may be due to the production of tissue factor by infected cells, compounded by the lack of production of regulators of coagulation due to tissue destruction in the liver.[38] In particular, multiple organ failure appears not to be due to extensive virus replication in many tissues, as many of the cell types that are susceptible to virus in culture do not show evidence of virus replication in intact animal hosts. Vascular endothelial cells are an interesting case in point. The pronounced bleeding disorders in many patients with filovirus infection led to the hypothesis that these viruses replicate in vascular endothelial cells and damage the vasculature due to viral cytopathic effects. Indeed, many endothelial cell lines and primary endothelial cells are highly permissive for filovirus replication in cell culture.[39] However, these cells are slow to develop cytopathic effects, and more importantly, vascular endothelial cells show little if any evidence of virus replication or cytopathic effects in intact nonhuman primate hosts.[39] The exceptions are cells of high endothelial venules in lymphoid organs and hepatic sinusoidal endothelial lining cells, which show evidence of virus infection (though not cytopathic effect) late in infection. Endothelial cells at these sites may differ from those in the rest of the body in terms of their susceptibility to virus, or they may simply be exposed to much higher virus titers, as these are the tissues that produce large amounts of infectious virus from the primary target cells.

Given the wide variety of cell types that support filovirus replication in cell culture, what prevents virus replication in these same cell types in intact animal hosts? The most likely candidates are antiviral cytokines that are generated as a result of filovirus infection. In fact, EBOV-infected macaques have very high levels of circulating IFNα and other cytokines.[37] While it is not feasible to perform the genetic experiment to ablate the response to type I IFNs in nonhuman primates, this experiment has been performed in mice. Interestingly, infection with wild-type EBOV or MARV does not cause disease in adult immunocompetent mice, although newborn mice are susceptible.[40] Sequential passage of the Zaire strain of EBOV in progressively older suckling mice has generated a mouse-adapted strain of EBOV that causes lethal infection of adult mice when inoculated intraperitoneally, though not when inoculated subcutaneously.[40] Similar to results with VSV, transgenic mice that lack the IFNα/β receptor or STAT1 rapidly succumb to lethal infection with many wild-type strains of EBOV and MARV, as well as the mouse-adapted EBOV regardless of the route of infection.[41] These results indicate that the interaction of filoviruses with the innate immune system is responsible in large part for the species specificity of the diseases they cause. By analogy, similar effects may be responsible for the restricted cell-type distribution of these viruses in intact human and nonhuman primate hosts.

3. Activators of Innate Antiviral Signaling

If innate antiviral responses serve to limit the invasion of viruses into most tissues in intact animal hosts, one of the critical questions is how innate antiviral signaling is activated in the host. A related question is what accounts for the ability of viruses to replicate in tissues that are susceptible to infection? The answer to this question is that viruses have specific mechanisms to suppress or evade antiviral responses in susceptible tissues. While most of these mechanisms of activation and suppression of host antiviral responses have been defined in cell cultures, in which the cells are usually susceptible to virus infection, an important future direction will be to determine how the differential responsiveness of cells in different tissues or the differential effectiveness of viral inhibitory mechanisms contributes to the specificity of viral tissue tropism and pathogenesis.

Most of the recent research in this area has focused on the pathways leading to the synthesis and response to type I IFNs as perhaps the most important antiviral cytokines, though the pathways associated with other cytokines and intracellular effectors that contribute to the host antiviral response are also extensively studied. Figure 2 summarizes the important elements of the pathways leading to the synthesis and response of type I IFNs and the viral inhibitors that lead to their suppression. In order for a cell to mount an antiviral response, viral products have to be recognized by sensors known as pathogen pattern recognition receptors (PRRs). For most cell types, the major PRR that initiates the response to rhabdoviruses and filoviruses appears to be a cytoplasmic RNA helicase, RIG-I (retinoic acid-inducible

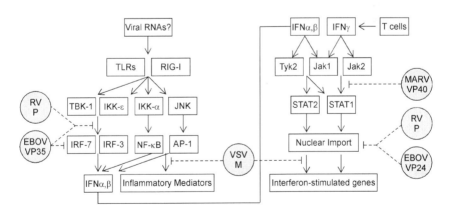

Fig. 2. Viral suppressors of innate antiviral signaling. Pathways leading to the synthesis of type I IFNs and other inflammatory mediators are shown on the left, and pathways that mediate the response to IFNs are shown on the right. Viral suppressors of antiviral signaling are indicated by shaded circles.

gene-I).[42–45] One of the principal ligands for RIG-I is 5′ phosphorylated RNA that is part of a short double-stranded RNA (dsRNA).[42,46] In the normal replication cycle of nonsegmented negative-strand RNA viruses, 5′ phosphorylated RNAs are produced during the process of transcription in the form of leader RNAs and during genome replication as either genomic or antigenomic RNAs.[4] Another potential source of such RNAs would be aberrant transcription products that are not capped by the viral polymerase. The 5′ ends of most of these RNAs as well as their complementary sequences are likely to be shielded from recognition by the viral nucleocapsid proteins. Processes that result in exposure of the free 5′ ends of viral RNA and formation of dsRNA, such as removal of the nucleocapsid protein, would appear to be necessary for recognition by RIG-I.[47] Thus, the origins of the signals that activate RIG-I are somewhat obscure, but are to a large extent coupled to the production of viral RNAs.

The other major PRRs that have been implicated in the host response to rhabdoviruses and filoviruses are toll-like receptors (TLRs). These PRRs act either at the plasma membrane or in the endocytic compartment to recognize a wide variety of molecules that may be associated with infection by bacteria, fungi, and protozoa as well as viruses. Unlike RIG-I-dependent signaling that is widely distributed among many cell types, the distribution of TLRs and the relative importance of their signaling pathways are often cell-type dependent. For example, TLR-7 is a major PRR in the response to VSV.[48] However, TLR-7 is expressed only in a limited subset of cells in the innate immune system, primarily plasmacytoid dendritic cells.[49] The signal that activates TLR-7 appears to be the presence of single-stranded RNA in the endocytic compartment.[50] Such single-stranded RNA can arise during virus penetration by degradation of virions or can be generated from viral RNA in the cytoplasm, which enters the endocytic compartment by autophagy.[51] In addition to TLR-7, TLR-4 has also been implicated in the innate response to VSV in murine macrophages.[52] The signal that activates TLR-4 in macrophages appears to be the VSV G protein, which interacts with TLR-4 during virus attachment and penetration.[52] Similar to VSV, the glycoprotein of EBOV can also activate TLR-4 during virus attachment and penetration.[53] In this case, the mucin-like domain of GP_1 is required for activation.

In the case of RV, the only TLR that has been implicated in the innate response is TLR-3.[54,55] TLR-3 responds to the presence of dsRNA in the endocytic compartment.[56] As pointed out above, nonsegmented negative-strand RNA viruses do not produce dsRNA as a part of their normal replication cycle. However, some dsRNA may arise from abnormal replication products in which negative-strand RNA is not completely encapsidated. While TLR-3 may play a role in the antiviral response to RV, on balance the presence of TLR-3 actually enhances RV neuropathogenesis, as TLR-3$^{-/-}$ mice are less susceptible to RV than their wild-type controls.[57] This appears to be due to a role for TLR-3 in the formation of Negri bodies, which are viral inclusions that may play a role in enhancing virus replication.[57,58] This is a good example of a virus taking advantage of a host response to virus infection to promote its own replication.

Activation of RIG-I or TLRs results in formation of signaling complexes through a variety of adapter proteins, which activate protein kinases that turn on the expression of antiviral genes.[59,60] The major protein kinases activated by these pathways that result in the production of type I IFNs are TANK-binding kinase-1 (TBK-1), inhibitor of NF-κB kinase (IKK)-α, IKK-ε, and Jun N-terminal kinase (JNK). TBK-1 and IKK-ε activate the transcription factors IRF-3 and IRF-7, which are essential for transcription of the IFN-β and most of the IFN-α genes, respectively, in many cell types. However, in macrophages and dendritic cells, IRF-5 and IRF-8 play a major role in the production of type I IFNs and other inflammatory cytokines.[61] In addition to cytokines, IRFs also activate many IFN-stimulated genes (ISGs), thus rendering cells in an antiviral state even in the absence of IFN signaling. RIG-I or TLR signaling also results in activation of IKK-α and JNK that in turn activate NF-κB and AP-1 transcription factors, respectively. These transcription factors are necessary in addition to IRFs for transcription of the genes for type I IFNs and many other inflammatory and antiviral cytokines.

4. Viral Suppressors of Innate Antiviral Signaling

The mechanisms that viruses have developed to suppress innate antiviral responses are quite diverse among different virus families. Many viruses encode proteins whose principal function is to suppress host antiviral responses. However, the proteins of rhabdoviruses and filoviruses that suppress host responses also serve major functions in the virus life cycle either as RNA polymerase subunits or as matrix proteins involved in virus assembly. Interestingly, the mechanisms by which RV and filoviruses suppress the production of type I IFNs are similar to each other, while those of VSV are quite distinct (Fig. 2). For RV and filoviruses, the P protein (known as VP35 in filoviruses) functions both as a subunit of the viral RNA-dependent RNA polymerase and as a suppressor of IFN production.[1,3] Both the RV P protein and filovirus VP35 inhibit IFN production by preventing the phosphorylation of IRF-3 by TBK1 and IKK-ε.[62,63] VP35 has been shown to interact directly with the kinase domains of TBK1 and IKK-ε and acts as a substrate for phosphorylation. This interaction also interferes with binding of the kinases to their normal substrate IRF-3. In addition, VP35 promotes the conjugation of IRF7 with SUMO, a small ubiquitin-like protein that represses the transcriptional activity of IRF7.[64] VP35 appears to promote SUMO conjugation by direct interaction with IRF7 and Ubc9 and PIAS1, which are the SUMO E2 and E3 enzymes, respectively. In addition to inhibiting the production of IFNs, VP35 also inhibits other antiviral mechanisms, such as the activation of protein kinase R (PKR).[65]

The functions of VP35 as a subunit of the viral RNA polymerase and as a suppressor of IFN production are genetically distinct. VP35 has an N-terminal domain that appears to be primarily involved in RNA synthesis and a C-terminal domain that is primarily involved in IFN inhibition.[1] The C-terminal domain binds dsRNA.[66] The structure of the C-terminal domain determined by X-ray crystallography consists of a unique fold that differs from other viral dsRNA-binding proteins and contains a cluster of basic residues that are important for binding dsRNA.[67] Mutations that abolish RNA-binding activity reduce, but do not completely eliminate the ability of VP35 to inhibit IFN production.[66] While optimal function of the C-terminal domain requires oligomerization mediated by the N-terminal domain,[68] point mutations in the C-terminal domain can render VP35 ineffective in the inhibition of IFN signaling, but fully functional in viral RNA synthesis.[66,69–71] Incorporation of such mutations into recombinant EBOVs dramatically reduces the virulence of these viruses in mice and guinea pigs.[69,72] Mutations in the RV P protein that inactivate its IFN inhibitory function without affecting its RNA synthesis function have not been identified. However, recombinant viruses have been generated that express either mutant P protein or lower levels of P protein than their wild-type controls. These viruses are able to replicate in cell types that are defective in their IFN responses, but are rapidly eliminated from IFN-competent cell types.[63,73]

In addition to inhibiting IFN production, both RV and filoviruses have specific mechanisms that inhibit signal transduction in response to IFN. The inhibition is due to the activity of viral proteins that interfere with formation of the transcription factors that activate interferon-stimulated genes ISGs. However, the different viruses inhibit different points in the pathways that lead from IFN receptors to transcription factor formation (Fig. 2). Type I IFNs bind to a common receptor that is coupled to two tyrosine kinases, Jak1 and Tyk2.[74] Receptor activation leads to autophosphorylation and activation of these kinases, which in turn phosphorylate two cytoplasmic proteins, STAT1 and STAT2. Phospho-STAT1 and -STAT2 are transported to the nucleus, where they associate with IRF9 to form the ISGF-3 transcription factor that activates expression of ISGs. Type II IFN (IFNγ) binds to a different receptor, which is coupled to Jak1 and Jak2 kinases. Activation of this receptor leads to phosphorylation of STAT1, but not STAT2. The phospho-STAT1 forms a homodimer, which is transported to the nucleus, where it activates a similar set of genes that respond to IFN-γ.

The RV P protein, in addition to serving as a polymerase subunit and as an inhibitor of IRF-3 phosphorylation, also functions to inhibit STAT1 signaling. P protein does not interfere with STAT phosphorylation. Instead, it binds to phosphorylated STAT1 and STAT2, and inhibits their translocation to the nucleus and binding to target DNAs.[75–78] This appears to be due to the association of the P protein-STAT complex with microtubules in the cytoplasm, which prevents transport to the nucleus.[76] In the event that some of the STAT1–STAT2 complex does get transported to the nucleus, association of P protein or its truncated derivatives also interferes with DNA-binding activity.[78]

Similar to RV, EBOV inhibits the transport of STAT1 to the nucleus. However, the viral protein that mediates this inhibition is not the P protein (VP35), but rather VP24, a protein encoded by filoviruses that does not have homologues among other negative-strand RNA viruses.[79] VP24 is a membrane-associated protein that appears to function in virus release, and thus may have activities similar to viral matrix proteins.[80] In addition to VP24, filoviruses also encode a matrix protein (VP40) that functions in virus assembly and is similar to the matrix proteins of other negative-strand RNA viruses, particularly those of the paramyxoviruses.[81,82] EBOV VP24 inhibits STAT1 nuclear import not by binding directly to STAT1, but rather to karyopherins $\alpha 1$, $\alpha 5$, and $\alpha 6$, which are the intracellular receptors that recognize the nuclear localization signal on phospho-STAT1.[79,83,84] VP24 binds to the same site as STAT1 on the karyopherins, and is likely a competitive inhibitor.

In the case of MARV, both the site of inhibition of IFN signaling and the viral protein responsible for the inhibition are distinct from that of EBOV. The VP40 (matrix) protein of MARV inhibits the tyrosine phosphorylation of STAT1 and STAT2 in response to IFN, as well as the autophosphorylation of the Jak1 and Tyk2 protein kinases.[85] The mechanism of this inhibition has not been determined, but Jak1 appears to be a more sensitive target than Tyk2, as MARV VP40 can inhibit the activity of overexpressed Jak1, but not overexpressed Tyk2.

In contrast to RV and the filoviruses, which target specific factors involved in the induction and response to IFNs, VSV suppresses the induction of IFNs and other antiviral factors through the global inhibition of host gene expression. This inhibition is mediated by the VSV matrix (M) protein.[3] The ability of M protein to inhibit host gene expression is genetically separable from its role in virus structure and assembly, and several M protein mutations have been identified that render M protein defective in the inhibition of host gene expression, but fully functional in virus assembly.[86–89] When expressed in transfected cells in the absence of other VSV components, M protein inhibits transcription by all three host RNA polymerases,[90] and also inhibits nuclear-cytoplasmic transport of host RNA.[91,92] In

VSV-infected cells, there is also an inhibition of translation of host mRNA that is due in part to the activity of M protein, but appears also to involve other viral components.[86,93] In principle, the global inhibition of host gene expression by VSV M protein inhibits both the synthesis and response of type I IFNs and other cytokines. Indeed, permissive cells infected with wild-type VSV produce very little IFN or other cytokines. Similarly, treatment of VSV-infected cells with IFN has relatively little effect on virus yield. However, once the antiviral state becomes established by pre-treating cells with IFN, VSV does not have a mechanism to suppress the response to IFN, making it one of the most sensitive viruses to the antiviral effects of IFN.

Similar to RABV P and the filovirus VP24, VP35, and VP40, VSV M protein does not have enzymatic activity, but instead it probably inhibits host antiviral responses by binding to host factors and interfering with their normal function. Thus far, the principal host protein whose binding to VSV M is correlated with the inhibition of host gene expression is Rae1.[94] Rae1 was originally implicated in mRNA transport, but more recent experiments suggest that its principal function is in mitotic spindle assembly and mitotic checkpoint regulation.[95–97] M protein and Rae1 form complexes with multiple proteins involved in mRNA transport and other cellular functions, such as Nup98, hnRNP-U, and E1B-AP5.[92,94,98] While binding of M protein to the Rae1-Nup98 complex is likely to be responsible for the inhibition of nuclear-cytoplasmic RNA transport, this is not due simply to the inhibition of Rae1 function, as Rae1 is not essential for mRNA transport.[95–97] Because Rae1 is distributed throughout the cytoplasm and the nucleus of the cell, it may be involved in multiple steps in host gene expression. Thus, interaction of VSV M protein with Rae1 might also be involved in the inhibition of transcription and translation, though it is also possible that M protein interacts with other host factors to inhibit these steps in host gene expression.

Many M protein mutants of VSV that are defective in their ability to inhibit host gene expression replicate as well or better than wt viruses in most cell types in culture. The principal defect in replication of these viruses is in multiple cycle infection of cells that are competent to produce and respond to IFNs.[86,99,100] Thus, the inhibition of host gene expression by M protein is not essential for virus replication in a single cycle infection in most cell types. However, such M protein mutant viruses are dramatically attenuated for virulence in intact animal hosts.[99,100] For example, following intranasal inoculation, M protein mutant VSV replicates in the nasal epithelium, but is rapidly cleared compared with virus with wt M protein, and does not invade the CNS.[101] The inability of the M protein mutant to spread through the CNS is due to the local production of type I IFNs, due to the inability to suppress the antiviral response in host cells.[14]

5. The Role of Specialized Cell Types in the Innate Immune System in Determining the Outcome of Virus Infection

Each of the viruses under consideration here has potent mechanisms to suppress the production of IFNs and other cytokines by infected cells, yet infection with these viruses is characterized in most cases by high levels of IFNs and other cytokines produced either systemically or locally at the site of infection. Indeed, in some cases the pathological effects of virus infection can be attributed to the inflammatory response rather than direct cytopathic effects of virus infection. What then are the sources of these cytokines? The general answer is that they are produced by specialized cell types in the innate immune system, which either are partially resistant to the viral inhibitory mechanisms or are largely nonpermissive for virus replication. Thus, the cells that are responsible for producing most of the cytokines in the host are not the same cells that are responsible for producing most of the infectious virus (Fig. 3).

An excellent example of this principle is the production of type I IFNs in mice infected with VSV. As pointed out earlier, intranasal inoculation of VSV results in spread of the virus from the nasal epithelium through the olfactory tract to the CNS over the course of about 6–7 days. The principal cells that produce virus during this process are the cells of the nasal epithelium and neurons.[13,15] Lesser amounts of virus can occasionally be detected in the lungs, the lymph nodes that serve the respiratory tract, and the spleen. In contrast, type I IFN is detected primarily in the serum, spleen, and lungs, but not in the CNS.[14] The principal source of this IFN appears to be a subset of plasmacytoid dendritic cells in the

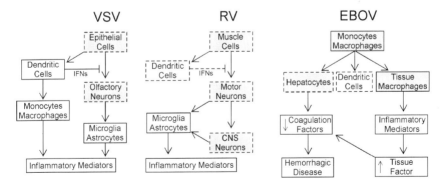

Fig. 3. Cell types that mediate innate responses to virus infection. Major cell types that generate innate antiviral responses are indicated by solid lines. Cells in which innate responses are suppressed (or in the case of RV are evaded by low levels of virus replication) are indicated by dashed lines. Major cell types that produce infectious virus are indicated by filled symbols.

spleen.[102] Plasmacytoid dendritic cells can be productively infected with VSV, though they produce less infectious virus than highly permissive cell types.[103] Importantly, both plasmacytoid dendritic cells and myeloid dendritic cells that express TLR7 are more resistant to the inhibition of host gene expression by VSV M protein than dendritic cells that do not express TLR7.[103,104] This leads to a delay in shut-off of host gene expression that provides these cells with the opportunity to respond to VSV infection by the production of type I IFN and other cytokines as well as costimulatory molecules involved in T cell activation. In general, the systemic production of type I IFNs is not sufficient to prevent invasion of VSV into the CNS of intranasally infected mice. This is likely due to the inefficient penetration of systemically produced cytokines into the CNS, as local production of IFN and other cytokines in mice infected with M protein mutant viruses prevents neuroinvasion.[14]

The extent of morbidity and mortality resulting from VSV infection in mice is subject to many variables, including the strain of VSV and the age, sex, and strain of the mice.[12,14,105] This is likely due to variability in the rate of virus spread in the CNS *versus* the antiviral defenses that prevent spread. In addition to the systemic production of type I IFNs, VSV infection induces local activation of innate antiviral responses in astrocytes and microglial cells.[106–108] Similar to dendritic cells, these cells can be productively infected with VSV, though the virus titers produced are low.[107] Importantly, infection of these cells leads to production of inflammatory mediators, implying that they too are partially resistant to the inhibitory effects of M protein. VSV infection also induces adaptive immune responses that serve to limit the spread of virus in the CNS. Mice that are deficient in T cells or B cells show a greater sensitivity to the VSV-induced mortality than their wild-type controls.[109–111] Thus, the outcome of VSV infection in mice is determined by a balance between the ability of the virus to spread through susceptible cells, primarily neurons in the CNS, and the ability of specialized cells of the innate and adaptive immune systems to mount an effective antiviral response.

Similar to VSV, infection of mice with RV by peripheral routes results in the systemic production of type I IFN and other cytokines, as well as production in the CNS.[112–114] While the major cellular sources of these cytokines have not been identified, infection of human dendritic cells and monocytes with RV results in the induction of type I IFN and other cytokine genes, indicating that these cell types are largely resistant to the inhibitory effects of the RV P protein and may be responsible for cytokine production *in vivo*.[115] Importantly, the difference in pathogenicity of attenuated RV *versus* field strains ("street RVs") is correlated with lower levels of viral gene expression by the more virulent strains.[116,117] This leads to correspondingly lower induction of antiviral responses by the more virulent viruses.

Thus, the key to the pathogenicity of RV is the combination of having a potent suppressive mechanism to inhibit IFN responses in susceptible cells together with a sufficiently low level of viral gene expression to reduce the responses generated in the cells of the innate immune system. In other words, RV is said to "[use] stealth to reach the brain."[118]

In contrast to RV, the pathogenesis of disease caused by filoviruses would not be considered to involve "stealth." Indeed, activation of cells of the innate immune system appears to play a critical role in viral pathogenicity. An important difference among the cells of the innate immune system appears to be the response of macrophages *versus* dendritic cells. EBOV is able to productively infect human dendritic cells.[119,120] However, EBOV infection results in expression of few if any cytokines or surface markers of maturation. This is not due to an inability to recognize EBOV-derived stimuli, as EBOV virus-like particles do induce maturation.[121] Instead, the lack of maturation is likely due to the inhibitory effects of VP35, as described in the previous section, as infected dendritic cells are refractory to other stimuli. In support of this idea, expression of VP35 in murine dendritic cells in the absence of other EBOV gene products inhibits the response to other stimuli.[122]

EBOV also establishes productive infection of human macrophages. However, in this case, the infected cells respond by producing a wide variety of inflammatory cytokines, such as IL-1β, TNF-α, and MIP-1α, as well as other inflammatory products, such as nitric oxide.[123–125] Most of these cytokines are present at high levels in infected humans or macaques. There is some inconsistency in the literature as to whether EBOV-infected macrophages produce type I IFNs.[123,124] However, high levels of type I IFNs are present in sera from infected humans and macaques,[124,126] indicating the existence of a cell type in the body that is resistant to the VP35-mediated suppression of IFN production. The ability of EBOV to replicate in susceptible cell types *in vivo* such as hepatocytes and macrophages in the presence of high levels of type I IFNs and other cytokines suggests that the inhibition of IFN signaling by VP24 is a major factor in allowing virus replication in the host. It is also likely that the high levels of inflammatory cytokines produced by EBOV-infected macrophages play a major role in pathogenesis of disease.[38] Early in the disease, the production of chemokines at the initial sites of infection may serve to recruit susceptible cells, such as macrophages and dendritic cells, which become infected and disseminate virus to other parts of the body. Later in the disease, the production of tissue factor by infected macrophages may contribute to the coagulopathy that is characteristic of this type of hemorrhagic fever. Also the production of apoptosis-inducing cytokines by infected macrophages may be responsible for the pronounced depletion of T cells observed during EBOV infection.[38]

The focus of this section has been on cells of the innate immune system that are infected by virus but are resistant to the inhibitory mechanisms that suppress antiviral responses in other susceptible cell types. There are likely to be additional cells of the innate immune system that are capable of responding to virus infection without being directly infected or by being abortively infected, and thus are not subject to viral inhibitory mechanisms. For example, natural killer cells and other lymphoid cells are largely nonpermissive for infection by some viruses, yet have mechanisms to recognize and respond to virus-infected cells. Another source of such cells would be cells that are normally permissive to virus infection, but are rendered nonpermissive by responding to type I IFN. For example, most conventional dendritic cells are susceptible to VSV infection but are unable to respond due to the inhibitory effects of M protein.[103,104,127,128] However, it has been proposed that dendritic cells that have been exposed to type I IFN are abortively infected, but are able to respond to the low levels of viral signals produced as a result of the abortive infection.[128] Such a response has recently been demonstrated in IFN-treated human dendritic cells infected with influenza virus.[129] Finally, there are likely to be cells of the innate immune system that are capable of responding to virus-induced tissue damage and other "danger signals" in addition to the viral products themselves.[130] A particularly interesting aspect of the innate immune system is that there is a tissue-specific distribution of cell types of the innate immune system that are capable of responding to infection and danger signals in different ways.[130] Returning to the theme put forth at the beginning, the role of tissue-specific responses to virus infection in viral tropism and pathogenesis is a key question and will be a fertile area for future research.

References

1. Basler, C.F. and G.K. Amarasinghe, Evasion of interferon responses by Ebola and Marburg viruses. *J Interferon Cytokine Res*, 2009. **29**(9): 511–20.
2. Dolnik, O., L. Kolesnikova and S. Becker, Filoviruses: Interactions with the host cell. *Cell Mol Life Sci*, 2008. **65**(5): 756–76.
3. Faul, E.J., D.S. Lyles and M.J. Schnell, Interferon response and viral evasion by members of the family Rhabdoviridae. *Viruses*, 2009. **1**(3): 832–51.
4. Lyles, D.S. and C.E. Rupprecht, Rhabdoviridae, in *Fields Virology*, D.M. Knipe and P.M. Howley, Editors. Lippincott Williams & Wilkins: Philadelphia, PA. 2007, pp. 1363–1408.
5. Sanchez, A., T.W. Geisbert and H. Feldmann, Filoviridae: Marburg and Ebola viruses, in *Fields Virology*, D.M. Knipe and P.M. Howley, Editors. Lippincott Williams & Wilkins: Philadelphia, PA. 2007, pp. 1409–48.
6. Towner, J.S., *et al.*, Newly discovered ebola virus associated with hemorrhagic fever outbreak in Uganda. *PLoS Pathog*, 2008. **4**(11): e1000212.

7. Groseth, A., H. Feldmann and J.E. Strong, The ecology of Ebola virus. *Trends Microbiol*, 2007. **15**(9): 408–16.

8. Pourrut, X., *et al.*, Large serological survey showing cocirculation of Ebola and Marburg viruses in Gabonese bat populations, and a high seroprevalence of both viruses in *Rousettus aegyptiacus*. *BMC Infect Dis*, 2009. **9**: 159.

9. Bailey, C.A., D.K. Miller and J. Lenard, Effects of DEAE-dextran on infection and hemolysis by VSV. Evidence that nonspecific electrostatic interactions mediate effective binding of VSV to cells. *Virology*, 1984. **133**(1): 111–8.

10. Carneiro, F.A., *et al.*, Membrane recognition by vesicular stomatitis virus involves enthalpy-driven protein–lipid interactions. *J Virol*, 2002. **76**(8): 3756–64.

11. Coil, D.A. and A.D. Miller, Phosphatidylserine is not the cell surface receptor for vesicular stomatitis virus. *J Virol*, 2004. **78**(20): 10920–6.

12. Durbin, R.K., *et al.*, PKR protection against intranasal vesicular stomatitis virus infection is mouse strain dependent. *Viral Immunol*, 2002. **15**(1): 41–51.

13. Huneycutt, B.S., *et al.*, Distribution of vesicular stomatitis virus proteins in the brains of BALB/c mice following intranasal inoculation: An immunohistochemical analysis. *Brain Res*, 1994. **635**(1–2): 81–95.

14. Trottier, M.D., D.S. Lyles and C.S. Reiss, Peripheral, but not central nervous system, type I interferon expression in mice in response to intranasal vesicular stomatitis virus infection. *J Neurovirol*, 2007. **13**(5): 433–45.

15. Plakhov, I.V., *et al.*, The earliest events in vesicular stomatitis virus infection of the murine olfactory neuroepithelium and entry of the central nervous system. *Virology*, 1995. **209**(1): 257–62.

16. Muller, U., *et al.*, Functional role of type I and type II interferons in antiviral defense. *Science*, 1994. **264**(5167): 1918–21.

17. Durbin, J.E., *et al.*, Targeted disruption of the mouse Stat1 gene results in compromised innate immunity to viral disease. *Cell*, 1996. **84**(3): 443–50.

18. Reagan, K.J. and W.H. Wunner, Rabies virus interaction with various cell lines is independent of the acetylcholine receptor. *Arch Virol*, 1985. **84**(3–4): 277–82.

19. Superti, F., M. Derer and H. Tsiang, Mechanism of rabies virus entry into CER cells. *J Gen Virol*, 1984. **65**(Pt 4): 781–9.

20. Wunner, W.H., K.J. Reagan and H. Koprowski, Characterization of saturable binding sites for rabies virus. *J Virol*, 1984. **50**(3): 691–7.

21. Gastka, M., J. Horvath and T.L. Lentz, Rabies virus binding to the nicotinic acetylcholine receptor alpha subunit demonstrated by virus overlay protein binding assay. *J Gen Virol*, 1996. **77**(Pt 10): 2437–40.

22. Lentz, T.L., *et al.*, Is the acetylcholine receptor a rabies virus receptor? *Science*, 1982. **215**(4529): 182–4.

23. Thoulouze, M.I., *et al.*, The neural cell adhesion molecule is a receptor for rabies virus. *J Virol*, 1998. **72**(9): 7181–90.

24. Tuffereau, C., *et al.*, Low-affinity nerve-growth factor receptor (P75NTR) can serve as a receptor for rabies virus. *EMBO J*, 1998. **17**(24): 7250–9.

25. Jackson, A.C. and H. Park, Experimental rabies virus infection of p75 neurotrophin receptor-deficient mice. *Acta Neuropathol*, 1999. **98**(6): 641–4.

26. Tuffereau, C., *et al.*, The rabies virus glycoprotein receptor p75NTR is not essential for rabies virus infection. *J Virol*, 2007. **81**(24): 13622–30.

27. Alvarez, C.P., *et al.*, C-type lectins DC-SIGN and L-SIGN mediate cellular entry by Ebola virus in *cis* and in trans. *J Virol*, 2002. **76**(13): 6841–4.

28. Marzi, A., *et al.*, DC-SIGN and DC-SIGNR interact with the glycoprotein of Marburg virus and the S protein of severe acute respiratory syndrome coronavirus. *J Virol*, 2004. **78**(21): 12090–5.

29. Simmons, G., *et al.*, DC-SIGN and DC-SIGNR bind ebola glycoproteins and enhance infection of macrophages and endothelial cells. *Virology*, 2003. **305**(1): 115–23.

30. Takada, A., *et al.*, Human macrophage C-type lectin specific for galactose and N-acetylgalactosamine promotes filovirus entry. *J Virol*, 2004. **78**(6): 2943–7.

31. Gramberg, T., *et al.*, LSECtin interacts with filovirus glycoproteins and the spike protein of SARS coronavirus. *Virology*, 2005. **340**(2): 224–36.

32. Becker, S., M. Spiess and H.D. Klenk, The asialoglycoprotein receptor is a potential liver-specific receptor for Marburg virus. *J Gen Virol*, 1995. **76**(Pt 2): 393–9.

33. Chan, S.Y., *et al.*, Folate receptor-alpha is a cofactor for cellular entry by Marburg and Ebola viruses. *Cell*, 2001. **106**(1): 117–26.

34. Takada, A., *et al.*, Downregulation of beta1 integrins by Ebola virus glycoprotein: Implication for virus entry. *Virology*, 2000. **278**(1): 20–6.

35. Shimojima, M., *et al.*, Tyro3 family-mediated cell entry of Ebola and Marburg viruses. *J Virol*, 2006. **80**(20): 10109–16.

36. Chan, S.Y., *et al.*, Distinct mechanisms of entry by envelope glycoproteins of Marburg and Ebola (Zaire) viruses. *J Virol*, 2000. **74**(10): 4933–7.

37. Geisbert, T.W., *et al.*, Pathogenesis of Ebola hemorrhagic fever in cynomolgus macaques: Evidence that dendritic cells are early and sustained targets of infection. *Am J Pathol*, 2003. **163**(6): 2347–70.

38. Bray, M. and T.W. Geisbert, Ebola virus: The role of macrophages and dendritic cells in the pathogenesis of Ebola hemorrhagic fever. *Int J Biochem Cell Biol*, 2005. **37**(8): 1560–6.

39. Geisbert, T.W., *et al.*, Pathogenesis of Ebola hemorrhagic fever in primate models: Evidence that hemorrhage is not a direct effect of virus-induced cytolysis of endothelial cells. *Am J Pathol*, 2003. **163**(6): 2371–82.

40. Bray, M., *et al.*, A mouse model for evaluation of prophylaxis and therapy of Ebola hemorrhagic fever. *J Infect Dis*, 1998. **178**(3): 651–61.

41. Bray, M., The role of the type I interferon response in the resistance of mice to filovirus infection. *J Gen Virol*, 2001. **82**(Pt 6): 1365–73.

42. Hornung, V., *et al.*, 5'-Triphosphate RNA is the ligand for RIG-I. *Science*, 2006. **314**(5801): 994–7.

43. Kato, H., *et al.*, Cell type-specific involvement of RIG-I in antiviral response. *Immunity*, 2005. **23**(1): 19–28.

44. Kato, H., *et al.*, Differential roles of MDA5 and RIG-I helicases in the recognition of RNA viruses. *Nature*, 2006. **441**(7089): 101–5.

45. Habjan, M., *et al.*, Processing of genome 5' termini as a strategy of negative-strand RNA viruses to avoid RIG-I-dependent interferon induction. *PLoS One*, 2008. **3**(4): e2032.

46. Schlee, M., *et al.*, Recognition of 5' triphosphate by RIG-I helicase requires short blunt double-stranded RNA as contained in panhandle of negative-strand virus. *Immunity*, 2009. **31**(1): 25–34.

47. Rehwinkel, J., *et al.*, RIG-I detects viral genomic RNA during negative-strand RNA virus infection. *Cell*, 2010. **140**(3): 397–408.

48. Lund, J.M., *et al.*, Recognition of single-stranded RNA viruses by toll-like receptor 7. *Proc Natl Acad Sci U S A*, 2004. **101**(15): 5598–603.

49. Cao, W. and Y.J. Liu, Innate immune functions of plasmacytoid dendritic cells. *Curr Opin Immunol*, 2007. **19**(1): 24–30.

50. Diebold, S.S., *et al.*, Innate antiviral responses by means of TLR7-mediated recognition of single-stranded RNA. *Science*, 2004. **303**(5663): 1529–31.

51. Lee, H.K., *et al.*, Autophagy-dependent viral recognition by plasmacytoid dendritic cells. *Science*, 2007. **315**(5817): 1398–401.

52. Georgel, P., *et al.*, Vesicular stomatitis virus glycoprotein G activates a specific antiviral toll-like receptor 4-dependent pathway. *Virology*, 2007. **362**(2): 304–13.

53. Okumura, A., *et al.*, Interaction between Ebola virus glycoprotein and host toll-like receptor 4 leads to induction of proinflammatory cytokines and SOCS1. *J Virol*, **84**(1): 27–33.

54. Jackson, A.C., J.P. Rossiter and M. Lafon, Expression of toll-like receptor 3 in the human cerebellar cortex in rabies, herpes simplex encephalitis, and other neurological diseases. *J Neurovirol*, 2006. **12**(3): 229–34.

55. Prehaud, C., *et al.*, Virus infection switches TLR-3-positive human neurons to become strong producers of beta interferon. *J Virol*, 2005. **79**(20): 12893–904.

56. Alexopoulou, L., *et al.*, Recognition of double-stranded RNA and activation of NF-kappaB by toll-like receptor 3. *Nature*, 2001. **413**(6857): 732–8.

57. Menager, P., *et al.*, Toll-like receptor 3 (TLR3) plays a major role in the formation of rabies virus Negri bodies. *PLoS Pathog*, 2009. **5**(2): e1000315.

58. Lahaye, X., *et al.*, Functional characterization of Negri bodies (NBs) in rabies virus-infected cells: Evidence that NBs are sites of viral transcription and replication. *J Virol*, 2009. **83**(16): 7948–58.

59. Nakhaei, P., *et al.*, RIG-I-like receptors: Sensing and responding to RNA virus infection. *Semin Immunol*, 2009. **21**(4): 215–22.

60. Takeuchi, O. and S. Akira, Pattern recognition receptors and inflammation. *Cell*, 2010. **140**(6): 805–20.

61. Savitsky, D., *et al.*, Regulation of immunity and oncogenesis by the IRF transcription factor family. *Cancer Immunol Immunother*, 2010. **59**(4): 489–510.

62. Prins, K.C., W.B. Cardenas and C.F. Basler, Ebola virus protein VP35 impairs the function of interferon regulatory factor-activating kinases IKKepsilon and TBK-1. *J Virol*, 2009. **83**(7): 3069–77.

63. Brzozka, K., S. Finke and K.K. Conzelmann, Identification of the rabies virus alpha/beta interferon antagonist: Phosphoprotein P interferes with phosphorylation of interferon regulatory factor 3. *J Virol*, 2005. **79**(12): 7673–81.

64. Chang, T.H., *et al.*, Ebola Zaire virus blocks type I interferon production by exploiting the host SUMO modification machinery. *PLoS Pathog*, 2009. **5**(6): e1000493.

65. Feng, Z., *et al.*, The VP35 protein of Ebola virus inhibits the antiviral effect mediated by double-stranded RNA-dependent protein kinase PKR. *J Virol*, 2007. **81**(1): 182–92.

66. Cardenas, W.B., *et al.*, Ebola virus VP35 protein binds double-stranded RNA and inhibits alpha/beta interferon production induced by RIG-I signaling. *J Virol*, 2006. **80**(11): 5168–78.

67. Leung, D.W., *et al.*, Structure of the Ebola VP35 interferon inhibitory domain. *Proc Natl Acad Sci U S A*, 2009. **106**(2): 411–6.

68. Reid, S.P., W.B. Cardenas and C.F. Basler, Homo-oligomerization facilitates the interferon-antagonist activity of the ebolavirus VP35 protein. *Virology*, 2005. **341**(2): 179–89.

69. Hartman, A.L., *et al.*, Inhibition of IRF-3 activation by VP35 is critical for the high level of virulence of ebola virus. *J Virol*, 2008. **82**(6): 2699–704.

70. Hartman, A.L., *et al.*, Reverse genetic generation of recombinant Zaire Ebola viruses containing disrupted IRF-3 inhibitory domains results in attenuated virus growth *in vitro* and higher levels of IRF-3 activation without inhibiting viral transcription or replication. *J Virol*, 2006. **80**(13): 6430–40.

71. Hartman, A.L., *et al.*, Whole-genome expression profiling reveals that inhibition of host innate immune response pathways by Ebola virus can be reversed by a single amino acid change in the VP35 protein. *J Virol*, 2008. **82**(11): 5348–58.

72. Prins, K.C., *et al.*, Mutations abrogating VP35 interaction with double-stranded RNA render Ebola virus avirulent in guinea pigs. *J Virol*, 2010. **84**(6): 3004–15.

73. Finke, S., K. Brzozka and K.K. Conzelmann, Tracking fluorescence-labeled rabies virus: Enhanced green fluorescent protein-tagged phosphoprotein P supports virus gene expression and formation of infectious particles. *J Virol*, 2004. **78**(22): 12333–43.

74. Randall, R.E. and S. Goodbourn, Interferons and viruses: An interplay between induction, signalling, antiviral responses and virus countermeasures. *J Gen Virol*, 2008. **89**(Pt 1): 1–47.

75. Brzozka, K., S. Finke and K.K. Conzelmann, Inhibition of interferon signaling by rabies virus phosphoprotein P: Activation-dependent binding of STAT1 and STAT2. *J Virol*, 2006. **80**(6): 2675–83.

76. Moseley, G.W., *et al.*, Nucleocytoplasmic distribution of rabies virus P-protein is regulated by phosphorylation adjacent to C-terminal nuclear import and export signals. *Biochemistry*, 2007. **46**(43): 12053–61.

77. Vidy, A., M. Chelbi-Alix and D. Blondel, Rabies virus P protein interacts with STAT1 and inhibits interferon signal transduction pathways. *J Virol*, 2005. **79**(22): 14411–20.

78. Vidy, A., *et al.*, The nucleocytoplasmic rabies virus P protein counteracts interferon signaling by inhibiting both nuclear accumulation and DNA binding of STAT1. *J Virol*, 2007. **81**(8): 4255–63.

79. Reid, S.P., *et al.*, Ebola virus VP24 binds karyopherin alpha1 and blocks STAT1 nuclear accumulation. *J Virol*, 2006. **80**(11): 5156–67.

80. Han, Z., *et al.*, Biochemical and functional characterization of the Ebola virus VP24 protein: Implications for a role in virus assembly and budding. *J Virol*, 2003. **77**(3): 1793–800.

81. Dessen, A., *et al.*, Crystal structure of the matrix protein VP40 from Ebola virus. *EMBO J*, 2000. **19**(16): 4228–36.

82. Money, V.A., *et al.*, Surface features of a Mononegavirales matrix protein indicate sites of membrane interaction. *Proc Natl Acad Sci U S A*, 2009. **106**(11): 4441–6.

83. Mateo, M., *et al.*, Ebolavirus VP24 binding to karyopherins is required for inhibition of interferon signaling. *J Virol*, 2010. **84**(2): 1169–75.

84. Reid, S.P., *et al.*, Ebola virus VP24 proteins inhibit the interaction of NPI-1 subfamily karyopherin alpha proteins with activated STAT1. *J Virol*, 2007. **81**(24): 13469–77.

85. Valmas, C., *et al.*, Marburg virus evades interferon responses by a mechanism distinct from ebola virus. *PLoS Pathog*, 2010. **6**(1): e1000721.

86. Ahmed, M., *et al.*, Ability of the matrix protein of vesicular stomatitis virus to suppress beta interferon gene expression is genetically correlated with the inhibition of host RNA and protein synthesis. *J Virol*, 2003. **77**(8): 4646–57.

87. Black, B.L., *et al.*, The role of vesicular stomatitis virus matrix protein in inhibition of host-directed gene expression is genetically separable from its function in virus assembly. *J Virol*, 1993. **67**(8): 4814–21.

88. Ferran, M.C. and J.M. Lucas-Lenard, The vesicular stomatitis virus matrix protein inhibits transcription from the human beta interferon promoter. *J Virol*, 1997. **71**(1): 371–7.

89. Francoeur, A.M., L. Poliquin and C.P. Stanners, The isolation of interferon-inducing mutants of vesicular stomatitis virus with altered viral P function for the inhibition of total protein synthesis. *Virology*, 1987. **160**(1): 236–45.

90. Ahmed, M. and D.S. Lyles, Effect of vesicular stomatitis virus matrix protein on transcription directed by host RNA polymerases I, II, and III. *J Virol*, 1998. **72**(10): 8413–9.

91. Her, L.S., E. Lund and J.E. Dahlberg, Inhibition of Ran guanosine triphosphatase-dependent nuclear transport by the matrix protein of vesicular stomatitis virus. *Science*, 1997. **276**(5320): 1845–8.

92. von Kobbe, C., *et al.*, Vesicular stomatitis virus matrix protein inhibits host cell gene expression by targeting the nucleoporin Nup98. *Mol Cell*, 2000. **6**(5): 1243–52.

93. Black, B.L., G. Brewer and D.S. Lyles, Effect of vesicular stomatitis virus matrix protein on host-directed translation *in vivo*. *J Virol*, 1994. **68**(1): 555–60.

94. Faria, P.A., *et al.*, VSV disrupts the Rae1/mrnp41 mRNA nuclear export pathway. *Mol Cell*, 2005. **17**(1): 93–102.

95. Babu, J.R., *et al.*, Rae1 is an essential mitotic checkpoint regulator that cooperates with Bub3 to prevent chromosome missegregation. *J Cell Biol*, 2003. **160**(3): 341–53. Epub 2003 Jan 27.

96. Blower, M.D., *et al.*, A Rae1-containing ribonucleoprotein complex is required for mitotic spindle assembly. *Cell*, 2005. **121**(2): 223–34.

97. Wong, R.W., G. Blobel and E. Coutavas, Rae1 interaction with NuMA is required for bipolar spindle formation. *Proc Natl Acad Sci U S A*, 2006. **103**(52): 19783–7. Epub 2006 Dec 15.

98. Chakraborty, P., *et al.*, Vesicular stomatitis virus inhibits mitotic progression and triggers cell death. *EMBO Rep*, 2009. **10**(10): 1154–60.

99. Ahmed, M., S.D. Cramer and D.S. Lyles, Sensitivity of prostate tumors to wild type and M protein mutant vesicular stomatitis viruses. *Virology*, 2004. **330**(1): 34–49.

100. Stojdl, D.F., *et al.*, VSV strains with defects in their ability to shutdown innate immunity are potent systemic anti-cancer agents. *Cancer Cell*, 2003. **4**(4): 263–75.

101. Ahmed, M., *et al.*, Immune response in the absence of neurovirulence in mice infected with m protein mutant vesicular stomatitis virus. *J Virol*, 2008. **82**(18): 9273–7.

102. Barchet, W., *et al.*, Virus-induced interferon alpha production by a dendritic cell subset in the absence of feedback signaling *in vivo*. *J Exp Med*, 2002. **195**(4): 507–16.

103. Ahmed, M., *et al.*, Vesicular stomatitis virus M protein mutant stimulates maturation of Toll-like receptor 7 (TLR7)-positive dendritic cells through TLR-dependent and -independent mechanisms. *J Virol*, 2009. **83**(7): 2962–75.

104. Waibler, Z., *et al.*, Matrix protein mediated shutdown of host cell metabolism limits vesicular stomatitis virus-induced interferon-alpha responses to plasmacytoid dendritic cells. *Immunobiology*, 2007. **212**(9–10): 887–94.

105. Barna, M., *et al.*, Sex differences in susceptibility to viral infection of the central nervous system. *J Neuroimmunol*, 1996. **67**(1): 31–9.

106. Bi, Z., *et al.*, Vesicular stomatitis virus infection of the central nervous system activates both innate and acquired immunity. *J Virol*, 1995. **69**(10): 6466–72.

107. Chauhan, V.S., *et al.*, Vesicular stomatitis virus infects resident cells of the central nervous system and induces replication-dependent inflammatory responses. *Virology*, 2010. **400**(2): 187–96.

108. Furr, S.R., *et al.*, Characterization of retinoic acid-inducible gene-I expression in primary murine glia following exposure to vesicular stomatitis virus. *J Neurovirol*, 2008. **14**(6): 503–13.

109. Brundler, M.A., *et al.*, Immunity to viruses in B cell-deficient mice: Influence of antibodies on virus persistence and on T cell memory. *Eur J Immunol*, 1996. **26**(9): 2257–62.

110. Gobet, R., *et al.*, The role of antibodies in natural and acquired resistance of mice to vesicular stomatitis virus. *Exp Cell Biol*, 1988. **56**(4): 175–80.

111. Huneycutt, B.S., *et al.*, Central neuropathogenesis of vesicular stomatitis virus infection of immunodeficient mice. *J Virol*, 1993. **67**(11): 6698–706.

112. Johnson, N., *et al.*, Inflammatory responses in the nervous system of mice infected with a street isolate of rabies virus. *Dev Biol (Basel)*, 2008. **131**: 65–72.

113. Johnson, N., *et al.*, Lyssavirus infection activates interferon gene expression in the brain. *J Gen Virol*, 2006. **87**(Pt 9): 2663–7.

114. Mendonca, R.Z. and C.A. Pereira, Relationship of interferon synthesis and the resistance of mice infected with street rabies virus. *Braz J Med Biol Res*, 1994. **27**(3): 691–5.

115. Li, J., *et al.*, Infection of monocytes or immature dendritic cells (DCs) with an attenuated rabies virus results in DC maturation and a strong activation of the NFkappaB signaling pathway. *Vaccine*, 2008. **26**(3): 419–26.

116. Morimoto, K., *et al.*, Reinvestigation of the role of the rabies virus glycoprotein in viral pathogenesis using a reverse genetics approach. *J Neurovirol*, 2000. **6**(5): 373–81.

117. Wang, Z.W., *et al.*, Attenuated rabies virus activates, while pathogenic rabies virus evades, the host innate immune responses in the central nervous system. *J Virol*, 2005. **79**(19): 12554–65.

118. Schnell, M.J., *et al.*, The cell biology of rabies virus: Using stealth to reach the brain. *Nat Rev Microbiol*, 2010. **8**(1): 51–61.

119. Bosio, C.M., *et al.*, Ebola and Marburg viruses replicate in monocyte-derived dendritic cells without inducing the production of cytokines and full maturation. *J Infect Dis*, 2003. **188**(11): 1630–8.

120. Mahanty, S., *et al.*, Cutting edge: Impairment of dendritic cells and adaptive immunity by Ebola and Lassa viruses. *J Immunol*, 2003. **170**(6): 2797–801.

121. Bosio, C.M., *et al.*, Ebola and Marburg virus-like particles activate human myeloid dendritic cells. *Virology*, 2004. **326**(2): 280–7.

122. Jin, H., *et al.*, The VP35 protein of Ebola virus impairs dendritic cell maturation induced by virus and lipopolysaccharide. *J Gen Virol*, 2010. **91**(Pt 2): 352–61.

123. Gupta, M., *et al.*, Monocyte-derived human macrophages and peripheral blood mononuclear cells infected with ebola virus secrete MIP-1alpha and TNF-alpha and inhibit poly-IC-induced IFN-alpha *in vitro*. *Virology*, 2001. **284**(1): 20–5.

124. Hensley, L.E., *et al.*, Proinflammatory response during Ebola virus infection of primate models: Possible involvement of the tumor necrosis factor receptor superfamily. *Immunol Lett*, 2002. **80**(3): 169–79.

125. Stroher, U., *et al.*, Infection and activation of monocytes by Marburg and Ebola viruses. *J Virol*, 2001. **75**(22): 11025–33.

126. Villinger, F., *et al.*, Markedly elevated levels of interferon (IFN)-gamma, IFN-alpha, interleukin (IL)-2, IL-10, and tumor necrosis factor-alpha associated with fatal Ebola virus infection. *J Infect Dis*, 1999. **179**(Suppl 1): S188–91.

127. Ahmed, M., K.L. Brzoza and E.M. Hiltbold, Matrix protein mutant of vesicular stomatitis virus stimulates maturation of myeloid dendritic cells. *J Virol*, 2006. **80**(5): 2194–205.

128. Ludewig, B., *et al.*, Induction of optimal anti-viral neutralizing B cell responses by dendritic cells requires transport and release of virus particles in secondary lymphoid organs. *Eur J Immunol*, 2000. **30**(1): 185–96.

129. Phipps-Yonas, H., *et al.*, Interferon-beta pretreatment of conventional and plasmacytoid human dendritic cells enhances their activation by influenza virus. *PLoS Pathog*, 2008. **4**(10): e1000193.

130. Matzinger, P., The danger model: A renewed sense of self. *Science*, 2002. **296**(5566): 301–5.

Chapter 12

Paramyxovirus and Rig-Like Helicases: A Complex Molecular Interplay Driving Innate Immunity

Denis Gerlier*

1. Introduction

The cytokine type I interferon (IFN) was discovered over half a century ago,[1] but the intimate molecular events initiating its activation upon a viral infection remained remarkably ill-defined for almost 50 years. Double-stranded RNA (dsRNA) was described as a strong inducer of IFN expression using total RNA extracted and purified from *Penicillium funiculosum* or reovirus or synthetic RNA injected in rabbits.[2] Further on, dsRNA has been considered as a general hallmark of any viral infection, particularly for the RNA viruses that require the obligate synthesis of complementary positive and negative RNA strands for replication. However, convincing evidence for the intracellular presence of such dsRNA intermediates *in situ* during the replication cycle of RNA viruses has not been so common. The main reason for this is that the popular proof for the presence of intracellular dsRNA comes from a biased biochemical approach still in use nowadays (see, for example, Refs. 3–5): it consists of making a protein-free total RNA extract from purified virions, infected cells or intracellular ribonucleoprotein (RNP) complexes purified from infected cells, i.e., giving a chance within the RNA mixture for "at random" annealing between all RNA initially present in the preparation, independently of their "true" ds or ssRNA status *in situ*. In the real life of the single negative-strand RNA viruses or Mononegavirales, the complementary genome and antigenome cannot anneal to each other or to the viral mRNAs, as both of them are tightly and regularly covered

*CNRS FRE3011, INSERM U758, Université de Lyon, Tour Cervi, 21 Avenue Tony Garnier, 69007 Lyon, France. Email: denis.gerlier@inserm.fr

with a helicoidal homopolymer of the nucleoprotein (N).[6–9] The RNA-dependent P1/eIF-2α protein kinase (PKR) was the first dsRNA sensor candidate to be identified. PKR binds to and is activated by dsRNA,[10] stimulates the NF-κB pathway,[11] but turns out to be unable to activate the IFN response and to be dispensable for its activation.[12] PKR appears to act as both an antiviral effector and a regulator of the IFN response.[10,13,14] Toll-like receptors were identified as sensors of extracellular RNAs and they deliver an IFN activation signalling only from the endosomal compartment.[15] In 2004, RIG-I was identified as one of the two intracytoplasmic cellular sensors of viral RNAs able to activate the IFN-β gene.[16] This seminal discovery has paved the way for the elucidation of the sensing of foreign RNA during a viral infection by the intracellular innate immune surveillance system.

The aim of this chapter is to propose a rational view on the molecular interplay between the RIG-I-like helicases (RLH) and paramyxovirus. The viral countermeasures targeting the signalling cascade downstream to the IFN receptor (IFNAR) have been the subject of recent reviews[17–19] and will not be detailed here.

2.   The RLH Family: Sensors of Viral RNAs and Key Players in the Activation of IFN Gene by Paramyxoviruses

In most cells, RNA viruses are detected mainly, if not exclusively, through the recognition of cytoplasmic viral RNAs by the RLH family. There are three members, RIG-I, MDA5 and LGP2.[20,21] The two formers are able to transduce a signal by recruiting MAVS (also known as IPS1, CARDIFF and VISA), which is anchored on the outside of the mitochondria membrane and mediates the activation of type I IFN and inflammatory cytokine genes. The downstream signalling includes the phosphorylation, dimerisation and nuclear import of the IFN responsive factors IRF-3 and IRF-7 to make a supramolecular complex called the enhanceosome, which drives the expression of IFN-β.[17] However, the signals downstream of RIG-I and MDA5 may not be equivalent,[22] and RIG-I also activates the inflammasome independently of MAVS.[23] LGP2 lacks the CARD domain connecting to MAVS and has regulatory function(s) yet to be clarified because of conflicting reports.[24,25] Paramyxoviruses, as other Mononegavirales, are detected by RIG-I[26] and MDA5.[27–29] MDA5 does not seem to be a major sensor of paramyxovirus infection. Indeed, the contribution of MDA5 in mediating the IFN activation induced by measles virus (MeV) is much lower than that of RIG-I, with little additional contribution of MDA5. Furthermore, only residual IFN stimulatory activity is detected when RIG-I is silenced even in the absence of virus V protein, a strong inhibitor of MDA5 (see below).[27] Similarly, in MDA5-disabled mice, the early onset of an

IFN response induced by Sendai virus (SeV) infection is normal but is decreased in a later phase, and this results in higher morbidity and mortality.[29]

3. 5′ppp RNA End and dsRNA as Key Features for Recognition by the RLH

The minimal RNA feature recognised by RIG-I is biochemically defined as a 5′triphosphate (5′ppp) end adjacent to a minimal dsRNA length of roughly 9–19 base pairs with tolerance of 1, 2 or 3 nucleotide (nt) mismatches and/or over-hangs[30,31] (Fig. 1). The dsRNA can be shortened to 7 base pairs to a maximal distance of 8 nt from the 5′ppp end in the case of an hairpin-folded ssRNA.[31] The 5′ppp (in the context of dsRNA?) binds to the C-terminal regulatory domain (RD) domain of RIG-I via an interaction with a large and positively-charged concave surface.[32] The downstream dsRNA is likely recognised by the central helicase domain according to a translocase mode[33] as revealed using helicase mutants,[34,35] limited proteolytic digestion of RIG-I in complex with dsRNA,[36] oligomerisation studies,[37] and *in vitro* measures of ATPase activity.[38] The recognition of the 5′ppp moiety can discriminate between (cellular) self and (pathogen) non-self RNA. Naturally present on any primary transcript because of the 5′ to 3′ condensation chemistry of the nucleic acid elongation, 5′ppp is normally excluded from the cytosol. Uncapped mRNA from mitochondria are not exported, and 5′ppp nuclear

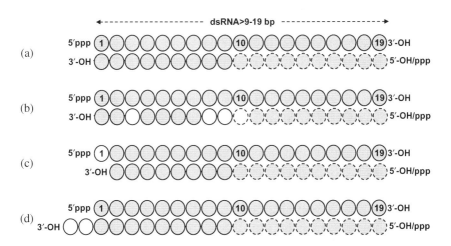

Fig. 1. Minimal RNA structures that bind and activate RIG-I with (a, b, c and d) possibly dispensable nt (dotted rings), tolerance to mismatches (b) or to 5′/3′overhang (c, d) (open rings) (according to Refs. 30 and 31).

transcripts are modified before their nuclear export by cleavage, capping or protein shielding (see Refs. 19 and 20 for review). The RNA feature recognised by MDA5 is less understood and is at best described as dsRNA of > 1 kb length lacking 5′ppp[21,39] associated with ssRNA into high-order RNA structures.[40] Similar to RIG-I, the COOH-terminal domain (CTD) of MDA5 exhibits a positively-charged area, but it is relatively flat with an open conformation of the RNA-binding loop that weakens the binding to dsRNA.[41] LGP2 recognises RNA agonists of both RIG-I and MDA5. Its CTD domain has a structure closer to that of RIG-I and binds to the termini of dsRNA rather devoid of 5′ppp end.[41–43] Which viral RNA(s) is(are) then recognised by the RLH and notably RIG-I? The answer to this simple question is not straight forward and direct convincing identification in infected cells is still lacking.

4. Which Viral RNA Species Could be Agonists of the RLH?

What can reasonably be the source of viral 5′ppp-ended RNA? The mRNAs from Paramyxoviruses, like others from Mononegavirales, are capped (Fig. 2a and b). When the capping is experimentally inhibited, the elongation of the 5′ppp transcript prematurely stops after 100 to 400 nt elongation.[44,45] To what extent such a capping failure occurs during natural infection is unknown. In light of the linear accumulation of full length ≃1.7 kb MeV N mRNA over 8 h by a constant number of incoming transcriptases,[46,47] one can estimate that such a failure should remain at least below ∼10% (i.e., within the precision limits of the measure).

Two small RNAs, *le*ader (*le*RNA) and *tr*ailer (*tr*RNA), are also produced by the viral polymerase (Fig. 2b). These RNAs remain 5′ppp[48] and are readily synthesised in viral transcription assays using permeabilised virions from some mononegavirales.[48,49] In cells infected with SeV, free *le*RNAs (31, 55 and 65 nt long) separated from encapsidated viral RNA by ultracentrifugation on CsCl gradient are detected, but in limited amounts. This suggests that they are unstable in the cells.[50,51] More abundant free short *le*RNA-N read-through transcripts (<300 nt long) amounting up to 40% of N mRNA are also detected.[50] These short *le*RNA-N read-throughs are distinct from the complete (∼2 kb long) *le*RNA-N read-troughs, which are encapsidated and thus likely due to some abortive replication as also found in the case of MeV.[47,50,52] As the abundance of free *le*RNA-N read-through RNAs remains unchanged when the replication is blocked by inhibition of protein synthesis, they represent transcripts that fail to be capped and have their elongation prematurely stopped.[50] The mechanism of such an elongation stop is unknown. Is it the presence of *le*RNA sequence and/or the lack of polymerase priming for capping? Indeed, polymerase priming for transcription requires the recognition of

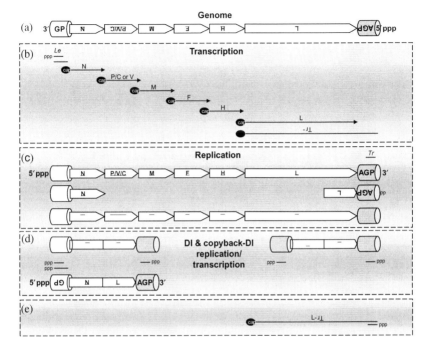

Fig. 2. Paramyxovirus RNA species produced during infection: (a) (morbillivirus) genomic organisation (negative strand). Abbreviations: GP, genomic promoter; N, nucleoprotein; P, phosphoprotein, V & C, V and C proteins; M, matrix; F, fusion glycoprotein; H, hemagglutinin; L, large (polymerase) protein; AGP, antigenomic promoter; *le*, leader; *tr*, trailer; DI, defective interferent (particles). Because the genome is fully encapsidated, all elements are represented as 2D drawings. (b) RNA synthesised during transcription from the genomic-NC template. Transcripts are not encapsidated and are represented as 1D lines with (or without) (cap) and polyA tails (arrow at 3′ end). (c) RNA synthesised during replication. Encapsidated RNAs are in 2D with encapsidated antigenome (positive strand, top) and genome (negative strand, bottom), respectively, and encapsidated abortive replication fragments in the middle. (d) Encapsidated (2D) and free RNA (1D) species produced from DI (left) and copy-back DI (right). (e) Predicted dsRNA from annealing of two viral RNA species, which have been biochemically characterised in infected cells until now. Other viral RNAs species and/or dsRNA annealing opportunities that can be hypothesised are not shown, but can be found in the text.

the unique bipartite transcriptase promoter (located within *le*RNA and N 5′UTR regions of the genome).[53,54] Free *tr*RNA is also 5′ppp-ended but would likely be synthesised only after the onset of replication.[55]

The last viral RNA species produced during an infection are the genome and antigenome, which are both 5′ppp-ended[56] (Fig. 2c). Indeed, protein-free and annealed genome+antigenome RNA mixtures are good inducers of RIG-I-mediated IFN response.[3,4] However, they are never found as free RNA, but exclusively as part of nucleocapsid (NC) in CsCl gradients.[50] Therefore, the intracellular (anti)genomes are fully shielded within the helicoidal homopolymeric nucleoprotein, and they resist nuclease digestion. Each N subunit covers exactly 6 nt, and the genome length has to be a multiple of 6 for every Paramyxovirinae so far tested.[57–61] The genome and antigenome are resistant to silencing by RNAi.[62–64] Finally, in the crystal structure of NC from respiratory syncytial virus (RSV), the phosphate backbone of the encapsidated RNA and 3 out of 7 contiguous nt are buried within a protein groove unless N subunits are opened.[6]

Several observations strongly argue for the recognition of a viral transcript by RIG-I during a paramyxovirus infection. In MeV, the kinetics of IFN-β gene activation parallels that of MeV transcription and not replication.[35] When replication-disabled MeV is delivered into the cytosol, it still transcribes the virus genes and activates the IFN-β response.[35] The parainfluenza virus 5 (PIV5) expressing a P mutant, unable to bind to or be phosphorylated by the PLK1 kinase, concomitantly exhibits an enhancement of the viral transcription and a stronger activation of the IFN-β and cytokine genes.[65,66] An SeV variant, which produces lower amounts of free leader and leader-N read-throughs, is less efficient in activating the IFN response.[50,67] Accordingly, the substitution of the genomic promoter with the stronger antigenomic promoter up-regulates both viral transcription and IFN response.[55,68] Upon their transfection, short viral transcripts made *in vitro* from permeabilised Mononegavirales virions and likely corresponding to *le*RNAs activate the IFN-β gene[35,69] as does *le*RNA transcribed from DNA in the cytosol but not in the nucleus.[35] However, SeV genomic RNA was claimed to be the RIG-I ligand in infected cells, as 211-nt long primer extension products encompassing the end of L gene and *tr*RNA are enriched in RIG-I immunoprecipitates.[5] The interpretation is, however, obscured by the lack of further characterisation of this RNA (full-length genomic sequence? and free or NC associated?) and poor analysis of short transcripts.

So far only the source of 5′ppp-ended RNA has been discussed. What about the source of the dsRNA moiety? RIG-I dsRNA agonist can be a single-strand RNA (ssRNA) with intramolecular annealing of stretches of complementary sequences or two annealed complementary ssRNAs. It should be stressed that infection with negative-strand RNA viruses does not produce much >40 bp-long dsRNA, whereas such dsRNAs are readily found in cells infected with positive-strand RNA viruses.[70] The dsRNA between two capped mRNAs produced by coinfection with two recombinant SeVs expressing the green fluorescent protein sense mRNA and its complementary antisense mRNA, respectively, strongly activates the IFN-β gene through a RIG-I-dependent pathway.[38,71] This suggests that long complementary dsRNA without 5′ppp ends may also be activators of RIG-I. Alternatively, one cannot exclude that the presence of one transcript may prevent the capping of its nascent complementary strand. While a carefully cloned SeV is a poor inducer of the IFN response, SeV stocks contaminated with defective interfering (DI) particles, i.e., genome with a large internal deletion, are good stimulators[72] (Fig. 2d). Notably, the IFN stimulatory activity is stronger for a copy-back DI, which contains the stronger antigenomic promoter on both negative and positive strands[72] (Fig. 2d). In addition, the ability of DIs to activate the IFN-β gene is inhibited by UV irradiation of the viral stock according to a dose–response relationship[72] as observed for the UV sensitivity of MeV N transcription,[35] whereas the replication of full-length SeV and MeV displays higher sensitivity to UV (note that SeV DI RNA and MeV N mRNA as well as their corresponding genomes are of comparable ∼1.6 kb and ∼16 kb size, respectively). Although this was interpreted as reflecting the levels of DI replication, the free or encapsidated status of the DI RNA produced from UV-irradiated virus stock was not reported. What are the RNA species that are synthesised from DI genome and antigenome? Fully encapsidated DI genomes and antigenomes are replicated, but their 5′ppp ends are shielded[73] and are unlikely available for recognition by RIG-I (Fig. 2d). Genomic and antigenomic DI RNAs may fail to be encapsidated, as ∼5% of DI (anti)genomes can pellet as free RNAs through CsCl gradients. Upon self-hybridation into panhandle RNA structures, because the 5′ and 3′ends of copy-back DI are complementary, or upon genome–antigenome hybridation, they could make RNA with both 5′ppp-end and dsRNA stretches.[72] *le*RNA and *tr*RNA (or complementary c*tr*RNA for copy-back) can be produced during transcription (*le*RNA from DI) or during replication (*tr*RNA from DI, *tr*RNA and c*tr*RNA from copy-back DI). *tr*RNA/c*tr*RNA hybrids would represent perfect RIG-I agonists, forming 5′ppp-ended >44 nt-long dsRNA (Fig. 2d and e). However, none of these three putative dsRNA forms were detected intracellularly by J2 antibody that is specific for >40 bp-long dsRNA indicating either their absence of their presence in undetectable amounts.[74] Whether *le* and/or *tr*RNA, or short *le*RNA-N read-throughs fold into secondary structures with long enough dsRNA regions at suitable distances from the 5′ppp end to act as RIG-I agonists remains to be determined. Alternative candidates are (i) read-through L-*tr*RNA mRNA[38,75–78] hybridised to 5′ppp *tr*RNA (Fig. 2e) and (ii) self-complementary NC-free uncapped RNA resulting from copy-choice mechanisms involving polymerase jumping during replication as occurs for the generation of copy-back DI.[79]

5. Paramyxovirus and RLH Cross-talk: A Piece of Molecular Choreography

By associating with nascent 5′ppp-ended genome and antigenome and covering of every single nt, the N protein acts as a major inhibitor of RIG-I recognition and IFN-β activation. It shields the 5′ppp end away from the RD of RIG-I and it prevents the annealing of genome–antigenome and/or mRNA(s)-genome complexes into dsRNA.

That the virus avoids production of dsRNA has been experimentally challenged with a recombinant ambisense SeV. A transcription chloramphenicol acetyltransferase (CAT) unit has been added at the 3′end of the antigenome under the control of the genomic replication and transcription promoter. This virus grew very poorly in IFN competent cells. It rapidly loses the expression of antisense transcript by being selected for debilitating mutations within the antisense-orientated transcription promoter. The likely explanation is that the plus sense L-CAT read-through and the minus sense CAT-L read-through transcripts annealed into dsRNA. This dsRNA strongly activates the antiviral response, which leads to select SeV with disabled antisense transcription promoter.[68]

Two proteins, T-cell-activated intracellular antigen related 1 (TIAR) and La autoantigen can bind to *tr*RNA and *le*RNA, respectively. While TIAR binding to SeV *tr*RNA is associated with modulation of apoptosis,[55,80] La protein can bind to *le*RNA from numerous Mononegavirales including RSV, PIV3 and rinderpest virus (RPV) *in vitro* and in infected cells.[69,81–85] During RSV and RPV infection, the nuclear La protein is redistributed to the cytoplasm,[69,83] and its expression is required for efficient transcription of RSV in an IFN-independent manner.[69] In addition to this transcriptional enhancing effect, La protein also seems to compete out with RIG-I for binding to RSV *le*RNA as, when La expression is silenced, there is both increased amounts of *le*RNA bound to RIG-I and enhanced IFN-β response. Thus, in the case of SeV, an IFN sensitive virus, the silencing of La expression strongly affects the production of SeV because of the additive effects of the IFN-independent decrease of viral transcription and the enhanced RIG-I-mediated activation of the cellular antiviral response[69] (Fig. 3). La autoantigen belongs to the ribonucleoprotein family of RNA recognition motif (RRM) with diverse roles in the metabolism of multiple cellular and viral RNAs.[86] Via its La domain, it binds to 3′OH-UUU motif and protects nascent pol III transcript from exonuclease digestion.[87] Via its adjacent RRM1 secondary RNA binding domain,[88] it binds to mRNA as part of an RNA regulon, which is involved in the coordinated ribosome biogenesis.[89] High-affinity binding of RNA to La protein usually requires the 3′OH-UUU motif,[87,88] which is absent from the mononegavirales *le*RNAs. Accordingly *le*RNA binds to the RRM1 La domain,[69] but with a surprisingly high affinity (80 nM range).[83] Interestingly, the presence of a 5′ppp end can contributes to enhance the RNA–La interaction.[90] That *le*RNA and La protein make stable complexes is illustrated by their stability in CsCl gradient,[69,83] and their similar buoyant density with that of viral NC.[69] Interestingly, a time-course analysis reveals that RSV *le*RNA is early associated with La in complexes free from N protein to become later associated

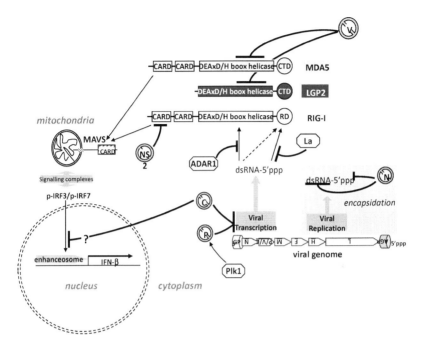

Fig. 3. The molecular choreography of the recognition of paramyxovirus RNA synthesis by intracellular innate immunity. This figure summarises all the known molecular mechanisms leading to dsRNA with 5′ppp end(s) or 5′ppp-ssRNA with self-annealing sequences into dsRNA (the distinction is not made), their down-regulation and/or shielding away from recognition by RLR, their cognate RLR, the viral inhibitors of the RLR and downstream signalling. Data for C proteins relate to SeV, RPV and MeV, for P to PIV5, for V to all tested paramyxovirinae, and for NS2 from RSV. Data on ADAR1 relate to MeV (see text for relevant references).

with this protein.[69] Moreover, La is able to unwind dsRNA.[91] Overall these data favour the following model. At early times, post-infection primary transcription results in the synthesis of *le*RNA, which is concurrently bound by La and RIG-I, with an advantage for La. In addition, La would also unwind the dsRNA feature, thus desensitising the viral RNA from activating RIG-I. At later times, concomitant replication and transcription result in preferential encapsidation of *le*RNA. In addition to La, PKR may also be a competitor for RNA agonist recognition by RIG-I because it binds to and is activated by dsRNA with a 5′ppp end.[92]

6. Viral Strategies to Counteract the RLH-Mediated Activation of IFN-β Gene

Every paramyxovirus expresses at least two proteins with potent anti-IFN activities. Pneumovirinae express NS1 and NS2 proteins from independent transcription units located at the very 3′ end of the genome. Paramyxovirinae express P, V (I, W, D) and C (Y) proteins from the second transcription P/V/C unit by mRNA editing (V protein) or by using an alternative reading frame (C protein).[93] NS1, P, V and C proteins block signalling downstream of the type IFN receptor (IFNAR). These proteins mostly target STAT-1, STAT-2 and/or Jak-1 proteins through various mechanisms including protein sequestration and degradation as reviewed elsewhere.[17–19] These activities are beyond the scope of this review and will not be detailed further. Instead, we will focus on the viral mechanisms that counteract the initiation of the IFN response mediated by the RLRs.

All paramyxovirus V proteins can inhibit MDA5.[94] They bind MDA5 and prevent MDA5-mediated activation of the IFN-β gene. PIV-5 V protein binds to the helicase domain of MDA5, competes with dsRNA for binding to this domain and prevents its oligomerisation[37] (Fig. 3). As V proteins from MeV, SeV and Hendra virus are also able to inhibit the oligomerisation of MDA5 induced by dsRNA, binding to the dsRNA binding site on the helicase domain appears a common property for paramyxovirus V proteins.[37] V also inhibits the ATPase activity of MDA5.[95] The high efficiency of MDA5 inhibition by V protein explains the low contribution of MDA5 in mediating the activation of the IFN-β gene, which is better evidenced after silencing of RIG-I expression.[27,29] V protein also binds to LGP2, blocks its ATPAse activity and is therefore predicted to inhibit its regulatory activities.[95]

So far, RSV NS2 protein is the only paramyxovirus gene product that binds to RIG-I (Fig. 3). NS2 acts as a competitive inhibitor of MAVS for binding to the N-terminal CARD domain of RIG-I.[96] This direct activity on this viral RNA sensor together with the later blockade of the IFNAR signalling by NS1 and NS2 proteins likely explains the particular resistance of RSV to the IFN system.[17,29] RSV NS1 also contributes to dampen the RIG-I-mediated activation of IFN-β, as NS1-deleted RSV induces a higher IFN response despite the expression of NS2.[96] Interestingly, NS1 expression is a viral transcription inhibitor in minigenome assay,[97] and NS1-deleted RSV grew faster and induced a stronger IRF-3 nuclear localisation, as if enhanced viral transcription has led to enhanced activation of RIG-I.[96]

The C protein from RPV,[98] MeV[99] and SeV[71,100,101] inhibits the virus-induced IFN-β gene activation mediated by RIG-I. C protein can also inhibit the activation of RIG-I by an artificial RNA agonist,[71,98,99] but the effect is moderate and variable according to the virus strains.[98,102] It can also be observed in cells devoid of the IFN/IFNAR amplification loop.[98] The underlying mechanism is unknown. The RPV C acts downstream of the phosphorylation, dimerisation and nuclear import of IRF-3, i.e., possibly at the level of the IFN enhanceosome (Fig. 3). This would require C shuttling into the nucleus and the MeV C protein is endowed with such shuttling properties.[103] Alternatively, and/or additionally, the paramyxovirus C proteins negatively regulate virus transcription, and, as such, limit the production of transcript agonists of RIG-I.[104–109] Indeed, when cells are infected with a C-deleted SeV, dsRNA detected by the J2 antibody accumulates with a concomitant activation of the PKR and phosphorylation of eIF2.[74] The infection with C-deleted MeV also induces PKR activation and eIF2 phosphorylation that results in decreased protein synthesis.[110–112] In this context, the adenosine deaminase ADAR1 seems to counteract both PKR-induced activation and IRF-3 activation by C-deleted MeV, pointing to the deamination of (5′ppp)dsRNA as a possible mechanism to lower both PKR and RIG-I activation.[113]

7. Viral Strategy to Sneak Through a Limited IFN Activation

Paramyxoviruses seem to intrinsically induce a limited amount of type I IFN early on after entry because they are transcribing in the cytosol. How then can they successfully propagate into adjacent cells that have activated their antiviral genes due to this local IFN priming? An elegant answer recently emerged from studies of IFN-primed cells infected by PIV5. Although the replication cycle is severely slowed down, a sneaking-through viral transcription and replication occur. This results in the progressive degradation of STAT1, one of the key transcription factors downstream of the IFNAR signalling. As a result, the IFN priming effect is progressively vanishing away within the infected cell. Then, this permits the delayed onset of efficient PIV5 replication. Following the infection, the incoming NCs initially localise within small cytoplasmic bodies that seem to act as viral factories shielding the virus replication away from the cellular antiviral response.[114]

8. Conclusion

RIG-I appears as the major sensor of paramyxovirus RNA. Most of the available data strongly argue for a transcription (i.e., made by a transcriptase P + L polymerase) and not replication (i.e., made by a replicase N + P + L polymerase) origin of the viral RNA agonist(s), which can bind to and activate RIG-I. This RNA should be 5′ppp-ended and have a stretch of hybrid (ds) RNA over a minimal length located close to the 5′ppp end.[30,31] The dsRNA can result from the folding of ssRNA into secondary structures or from the partial annealing of two RNAs, which exhibit a stretch of complementary sequences. Similar to miRNA targeting mRNA, any small 5′ppp transcript could potentially anneal with another free RNA of viral or even cellular origin. More work will be needed to identify the true physiologic viral RNA species that act as RIG-I agonist(s). The knowledge of this(ese) viral RNA agonist(s) will tell us more on several aspects of paramyx-ovirus molecular biology. To what extent does the polymerase fail to cap a nascent mRNA? To what extent does the polymerase fail to encapsidate a nascent genome or antigenome particularly from DI minigenomes? Are the short *le*RNA transcrip-tion and/or *tr*RNA (transcription or replication?) products obligatory by-products of the transcription (or the replication?)? Indeed, current models[53,54] favour the hypothesis whereby the polymerase enters at the very 3′end of the tightly packed NC. There, it immediately starts RNA synthesis until it either recognises the tran-scription promoter for the synthesis of capped mRNA, or it pursues the elongation because it "senses" the ongoing encapsidation over the nascent *le*RNA or *tr*RNA. How often does the polymerase jump from one template to an adjacent template and produce aberrant RNAs? This knowledge, as well progress in our understanding on the role of V-mediated inactivation of MDA5 and on the viral counter-measures, which dampen the RLH recognition and/or signalling, will provide us with novel strategies to design new antivirals and/or attenuated vaccines.

Acknowledgments

The author thanks Jade Louber for her contribution in some literature analysis, P. Lawrence and B. Horvat for their helpful comments. Part of the work cited in this chapter has been supported by an ANR grant ANR-08-PCVI-0020 (AFF-IDP).

References

1. Isaacs, A. and J. Lindenmann, Virus interference. I. The interferon. *Proc R Soc Lond B Biol Sci*, 1957. **147**(927): 258–67.
2. Lampson, G.P., *et al.*, Inducers of interferon and host resistance. I. Double-stranded RNA from extracts of Penicillium funiculosum. *Proc Natl Acad Sci U S A*, 1967. **58**(2): 782–9.
3. Habjan, M., *et al.*, Processing of genome 5′ termini as a strategy of negative-strand RNA viruses to avoid RIG-I-dependent interferon induction. *PLoS One*, 2008. **3**(4): e2032.
4. Hornung, V., *et al.*, 5′-Triphosphate RNA is the ligand for RIG-I. *Science*, 2006. **314**(5801): 994–7.
5. Rehwinkel, J., *et al.*, RIG-I detects viral genomic RNA during negative-strand RNA virus infection. *Cell*, 2010. **140**(3): 397–408.
6. Tawar, R.G., *et al.*, Crystal structure of a nucleocapsid-like nucleoprotein–RNA com-plex of respiratory syncytial virus. *Science*, 2009. **326**(5957): 1279–83.
7. Albertini, A.A., *et al.*, Crystal structure of the rabies virus nucleoprotein–RNA com-plex. *Science*, 2006. **313**(5785): 360–3.

8. Green, T.J., *et al.*, Structure of the vesicular stomatitis virus nucleoprotein–RNA complex. *Science*, 2006. **313**(5785): 357–60.

9. Lamb, R.A. and D. Kolakofsky, Paramyxoviridae: The Viruses and Their Replication, *in Fields Virology* B.N. Fields, D.M. Knipe and P.M. Howley, Editors. Lippincott-Raven: (4th edition), Philadelphia, PA., 2001, pp. 1305–40.

10. Sadler, A.J. and B.R. Williams, Structure and function of the protein kinase R. *Curr Top Microbiol Immunol*, 2007. **316**: 253–92.

11. Kumar, A., *et al.*, Double-stranded RNA-dependent protein kinase activates transcription factor NF-kappa B by phosphorylating I kappa B. *Proc Natl Acad Sci U S A*, 1994. **91**(14): 6288–92.

12. Yang, Y.L., *et al.*, Deficient signaling in mice devoid of double-stranded RNA-dependent protein kinase. *EMBO J*, 1995. **14**(24): 6095–106.

13. McAllister, C.S. and C.E. Samuel, The RNA-activated protein kinase enhances the induction of interferon-beta and apoptosis mediated by cytoplasmic RNA sensors. *J Biol Chem*, 2009. **284**(3): 1644–51.

14. Garcia, M.A., E.F. Meurs and M. Esteban, The dsRNA protein kinase PKR: Virus and cell control. *Biochimie*, 2007. **89**(6–7): 799–811.

15. Barton, G.M. and J.C. Kagan, A cell biological view of toll-like receptor function: Regulation through compartmentalization. *Nat Rev Immunol*, 2009. **9**(8): 535–42.

16. Yoneyama, M., *et al.*, The RNA helicase RIG-I has an essential function in double-stranded RNA-induced innate antiviral responses. *Nat Immunol*, 2004. **5**(7): 730–7.

17. Goodbourn, S. and R.E. Randall, The regulation of type I interferon production by paramyxoviruses. *J Interferon Cytokine Res*, 2009. **29**(9): 539–47.

18. Randall, R.E. and S. Goodbourn, Interferons and viruses: An interplay between induction, signalling, antiviral responses and virus countermeasures. *J Gen Virol*, 2008. **89**(Pt 1): 1–47.

19. Gerlier, D. and H. Valentin, Measles virus interaction with host cells and impact on innate immunity. *Curr Top Microbiol Immunol*, 2009. **329**: 163–91.

20. Takeuchi, O. and S. Akira, MDA5/RIG-I and virus recognition. *Curr Opin Immunol*, 2008. **20**(1): 17–22.

21. Yoneyama, M. and T. Fujita, RNA recognition and signal transduction by RIG-I-like receptors. *Immunol Rev*, 2009. **227**(1): 54–65.

22. Loo, Y.M., *et al.*, Distinct RIG-I and MDA5 signaling by RNA viruses in innate immunity. *J Virol*, 2008. **82**(1): 335–45.

23. Poeck, H., *et al.*, Recognition of RNA virus by RIG-I results in activation of CARD9 and inflammasome signaling for interleukin 1 beta production. *Nat Immunol*, 2010. **11**(1): 63–9.

24. Satoh, T., *et al.*, LGP2 is a positive regulator of RIG-I- and MDA5-mediated antiviral responses. *Proc Natl Acad Sci U S A*, 2010. **107**(4): 1512–7.

25. Venkataraman, T., *et al.*, Loss of DExD/H box RNA helicase LGP2 manifests disparate antiviral responses. *J Immunol*, 2007. **178**(10): 6444–55.

26. Wilkins, C. and M. Gale Jr., Recognition of viruses by cytoplasmic sensors. *Curr Opin Immunol*, 2010. **22**(1): 41–7.

27. Ikegame, S., *et al.*, Both RIG-I and MDA5 RNA helicases contribute to the induction of alpha/beta interferon in measles virus-infected human cells. *J Virol*, 2010. **84**(1): 372–9.

28. Yount, J.S., *et al.*, MDA5 participates in the detection of paramyxovirus infection and is essential for the early activation of dendritic cells in response to Sendai Virus defective interfering particles. *J Immunol*, 2008. **180**(7): 4910–8.

29. Gitlin, L., *et al.*, Melanoma differentiation-associated gene 5 (MDA5) is involved in the innate immune response to Paramyxoviridae infection *in vivo*. *PLoS Pathog*, 2010. **6**(1): e1000734.

30. Schlee, M., *et al.*, Recognition of 5′ triphosphate by RIG-I helicase requires short blunt double-stranded RNA as contained in panhandle of negative-strand virus. *Immunity*, 2009. **31**(1): 25–34.

31. Schmidt, A., *et al.*, 5′-triphosphate RNA requires base-paired structures to activate antiviral signaling via RIG-I. *Proc Natl Acad Sci U S A*, 2009. **106**(29): 12067–72.

32. Cui, S., *et al.*, The C-terminal regulatory domain is the RNA 5′-triphosphate sensor of RIG-I. *Mol Cell*, 2008. **29**(2): 169–79.

33. Myong, S., *et al.*, Cytosolic viral sensor RIG-I is a 5′-triphosphate-dependent translocase on double-stranded RNA. *Science*, 2009. **323**(5917): 1070–4.

34. Bamming, D. and C.M. Horvath, Regulation of signal transduction by enzymatically inactive antiviral RNA helicase proteins MDA5, RIG-I, and LGP2. *J Biol Chem*, 2009. **284**(15): 9700–12.

35. Plumet, S., *et al.*, Cytosolic 5′-triphosphate ended viral leader transcript of measles virus as activator of the RIG I-mediated interferon response. *PLoS One*, 2007. **2**: e279.

36. Takahasi, K., *et al.*, Nonself RNA-sensing mechanism of RIG-I helicase and activation of antiviral immune responses. *Mol Cell*, 2008. **29**(4): 428–40.

37. Childs, K.S., *et al.*, Mechanism of mda-5 inhibition by paramyxovirus V proteins. *J Virol*, 2009. **83**(3): 1465–73.

38. Hausmann, S., *et al.*, RIG-I and dsRNA-induced IFNbeta activation. *PLoS One*, 2008. **3**(12): e3965.

39. Kato, H., *et al.*, Length-dependent recognition of double-stranded ribonucleic acids by retinoic acid-inducible gene-I and melanoma differentiation-associated gene 5. *J Exp Med*, 2008. **205**(7): 1601–10.

40. Pichlmair, A., *et al.*, Activation of MDA5 requires higher-order RNA structures generated during virus infection. *J Virol*, 2009. **83**(20): 10761–9.

41. Takahasi, K., *et al.*, Solution structures of cytosolic RNA sensor MDA5 and LGP2 C-terminal domains: Identification of the RNA recognition loop in RIG-I-like receptors. *J Biol Chem*, 2009. **284**(26): 17465–74.

42. Li, X., *et al.*, The RIG-I-like receptor LGP2 recognizes the termini of double-stranded RNA. *J Biol Chem*, 2009. **284**(20): 13881–91.

43. Pippig, D.A., *et al.*, The regulatory domain of the RIG-I family ATPase LGP2 senses double-stranded RNA. *Nucleic Acids Res*, 2009. **37**(6): 2014–25.

44. Li, J., J.T. Wang and S.P. Whelan, A unique strategy for mRNA cap methylation used by vesicular stomatitis virus. *Proc Natl Acad Sci U S A*, 2006. **103**(22): 8493–8.

45. Liuzzi, M., *et al.*, Inhibitors of respiratory syncytial virus replication target cotranscriptional mRNA guanylylation by viral RNA-dependent RNA polymerase. *J Virol*, 2005. **79**(20): 13105–15.

46. Plumet, S. and D. Gerlier, Optimized SYBR green real-time PCR assay to quantify the absolute copy number of measles virus RNAs using gene specific primers. *J Virol Methods*, 2005. **128**(1–2): 79–87.

47. Plumet, S., W.P. Duprex and D. Gerlier, Dynamics of viral RNA synthesis during measles virus infection. *J Virol*, 2005. **79**(11): 6900–8.

48. Colonno, R.J. and A.K. Banerjee, A unique RNA species involved in initiation of vesicular stomatitis virus RNA transcription *in vitro. Cell*, 1976. **8**(2): 197–204.

49. Horikami, S.M. and S.A. Moyer, Synthesis of leader RNA and editing of the P mRNA during transcription by purified measles virus. *J Virol*, 1991. **65**(10): 5342–7.

50. Vidal, S. and D. Kolakofsky, Modified model for the switch from Sendai virus transcription to replication. *J Virol*, 1989. **63**(5): 1951–8.

51. Leppert, M., *et al.*, Plus and minus strand leader RNAs in negative strand virus-infected cells. *Cell*, 1979. **18**(3): 735–47.

52. Castaneda, S.J. and T.C. Wong, Leader sequence distinguishes between translatable and encapsidated measles virus RNAs. *J Virol*, 1990. **64**(1): 222–30.

53. Whelan, S.P., J.N. Barr and G.W. Wertz, Transcription and replication of nonsegmented negative-strand RNA viruses. *Curr Top Microbiol Immunol*, 2004. **283**: 61–119.

54. Kolakofsky, D., *et al.*, Viral RNA polymerase scanning and the gymnastics of Sendai virus RNA synthesis. *Virology*, 2004. **318**(2): 463–73.

55. Iseni, F., *et al.*, Sendai virus trailer RNA binds TIAR, a cellular protein involved in virus-induced apoptosis. *EMBO J*, 2002. **21**(19): 5141–50.

56. Leppert, M. and D. Kolakofsky, 5' Terminus of defective and nondefective Sendai viral genomes is ppp Ap. *J Virol*, 1978. **25**(1): 427–32.

57. Calain, P. and L. Roux, The rule of six, a basic feature for efficient replication of Sendai virus defective interfering RNA. *J Virol*, 1993. **67**(8): 4822–30.

58. Kolakofsky, D., *et al.*, Paramyxovirus RNA synthesis and the requirement for hexamer genome length: The rule of six revisited. *J Virol*, 1998. **72**(2): 891–9.

59. Peeters, B.P., *et al.*, Genome replication of Newcastle disease virus: Involvement of the rule-of-six. *Arch Virol*, 2000. **145**(9): 1829–45.

60. Halpin, K., *et al.*, Nipah virus conforms to the rule of six in a minigenome replication assay. *J Gen Virol*, 2004. **85**(Pt 3): 701–7.

61. Skiadopoulos, M.H., *et al.*, The genome length of human parainfluenza virus type 2 follows the rule of six, and recombinant viruses recovered from non-polyhexameric-length antigenomic cDNAs contain a biased distribution of correcting mutations. *J Virol*, 2003. **77**(1): 270–9.

62. Bitko, V. and S. Barik, Phenotypic silencing of cytoplasmic genes using sequence-specific double-stranded short interfering RNA and its application in the reverse genetics of wild type negative-strand RNA viruses. *BMC Microbiol*, 2001. **1**(1): 34.

63. Mottet-Osman, G., *et al.*, Suppression of the Sendai virus M protein through a novel short interfering RNA approach inhibits viral particle production but does not affect viral RNA synthesis. *J Virol*, 2007. **81**(6): 2861–8.

64. Reuter, T., *et al.*, RNA interference with measles virus N, P, and L mRNAs efficiently prevents and with matrix protein mRNA enhances viral transcription. *J Virol*, 2006. **80**(12): 5951–7.

65. Timani, K.A., *et al.*, A single amino acid residue change in the P protein of parainfluenza virus 5 elevates viral gene expression. *J Virol*, 2008. **82**(18): 9123–33.

66. Sun, D., *et al.*, PLK1 down-regulates parainfluenza virus 5 gene expression. *PLoS Pathog*, 2009. **5**(7): e1000525.

67. Strahle, L., *et al.*, Sendai virus targets inflammatory responses, as well as the interferon-induced antiviral state, in a multifaceted manner. *J Virol*, 2003. **77**(14): 7903–13.

68. Le Mercier, P., *et al.*, Ambisense sendai viruses are inherently unstable but are useful to study viral RNA synthesis. *J Virol*, 2002. **76**(11): 5492–502.

69. Bitko, V., *et al.*, Cellular La protein shields nonsegmented negative-strand RNA viral leader RNA from RIG-I and enhances virus growth by diverse mechanisms. *J Virol*, 2008. **82**(16): 7977–87.

70. Weber, F., *et al.*, Double-stranded RNA is produced by positive-strand RNA viruses and DNA viruses but not in detectable amounts by negative-strand RNA viruses. *J Virol*, 2006. **80**(10): 5059–64.

71. Strahle, L., *et al.*, Activation of the beta interferon promoter by unnatural Sendai virus infection requires RIG-I and is inhibited by viral C proteins. *J Virol*, 2007. **81**(22): 12227–37.

72. Strahle, L., D. Garcin and D. Kolakofsky, Sendai virus defective-interfering genomes and the activation of interferon-beta. *Virology*, 2006. **351**(1): 101–11.

73. Lynch, S. and D. Kolakofsky, Ends of the RNA within Sendai virus defective interfering nucleocapsids are not free. *J Virol*, 1978. **28**(2): 584–9.

74. Takeuchi, K., *et al.*, Sendai virus C protein plays a role in restricting PKR activation by limiting the generation of intracellular double-stranded RNA. *J Virol*, 2008. **82**(20): 10102–10.

75. Wilde, A., C. McQuain and T. Morrison, Identification of the sequence content of four polycistronic transcripts synthesized in Newcastle disease virus infected cells. *Virus Res*, 1986. **5**(1): 77–95.

76. Gupta, K.C. and D.W. Kingsbury, Polytranscripts of Sendai virus do not contain intervening polyadenylate sequences. *Virology*, 1985. **141**(1): 102–9.

77. Masters, P.S. and C.E. Samuel, Detection of *in vivo* synthesis of polycistronic mRNAs of vesicular stomatitis virus. *Virology*, 1984. **134**(2): 277–86.

78. Cattaneo, R., *et al.*, Altered transcription of a defective measles virus genome derived from a diseased human brain. *EMBO J*, 1987. **6**(3): 681–8.

79. Re, G.G., E.M. Morgan and D.W. Kingsbury, Nucleotide sequences responsible for generation of internally deleted Sendai virus defective interfering genomes. *Virology*, 1985. **146**(1): 27–37.

80. Wiegand, M., S. Bossow and W.J. Neubert, Sendai virus trailer RNA simultaneously blocks two apoptosis-inducing mechanisms in a cell type-dependent manner. *J Gen Virol*, 2005. **86**(Pt 8): 2305–14.

81. Kurilla, M.G., *et al.*, Nucleotide sequence and host La protein interactions of rabies virus leader RNA. *J Virol*, 1984. **50**(3): 773–8.

82. Kurilla, M.G. and J.D. Keene, The leader RNA of vesicular stomatitis virus is bound by a cellular protein reactive with anti-La lupus antibodies. *Cell*, 1983. **34**(3): 837–45.

83. Raha, T., *et al.*, Leader RNA of Rinderpest virus binds specifically with cellular La protein: A possible role in virus replication. *Virus Res*, 2004. **104**(2): 101–9.

84. Wilusz, J., M.G. Kurilla and J.D. Keene, A host protein (La) binds to a unique species of minus-sense leader RNA during replication of vesicular stomatitis virus. *Proc Natl Acad Sci U S A*, 1983. **80**(19): 5827–31.

85. De, B.P., *et al.*, Specific interaction *in vitro* and *in vivo* of glyceraldehyde-3-phosphate dehydrogenase and LA protein with *cis*-acting RNAs of human parainfluenza virus type 3. *J Biol Chem*, 1996. **271**(40): 24728–35.

86. Bousquet-Antonelli, C. and J.M. Deragon, A comprehensive analysis of the La-motif protein superfamily. *RNA*, 2009. **15**(5): 750–64.

87. Teplova, M., *et al.*, Structural basis for recognition and sequestration of UUU(OH) 3′ temini of nascent RNA polymerase III transcripts by La, a rheumatic disease autoantigen. *Mol Cell*, 2006. **21**(1): 75–85.

88. Huang, Y., *et al.*, Separate RNA-binding surfaces on the multifunctional La protein mediate distinguishable activities in tRNA maturation. *Nat Struct Mol Biol*, 2006. **13**(7): 611–8.

89. Keene, J.D., RNA regulons: Coordination of post-transcriptional events. *Nat Rev Genet*, 2007. **8**(7): 533–43.

90. Fan, H., *et al.*, 5′ processing of tRNA precursors can be modulated by the human La antigen phosphoprotein. *Mol Cell Biol*, 1998. **18**(6): 3201–11.

91. Huhn, P., *et al.*, Characterization of the autoantigen La (SS-B) as a dsRNA unwinding enzyme. *Nucleic Acids Res*, 1997. **25**(2): 410–6.

92. Nallagatla, S.R., *et al.*, 5′-Triphosphate-dependent activation of PKR by RNAs with short stem-loops. *Science*, 2007. **318**(5855): 1455–8.

93. Nagai, Y. and A. Kato, Accessory genes of the paramyxoviridae, a large family of non-segmented negative-strand RNA viruses, as a focus of active investigation by reverse genetics. *Curr Top Microbiol Immunol*, 2004. **283**: 197–248.

94. Childs, K., *et al.*, Mda-5, but not RIG-I, is a common target for paramyxovirus V proteins. *Virology*, 2007. **359**(1): 190–200.

95. Parisien, J.P., *et al.*, A shared interface mediates paramyxovirus interference with antiviral RNA helicases MDA5 and LGP2. *J Virol*, 2009. **83**(14): 7252–60.

96. Ling, Z., K.C. Tran and M.N. Teng, Human respiratory syncytial virus nonstructural protein NS2 antagonizes the activation of beta interferon transcription by interacting with RIG-I. *J Virol*, 2009. **83**(8): 3734–42.

97. Atreya, P.L., M.E. Peeples and P.L. Collins, The NS1 protein of human respiratory syncytial virus is a potent inhibitor of minigenome transcription and RNA replication. *J Virol*, 1998. **72**(2): 1452–61.

98. Boxer, E.L., S.K. Nanda and M.D. Baron, The rinderpest virus non-structural C protein blocks the induction of type 1 interferon. Virology, 2009. **385**(1): 134–42.

99. Shaffer, J.A., W.J. Bellini and P.A. Rota, The C protein of measles virus inhibits the type I interferon response. *Virology*, 2003. **315**(2): 389–97.

100. Garcin, D., P. Latorre and D. Kolakofsky, Sendai virus C proteins counteract the interferon-mediated induction of an antiviral state. *J Virol*, 1999. **73**(8): 6559–65.

101. Gotoh, B., *et al.*, Knockout of the Sendai virus C gene eliminates the viral ability to prevent the interferon-alpha/beta-mediated responses. *FEBS Lett*, 1999. **459**(2): 205–10.

102. Fontana, J.M., *et al.*, Regulation of interferon signaling by the C and V proteins from attenuated and wild-type strains of measles virus. *Virology*, 2008. **374**(1): 71–81.

103. Nishie, T., K. Nagata and K. Takeuchi, The C protein of wild-type measles virus has the ability to shuttle between the nucleus and the cytoplasm. *Microbes Infect*, 2007. **9**(3): 344–54.

104. Malur, A.G., M.A. Hoffman and A.K. Banerjee, The human parainfluenza virus type 3 (HPIV 3) C protein inhibits viral transcription. *Virus Res*, 2004. **99**(2): 199–204.

105. Reutter, G.L., *et al.*, Mutations in the measles virus C protein that up regulate viral RNA synthesis. *Virology*, 2001. **285**(1): 100–9.

106. Baron, M.D. and T. Barrett, Rinderpest viruses lacking the C and V proteins show specific defects in growth and transcription of viral RNAs. *J Virol*, 2000. **74**(6): 2603–11.

107. Horikami, S.M., *et al.*, The Sendai virus C protein binds the L polymerase protein to inhibit viral RNA synthesis. *Virology*, 1997. **235**(2): 261–70.

108. Cadd, T., *et al.*, The Sendai paramyxovirus accessory C proteins inhibit viral genome amplification in a promoter-specific fashion. *J Virol*, 1996. **70**(8): 5067–74.

109. Curran, J., J.B. Marq and D. Kolakofsky, The Sendai virus nonstructural C proteins specifically inhibit viral mRNA synthesis. *Virology*, 1992. **189**(2): 647–56.

110. Nakatsu, Y., *et al.*, Translational inhibition and increased interferon induction in cells infected with C protein-deficient measles virus. *J Virol*, 2006. **80**(23): 11861–7.

111. McAllister, C.S., *et al.*, Mechanisms of protein kinase PKR-mediated amplification of beta interferon induction by C protein-deficient measles virus. *J Virol*, 2010. **84**(1): 380–6.

112. Toth, A.M., *et al.*, Protein kinase PKR mediates the apoptosis induction and growth restriction phenotypes of C protein-deficient measles virus. *J Virol*, 2009. **83**(2): 961–8.

113. Toth, A.M., *et al.*, RNA-specific adenosine deaminase ADAR1 suppresses measles virus-induced apoptosis and activation of protein kinase PKR. *J Biol Chem*, 2009. **284**(43): 29350–6.

114. Carlos, T.S., *et al.*, Parainfluenza virus 5 genomes are located in viral cytoplasmic bodies whilst the virus dismantles the interferon-induced antiviral state of cells. *J Gen Virol*, 2009. **90**(Pt 9): 2147–56.

Chapter 13

The Molecular and Cellular Biology of Emerging Bunyaviruses

John N. Barr*

1. Bunyavirus Classification

The *Bunyaviridae* family of segmented negative-stranded RNA viruses includes over 350 named species, classified into five genera namely *Orthobunyavirus*, *Hantavirus*, *Nairovirus*, *Phlebovirus* and *Tospovirus*. The defining features of these viruses are the possession of a three-segment RNA genome of negative or ambi-sense polarity, a spherical virion morphology with a diameter of between 90 and 120 nm and a replication strategy that occurs in the cytoplasm of infected cells. However, more recent molecular studies have revealed that in addition the bunyaviruses also share both fundamental mechanisms of RNA synthesis and common principles of virion architecture.

Classification of individual virus species into the five genera was originally based on the detection of serological markers and now is generally determined by multiple factors: the presence of genus-specific characteristics of individual proteins, the segment coding strategy used to encode these proteins and the presence of conserved sequence elements within the segment termini. In addition, the separate classification of the currently few members of the *Tospovirus* genera also depends on the fact that they are the only bunyaviruses able to infect plants. Prototypic members of each genus within the *Bunyaviridae* family are listed in Table 1, as well as selected members that hold particular significance in their pathogenicity, their economic importance, or their molecular and cellular biology. The reader is

---

*Institute of Molecular and Cellular Biology, Faculty of Biological Sciences, University of Leeds, Leeds, West Yorkshire LS2 9JT, UK. E-mail: j.n.barr@leeds.ac.uk

Table 1.   Prototypic and notable viruses of the family Bunyaviridae.

Genus	Current members*	Prototype	Notable members	Principle vectors
Orthobunyavirus	164	*Bunyamwera virus*	*La Crosse virus, California encephalitis virus, Jamestown Canyon virus, Oropouche virus*	Mosquitos, ticks, culicoid flies
Phlebovirus	37	*Rift Valley fever virus*	*Uukuniemi virus, Toscana virus, Punta Toro virus, Sandfly fever virus*	Phlebotomine flies, mosquitos
Nairovirus	34	*Dugbe virus*	*Crimean-Congo hemorrhagic fever virus, Hazara virus*	Ticks, mosquitos, culicoid flies
Hantavirus	42	*Hantaan virus*	*Sin Nombre virus, Puumala virus, Andes virus, Seoul virus*	Rodents
Tospovirus	8	*Tomato spotted wilt virus*	*Impatiens necrotic spot virus, Groundnut bud necrosis virus*	Thrips

*Current members as listed in the most recent publication of the International Committee on Taxonomy of Viruses (2006). The total number of named bunyaviruses, including those currently unclassified, exceeds 350.

directed to the website of the International Committee on Taxonomy of Viruses (ICTV) for all updates regarding bunyavirus classification.

2.   Bunyavirus Epidemiology and Pathogenesis

Bunyaviruses within the *Orthobunyavirus*, *Hantavirus*, *Nairovirus* and *Phlebovirus* genera are able to infect animals, including humans, often with fatal conse-quences. These viruses generally amplify in non-human vertebrates, although in some instances, humans sustain sufficient viremia also to act as amplifying reser-voirs, particularly during urban epidemics.[1,2] Reports of horizontal human-to-human transmission are rare, the only currently documented case being for the Andes hantavirus[3] although Crimean-Congo hemorrhagic fever virus (CCHFV) has been reported to be transmitted nosocominally, often during surgical pro-cedures on infected individuals.[4,5] Confirmed cases of vertical transmission are also rare,[6] and so in general, human infection appears to be a dead-end event.

With the exception of the hantaviruses, which are predominantly spread by rodents, bunyaviruses are transmitted by arthropods, including mosquitos, midges, sandflies and ticks. The plant-infecting tospoviruses are also transmitted by arthropods, predominantly thrips. Within these insect vectors, bunyaviruses are able to multiply and are transmitted both transovarially[7,8] and venereally.[9] However, it is clear from *in vivo* studies that the outcome of the bunyavirus infectious cycle in the insect vector is markedly different to that in the vertebrate host, being predominantly persistent rather than lytic, and with prolonged shedding of virus over several months following infection.

Hantaviruses maintain a long-term infection within the rodent reservoir that can persist for several years with no overt disease symptoms. Virus is shed in urine, feces and saliva, and contact with this material can transmit the virus horizontally to other rodents. Human infections are most often associated with contact with aerosolized rodent urine or feces, and result in a radically different disease outcome; hemorrhagic fever with renal syndrome or hantavirus pulmonary syndrome, the latter resulting in death in approximately 50% of infected individuals.[10]

The three-segment coding strategy of the bunyaviruses allows the opportunity for natural evolution through segment re-assortment, which has been detected both experimentally, and in several cases in nature.[11-14] This has led to generation of new viruses with altered pathogenesis and is a potent mechanism for disease emergence.

3. Bunyavirus Genome Structure

The three bunyavirus RNA segments are named small (S), medium (M), and large (L) reflecting their relative nucleotide length (Table 2). Viruses within each genus share similar overall segment length and a generally common expression strategy for their encoded protein products.[15] The organization of the three bunyavirus RNA segments is similar across all genera; each template strand possesses non-translated regions (NTRs) located at the 3′ and 5′ termini (Fig. 1) that surround a single open-reading frame (ORF). Reverse genetics analysis of members of the four animal-infecting bunyavirus genera has shown that these NTRs contain *cis*-acting signals involved in RNA synthesis and segment packaging.[16-25] The extreme terminal nucleotides within the NTRs are generally highly conserved within each genus; for example, all members of the *Orthobunyavirus* genus have the same 11 nucleotides at the 3′ and 5′ termini for all three of their segments (Table 2). This level of conservation indicates strong dependence on an invariant sequence for critical viral activities. The 3′ and 5′ NTRs of each segment display extensive nucleotide complementarity, which is often broken by a single conserved nucleotide mismatch, allowing the possibility that the NTRs may interact through canonical

Table 2. Conserved nucleotide sequence of the terminal 3′ and 5′ nucleotides of S, M and L segments of prototypic bunyaviruses and overall segment length.

Genus (prototype)	Consensus terminal sequences	Segment length (nts)		
		S	M	L
Orthobunyavirus (*Bunyamwera virus*)	3′-UCAUACCAUG... 5′-AGUAGUGUGC...	961	4458	6875
Phlebovirus (*Rift Valley fever virus*)	3′-UGUGUUUC... 5′-ACACAAAG...	1690	3885	6404
Nairovirus (*Dugbe virus*)	3′-AGAGUUUCU... 5′-UCUCAAAGA...	1712	4888	12 255
Hantavirus (*Hantaan virus*)	3′-AUCAUCAUCUG... 5′-UAGUAGUAUGC...	1696	3616	6533
Tospovirus (*Tomato spotted wilt virus*)	3′-UCUCGUUA... 5′-AGAGCAAU...	2916	4821	8897

Watson-Crick base pairing. In line with this, there is considerable evidence from both biochemical analyses and direct observation to suggest that bunyavirus RNA segments exist as circular molecules within infected cells and virus particles.[26–28] The finding that a high degree of nucleotide complementarity is common to all known bunyavirus segments suggests that NTR interaction is a conserved requirement of segment function.

While the respective NTRs of each segment possess strictly conserved nucleotides at their extreme termini, the internal nucleotides show much reduced conservation, indicative of more relaxed functional significance. These internal regions show considerable variation both between segments of the same virus and between members within the same genus, and these segment-specific sequences have been shown to play roles in the regulation of RNA synthesis for both *Bunyamwera virus* (BUNV) and *Rift Valley fever virus* (RVFV).[29,30] Analysis of recombinant BUNV bearing altered S segments shows that much of these segment-specific NTR sequences are dispensable for virus multiplication; just 22 nucleotides at the genomic 3′ end and 113 nucleotides at the 5′ end being absolutely required for infectious virus multiplication.[31] However, these internal NTR elements make a major contribution to virus fitness: In general, the segment-specific 3′ sequences were found to affect the balance of replication and transcription, whereas 5′ sequences predominantly affected overall virus growth, possibly due to effects relating to shut-off of host cell protein synthesis.

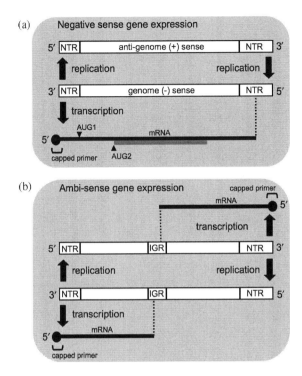

Fig. 1. Schematic representation of the negative sense and ambi-sense gene expression strategies of bunyavirus genome segments. All bunyavirus RNA segments include terminal non-translated regions (NTRs) that signal the viral RdRp to perform transcription and replication. RNA replication yields an exact complementary copy of the template, which is encapsidated in the virus-encoded N protein. (a) In the case of negative sense gene expression, only the genomic RNA possesses a transcription promoter, directing the synthesis of a single mRNA that is initiated using a capped oligoribonucleotide stolen from host cell mRNAs. This mRNA may include either one or two ORFs, accessed via alternative AUG initiation codons by scanning ribosomes. (b) In ambi-sense transcription, both replication products possess transcription promoters, and each yields a single mRNA product. The 3' end of these mRNAs is formed as a consequence of transcription termination signals within the intergenic region (IGR).

The possibility that additional signal elements are located outside of the segment NTRs cannot be ruled out. For BUNV, it has been shown that most segments in which the cognate pairings of NTR and ORF sequences have been rearranged cannot be rescued into viable infectious virus[32] suggesting some kind of interplay between the two regions. Only the MLM combination (in which the M NTRs flanked the L ORF) was recoverable, which was found to replicate to 100-fold reduced titres, and exhibited greatly increased particle to plaque-forming unit ratios due to reduced incorporation of the MLM segment within progeny virus particles. This finding suggested that the NTR/ORF boundaries might contain sequences involved in segment packaging, a scenario that has been firmly established for the eight-segment influenza virus.[33,34]

4. Bunyavirus Gene Products

4.1. *Proteins encoded by the S segment*

The S segment of all bunyaviruses encodes the nucleocapsid (N) protein, whose primary role is to encapsidate the viral RNA replication products to form the nucleocapsid as described below. Most members of the *Orthobunyavirus*, *Tospovirus* and *Phlebovirus* genera also encode a non-structural S (NSs) protein whose primary role is in modulating the host cell anti-viral response through diverse innate immunity pathways, described in more detail below. The N and NSs proteins of orthobunyaviruses are translated from the same mRNA encoded by the S segment genome, whereas the phlebovirus and tospovirus N and NSs proteins are translated from separate mRNAs transcribed from the genomic and anti-genomic strands, respectively (Fig. 1). While these coding strategies are generally conserved, and aid in bunyavirus classification, there are exceptions: Recent work has shown that three of the 18 serotypes within the *Orthobunyavirus* genus do not express an NSs protein, neither from overlapping reading frame nor as a separate mRNA.[35] In addition, several serotypes of New World and Northern Hemisphere hantaviruses also express an NSs protein, from an overlapping ORF on the N mRNA,[36] which is functional.[37] It is probable that additional exceptions to the established rules will emerge in the future.

The N protein is a highly abundant structural component of the infectious virion, and its primary role is to encapsidate the viral RNA to form the nucleocapsid, also known as the ribonucleoprotein (RNP) complex. As with all negative-stranded RNA viruses, the bunyavirus genome is only able to participate in either RNA synthesis or segment packaging when in the form of the RNP, and consequently the RNA encapsidation by the N protein is critical for virus viability. Formation of the RNP depends on the association of viral RNA with multiple copies of the N protein such that the RNA becomes encapsidated along its entire length, an activity that also depends on homotypic interactions between adjacent N protein monomers. The stoichiometry of the N:RNA in the nucleocapsid has been determined only for one bunyavirus, namely BUNV, in which each N protein monomer covers precisely 12 nucleotides.[38] This stoichiometry is distinct from all other negative-stranded RNA viruses studied to date and suggests that the number of N protein molecules associated with the BUNV S, M and L segments is 80, 372 and 565, respectively. BUNV N protein oligomerization results in monomers being stacked in 'head-to-head' and 'tail-to-tail' conformation,[39] and is able to form a range of different oligomeric states, consistent with its role in RNP formation.[39,40] While the primary role of the N protein is considered to be in providing structural uniformity to the RNA genome, it also possesses additional roles that are equally critical to the virus life cycle including interactions with membrane glycoproteins,[41–43] and interactions with the viral RNA-dependent RNA polymerase (RdRp) to allow access to the RNP during RNA synthesis. A recent analysis of BUNV N protein function has allowed a preliminary functional map to be constructed showing regions involved in oligomerization, polymerase binding and RNP assembly into virus particles.[40] It is apparent that the BUNV N protein is able to differentially affect replication and transcription activities of assembled RNPs, suggesting it plays a key role in allowing the correct recognition of the bunyavirus template by the RdRp,[40] although how it achieves this is presently unclear.

The recombinant hantavirus N protein forms predominantly trimers, and interacting domains have been mapped to both N-terminal and C-terminal regions.[44–48] In common with the BUNV N protein, homotypic oligomerization was proposed to involve interactions of adjacent monomers in 'head-to-head' and 'tail-to-tail' arrangements, and encapsidation was thought to proceed by successive addition of preformed trimers to the growing RNA strand.[48] NMR and X-ray crystallography data show that the hantavirus N-terminal region forms a coiled coil domain[49,50] and show that by itself, the N-terminal regions are insufficient for trimerization, implying important involvement of the C-terminal regions. The N protein trimer may not be a common assembly intermediate, as the RVFV N protein has been shown to predominantly dimerize,[51] whereas the BUNV N protein forms a variety of N protein multimers, including tetramers.[38,39] Multimerization of the *Tomato spotted wilt virus* (TSWV) N protein was proposed to involve a different head-to-tail dimer arrangement, with hydrophobic interacting domains restricted to the extreme terminal regions, although the subsequent identification of additional interacting residues within central regions of the N ORF suggests that dimer formation may be more complex.[52,53]

The only complete bunyavirus N protein for which high resolution structural information is available is RVFV, with the N protein structure being recently determined to a resolution of 1.93 Å.[54] The RVFV N protein was shown to consist of N-terminal and C-terminal helical lobes, separated by a short linker helix. Interestingly, the folds of both lobes appeared to be novel, with neither showing structural homology with any of the other negative-stranded RNA virus N proteins previously solved. Furthermore, while these other N proteins all possessed both a positively-charged RNA binding groove and also extended terminal loops involved in interacting with adjacent monomers, the RVFV N protein possessed neither of these features. These findings suggest that RVFV might utilize a novel strategy of RNP organization, and consistent with this conclusion, the RNPs harvested from infectious virus particles show little evidence of the compact helical RNPs that are characteristic of members of the order *Mononegavirales*. The structure also revealed the presence of a hydrophobic pocket at the interface between the two lobes, which was highly conserved among other phleboviruses, and was proposed to represent the site of the previously-described interaction with the viral glycoprotein Gn.[55]

For members of both *Hantavirus* and *Orthobunyavirus* genera, several studies have shown that their respective N proteins exhibit some degree of sequence specificity for the RNAs it encapsidates, and taken together, they indicate that while there is no obligatory encapsidation sequence, there is evidence for preferred binding for sequences or structural elements represented within the viral genome.[48,56–59] This has been interpreted as playing a contributing role in the exclusive packaging of viral replication products within virus particles, over either viral mRNAs or host RNAs in general.

In addition to its RNA encapsidation and multimerization abilities, the hantavirus N protein is also reported to possess additional functions relating to viral gene expression and also in manipulating the innate immune response within host cells as described below. Remarkably, the *Sin Nombre virus* N protein is thought to be able to functionally replace the entire cellular eIF4F complex, responsible for binding the mRNA 5′ cap, delivering the capped mRNA to the 43S pre-initiation complex, and acting as a helicase to unwind the mRNA during translation. The preferential association of N with 5′ capped RNAs is thought to occur in cellular processing (P) bodies, which store cellular mRNAs targeted for degradation.[60] The advantage to the virus of replacing eIF4F function is thought to be in allowing translation of viral mRNAs to be maintained despite efforts of the cell to shut down the translation machinery. How the hantavirus N protein manages to perform all these competing roles, in addition to its viral roles of RNA binding, formation of the helical RNP and association with the viral glycoproteins, is a remarkable feat of structural economy, especially given the size of the N protein, at just 48 kDa. It is also unknown how the hantavirus N protein manages to coordinate and regulate all these activities, which must presumably compete for interaction sites at various stages of the virus life cycle, although it has been suggested to depend in part on its oligomeric state.[61]

4.2. *Proteins encoded by the M segment*

The M segment of all members of the *Bunyaviridae* family encodes a polyprotein precursor (Fig. 2) that is co-translationally inserted into the membrane of the

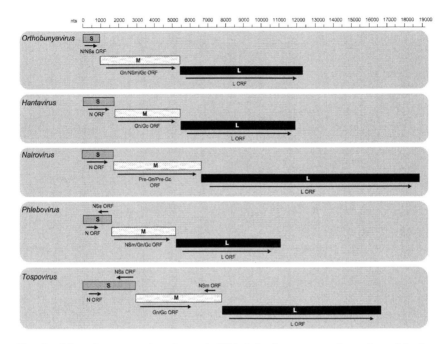

Fig. 2. Schematic representation of genomic RNAs belonging to prototypic members of the five genera classified within the *Bunyaviridae* family. All bunyaviruses possess three RNA segments named small (S), medium (M) and large (L). Arrows below each segment indicate ORFs expressed using a negative sense coding strategy, whereas arrows above the segments denote ORFs transcribed as mRNAs from positive sense templates. As described in the text, these schematics are generalizations, and several exceptions to these coding strategies have been identified.

endoplasmic reticulum, where it is cleaved into Gn and Gc components by host cell proteases. These cleaved polypeptides form a di-sulphide bond linked heterodimer, which is also modified by carbohydrate addition, and is transported to, and retained in the Golgi apparatus. The Gn/Gc heterodimer performs critical roles in mediating virus assembly, formation of the virus particle and attachment to new target cells, as described below.

The Gn/Gc heterodimer is retained in the Golgi by the virtue of a retention signal, where it associates with RNPs to mediate assembly and budding of mature virus particles. For most bunyavirus Gn/Gc heterodimers, the Golgi retention signal is located within the Gn component, and for BUNV has been precisely mapped to within its transmembrane domain (TMD).[62] In contrast, the Golgi retention signal for the phlebovirus *Uukuniemi virus* (UUKV) resides within the C-terminal tail,[63] whereas for *Punta Toro virus* and RVFV, the signal includes residues from both the TMD and the C-terminal tail.[64,65] The Golgi retention signal for the tospovirus TSWV Gn/Gc is also within the Gn cytoplasmic domain and has been mapped to a 10 residue region proximal to, but not including, the TMD.[66] In another case, the Golgi targeting signal for CCHFV nairovirus was found to include residues localized within the Gn ectodomain.[67,68] Interestingly, Golgi retention of the Hantaan hantavirus Gn/Gc heterodimer was proposed to depend on its correct overall conformation, rather than on the possession of a discrete amino acid sequence,[69] and similar results have been described for Sin Nombre and Andes hantaviruses.[70]

Most members of the *Orthobunyavirus, Phlebovirus* and *Tospovirus* genera also encode an NSm protein (Fig. 2), and recent evidence suggests that CCHFV nairovirus also generates an NSm protein through novel proteolytic processing.[71,72] The tospovirus NSm protein is translated from a separate mRNA encoded by the anti-genome, whereas the orthobunyavirus and phlebovirus NSm is cleaved by cellular proteases from the same polyprotein precursor that yields the Gn and Gc proteins. The BUNV NSm protein is thought to play a role in virus assembly[73] and consistent with this role, NSm is found within specific Golgi-associated tubular structures within infected cells where virus morphogenesis takes place.[74] In contrast, the NSm protein of TSWV is the putative movement protein, involved in inter-cell virus transmission within the infected plant,[75] whereas the phlebovirus NSm has been shown to be non-essential[76] but may play accessory roles in the regulation of cellular apoptosis.[77]

4.3. *Proteins encoded by the L segment*

For all bunyaviruses, the L segment encodes the viral component of the RdRp, which is the sole protein product of this segment (Fig. 2). As suggested by their name, the bunyavirus L proteins are large molecules in excess of 200 kDa that perform several complex functions that together result in the generation of RNA replication and mRNA transcription products from their respective viral templates. Comparison and alignment of the bunyavirus RdRps with all other viral RdRps reveals that they share the well-characterized polymerase module (region 3), including the defined motifs known as pre-A, A, B, C and D.[78,79] An additional motif-specific for the segmented negative-stranded RNA virus RdRps was also identified within this region, namely region E,[78] in the N-terminal part of the polymerase module. Furthermore, bunyavirus L proteins were shown to possess conserved N-terminal regions 1 and 2 that are also featured in the L proteins of the closely-related arenaviruses, as well as region 4 that may also be specific to segmented negative-stranded RNA viruses.[78,80] The location of domains involved in bunyavirus transcription activities such as cap binding and endonuclease cleavage of host mRNAs is currently undefined, despite being well characterized in the hetero-oligomeric RdRp of the related orthomyxovirus, *Influenza A virus*.[81–84] However, conserved motif E within region 3,[78] as well as the bunyavirus-specific region 4, has been implicated in these roles.[80]

Consistent with the concept that the bunyavirus RdRp is composed of independent functional modules, the BUNV RdRp was shown to maintain its function following insertion of epitope tags at two locations within its ORF, in common with the RdRp of several other non-segmented negative-stranded RNA viruses.[85–87] These altered polymerases were rescued into infectious viruses with little change in virus growth in BHK-21 cells,[88] although titres were considerably reduced in both Vero E6 and insect cells, suggesting a cell-type dependence that may depend on interaction of the polymerase with cell-specific components.

The RdRps belonging to members of the *Tospovirus* and *Nairovirus* genera are considerably larger than those of other bunyaviruses, with predicted molecular weights of approximately 330 kDa for the tospoviruses TSWV and *Impatiens necrotic spot virus* and approximately 450 kDa for the nairoviruses *Dugbe virus* and CCHFV.[89–93] Sequence alignment shows that these RdRps share the familiar pre-A through D motif structure of the polymerase module of other RdRps, and motif E specifically found in segmented negative-stranded RNA viruses. However, the nairovirus L proteins have acquired additional sequences predominantly at their N-termini that perform as yet uncharacterized functions. In the case of CCHFV, the additional sequences show strong similarity to proteases within the ovarian tumor (OTU) super-family, along with regions showing homology with known transcription factor, gyrase, helicase and topoisomerase domains.[90,93] It has been proposed that the L protein may represent a polyprotein that is autoproteolytically processed by the OTU-like protease domain to yield a polymerase and a range of accessory factors. The roles played by these additional domains in the virus life cycle are unknown, although recent evidence suggests that they are dispensable for RNA polymerization functions in the context of a model RNA segment.[16] Whether the same holds true for infectious virus remains to be tested.

Whether the virally-encoded L protein is the sole component of the active RdRp is currently a poorly explored concept, although it has been recently established that the RdRp of TSWV requires a host cell factor for replication but not transcription activity in its primary insect vector, the western flower thrip *Frankliniella occidentalis*.[94] Remarkably, expression of this cell component in normally non-permissive human cells made them able to support virus propagation.

The RdRp of RVFV has also been suggested to form biologically-active oligomers,[95–99] consistent with the idea that viral RdRps congregate within specialized compartments to form virus factories, described below.

5. Bunyavirus RNA Synthesis

As with all other negative-stranded RNA viruses, each bunyavirus segment acts as the template for two different RNA synthesis activities; mRNA transcription and RNA replication (Fig. 1). Genomic and anti-genomic replication products are full-length encapsidated complementary copies of each other, whereas transcription products are extended at their 5′ ends by a capped oligoribonucleotide stolen from host cell mRNAs and are generally truncated at their 3′ ends relative to the genome template (Fig. 1). Each bunyavirus RNA strand encodes a maximum of one transcriptional unit. This is different to the non-segmented RNA viruses, which possess multiple transcription units, and these differences are likely consequences of fundamental properties of the respective RdRps: The bunyavirus RdRp can only initiate mRNA synthesis at the segment termini, whereas in the non-segmented RNA viruses, transcription can be signaled by discrete linear signals at internal sites throughout the genome. The reason for this is unknown, but may reflect a requirement for proximity of 3′ and 5′ terminal sequences that together build-up the promoter required for transcription initiation. An important functional consequence of this is that the bunyavirus RdRp cannot reinitiate RNA synthesis following termination of a prior strand. It maybe that the ambi-sense arrangement of ORFs on RNA segments of phlebo-, tospo- and nairoviruses is a means of increasing coding capacity of each segment within the functional confines of an RdRp that cannot re-initiate.

The currently accepted model describing bunyavirus RNA synthesis proposes that upon entry to an infected cell, the template-associated RdRp performs primary transcription on the input RNA strand yielding 5′ capped mRNAs. Bunyavirus transcription is distinct from replication in that the mRNAs are not initiated *de novo*, but instead rely on primers generated from capped host cell mRNAs, in a mechanism that appears similar to that used by *Influenza virus*, which snatches caps in the nucleus. There is a direct evidence to suggest that the capped mRNAs are selected, possibly determined by complementarity of as little as one nucleotide between the mRNA and the 3′ template sequence utilizing a 'prime-and-realign' mechanism.[96–99] This involves annealing of the capped primer at an internal location at the 3′ NTR within a repeated tri-nucleotide sequence (3′-AUCAUCAU-5′). Following limited polymerization, the nascent strand slips backwards to realign with the template by virtue of the 3′ proximal triplet repeat. The position of RdRp-mediated endonuclease cleavage determines the length of the 5′ capped extension, which are genus-specific and typically between 10 and 21 nucleotides in length.

As primary transcription occurs on input segments prior to the first round of replication, expression of genes on ambi-sense segments would be predicted to be restricted to only the gene in the negative sense orientation. However, in the case of RVFV, both the ambi-sense N and NSs genes are transcribed during initial rounds of primary transcription due to incorporation of complementary copies of the three RVFV segments within infecting virus particles.[100]

At some point in time following the onset of primary transcription, the RdRp is able to replicate the input template. The 5′ of the replication product is fundamentally different from that of mRNAs, which implies that they are the products of fundamentally different pathways that differ at their onset. The reason for this apparent change in template activity is poorly characterized and may reflect a switch in RdRp function through either polymerase modification, association with host cell components, or alternatively may reflect increased stability of nascent replication products through encapsidation.

5.1. *Signals for initiation of transcription and replication*

As described above, functional analyses have shown that the sequence signals responsible for directing the bunyavirus RdRp to generate transcription and replication products are contained within the NTRs. For the phlebovirus UUKV, nucleotides responsible for reporter gene transcription were identified, and implied the existence of two functionally important modules linked by a hinge region.[21] For BUNV, nucleotides that are composed of the individual transcription and replication promoters have been mapped, and shown to comprise distinct sets of nucleotides at both ends of the template strands.[29,101–104] These nucleotides are arranged such that the NTRs of the genomic strand possess signals for both replication and transcription, whereas the anti-genomic NTRs only possess the replication signal. Transfer of transcription activity to the anti-genomic strand was possible, along with the ability to perform ambi-sense transcription. This suggests that the ambi-sense transcription strategy is solely dependent on positioning of a transcription promoter within both genomic and anti-genomic strands, and would predict that bunyavirus segments that perform ambi-sense transcription would naturally have this promoter arrangement. For both UUK and BUNV, overall RNA synthesis activity depends on the ability of the 3′ and 5′ NTRs to interact through the base-pairing potential of their complementary termini.[20,104] In the case of BUNV, the formation of best-replicating model segments depended on inter-terminal complementarity with no apparent sequence specificity, and in addition, the BUNV transcription promoter was found to comprise nucleotides located at both ends of the RNA template.[101,103,104] While S, M and L segments of each virus share common conserved nucleotides, they also exhibit considerable sequence differences, and these segment-specific sequences have been shown to play important roles in

RNA synthesis activities. For both BUNV[29] and UUKV,[20] the evidence suggests that the M segment has the highest activity, whereas for RVFV, S segment activities were highest.[30] The extent of nucleotide complementarity was shown to be important for RNA synthesis activity, but did not always directly correlate with overall levels of RNA products, and indeed in some instances nucleotide changes that increased complementarity decreased replication ability. Taken together, the available evidence indicates that the bunyavirus RNA synthesis promoters are composed of sequence-independent structural elements formed by inter-terminal base pairing, and also the specific identity of both paired and unpaired nucleotides. As described above, this conclusion is supported by both biochemical analysis and direct observation of bunyavirus segments.[26–28]

5.2. *Transcription termination*

Analysis of the 3' ends of transcription products generated from S, M and L segments of a range of bunyaviruses has revealed that in many cases the 3' mRNA ends are consistently truncated in relation to the corresponding templates, suggesting the presence of specific transcription termination signals; the S mRNAs are truncated by approximately 100 nts, whereas M and L mRNAs are truncated by about 40 nts.[105–109] Mutagenic analysis of the transcription termination ability of the BUNV S segment 5' NTR revealed the presence of a 33 nucleotide signal that included a critical hexanucleotide 3'-GUCGAC-5' and that mapped closely to the S mRNAs 3' end.[110] Consistent with the previously mapped 3' end of L mRNAs, a related sequence was also identified in the corresponding 5' NTR. In contrast, no such signal was identified in the M segment, raising the possibility that the M mRNA 3' end was formed by RdRp run-off.

In the case of the phlebovirus mRNAs transcribed from ambi-sense S and M segments, the 3' ends map to locations in between the segment ORFs, called intergenic regions. For RVFV, mutagenic analysis of the intergenic region identified a conserved sequence $3'-C_{1-3}GUCG-5'$ involved in transcription termination activity.[111] This sequence is shared by other phlebovirus ambi-sense segments and interestingly, is also related to that identified for BUNV transcription termination, despite these viruses being classified into different genera.

The lack of poly(A) tails at these mRNA 3' termini implied that the termination mechanism was likely different to that employed by non-segmented negative-stranded RNA viruses, which involve polymerase slippage on poly-(U) tracts,[112] although the Sin Nombre hantavirus M segment mRNA may be an exception to this.[113] There is some evidence to suggest that the termination mechanism may involve the secondary structure in either the template or the nascent strand,[114,115] and in support of this possibility, the BUNV termination signal includes a potential stem-loop structure,[110] and secondary structure elements within the intergenic regions of related arenavirus segments are required for 3′ end formation of corresponding mRNAs.[116,117] The general lack of poly(A) tails on bunyavirus mRNAs would be predicted to impact viral protein expression within the cytoplasm of infected cells, given the important role of this element in translation enhancement and mRNA stability. In the case of BUNV, it has been suggested that sequences towards the 3′ end of the newly terminated mRNA may contain signals that functionally replace the poly(A) tail. Mutagenesis of this 3′ proximal region indicates that the enhancing element may include a strong stem-loop structure[118] that interestingly is conserved within the orthobunyaviruses,[110] and a similar structure has also been identified within the 3′ region of tospovirus mRNAs.[119] Whether this structure enhances translation through interaction with a cellular or viral protein is unknown. Intriguingly, the BUNV 3′ stem loop displays both structural and sequence similarities to the stem loop present at the 3′ end of histone mRNAs that are also poly(A) deficient,[120] raising the possibility that BUNV may have hijacked a cellular poly(A)-independent translation process for its own ends. The strategy of adopting a translation enhancement mechanism that is distinct from the cellular process may also allow the virus to escape host cell translation shut-off.

A peculiarity of orthobunyavirus transcription is that it requires concurrent protein synthesis, and this requirement is known to be due to the presence of translocating ribosomes on nascent transcripts, rather than any protein product.[114,115] In support of the involvement of secondary structure in bunyavirus transcription termination as described above, the role of the translocating ribosome is thought to be in preventing the RdRp from recognizing spurious transcription termination signals within the bunyavirus template, through disruption of secondary structural elements in the nascent or template strands, or possibly between these RNA molecules. Interestingly, this coupling of transcription and translation is not required in insect cells,[121] a phenomenon that can be reproduced *in vitro*, using ribosomes supplied in reticulocyte lysates. Exploitation of this methodology showed that the cell-type dependence of the translational requirement was due to a host cell factor found in mammalian cells, which possibly facilitated RNA interactions. Interestingly, *in vitro* transcription from tospovirus RNPs also shows a requirement for reticulocyte lysates, although is resistant to inhibitors that prevent ribosome translocation,[122] hinting that tospovirus transcription requires a host cell factor other than translocating 40S subunits.

Less is known about the RNA synthesis activities of RNPs from either *Hantavirus*, *Nairovirus* or *Tospovirus* genera, which stems from the lack of efficient reverse genetics systems for any of their members, although a system for efficient generation and an analysis of CCHFV model genomes have recently been reported.[16]

6. Bunyavirus Assembly

6.1. *Bunyavirus assembly and budding*

Bunyaviruses are distinct from most other negative-stranded RNA viruses in that they are generally thought to assemble and bud from peri-nuclear viral factories that are built around the Golgi complex.[123–128] This may be driven by the specific retention of the Gn/Gc heterodimer within the Golgi compartment as described above, where the heterodimer accumulates with both Gn and Gc cytoplasmic tails likely protruding through the Golgi membrane. Both tails are required for assembly, and interact directly with the N protein component of viral RNPs[41,43,129,130] that are generated in large numbers by replication complexes within tubular structures surrounding the Golgi complex.[74,127,131] These structures are connected to both mitochondria and the rough endoplasmic reticulum, and are composed of both cellular and viral components, most notably NSm (Fig. 3). Details of how the RNPs are transported to the budding compartment are currently obscure, though the involvement of cytoskeletal components has been suggested to play a role in the assembly of several bunyaviruses, particularly in transport of the N protein.[74,132–135] The association between the RNP and the Gn/Gc heterodimer brings together all the major structural proteins of the virus particle in one compact location within a budding compartment, depicted in Fig. 3. This assembly pathway is generally well accepted for most bunyaviruses; however, there are exceptions: For example, Black Creek Canal hantavirus has been visualized budding at the plasma membrane of polarized epithelial cells[136] and a similar finding was reported for Sin Nombre hantavirus,[137] raising the possibility that hantaviruses in general may utilize alternative assembly pathways. In addition, the RVFV phlebovirus has been shown to bud from the plasma membrane in primary liver cells, which raises the interesting and important possibility that assembly may be cell-type dependant. The site of nairovirus assembly and budding is poorly characterized, though the retention of CCHFV glycoproteins in the Golgi complex strongly implicates this as the assembly site.[67,68]

6.2. *Bunyavirus morphogenesis and structure of the infectious particle*

Bunyavirus morphogenesis has been best studied for BUNV, for which generation of infectious virions is thought to involve the transition of virus particles through distinct structural stages, and currently three forms have been identified.[127] The initial forms are annular structures that are thought to represent immature precursors and are named type 1 intracellular viruses. Also detected within infected

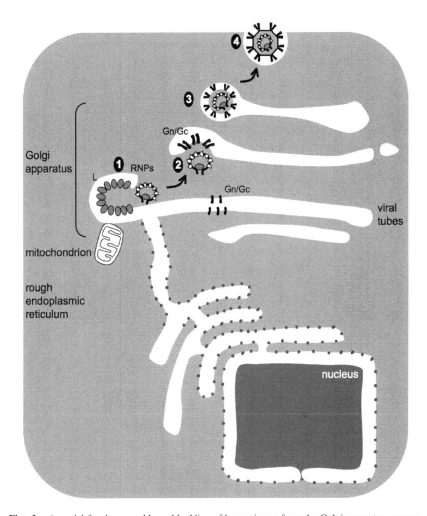

Fig. 3. A model for the assembly and budding of bunyaviruses from the Golgi apparatus, summarizing work performed in the laboratory of Cristina Risco and adapted from Refs. 74, 127 and 131. Viral tubes composing of both cellular (Golgi actin) and viral (NSm) components form in association with Golgi stacks and are linked to both mitochondria and the rough endoplasmic reticulum. (1) The globular domain of the tube forms a protected site of viral polymerase (L) activity to generate viral mRNA transcripts and newly-replicated S, M and L RNP complexes. (2) RNPs associate with Gn/Gc glycoprotein heterodimers that accumulate in Golgi stacks due to a retention signal on the Gn moiety. (3) Gn/Gc heterodimers that pass through the *trans*-Golgi compartment are modified by the addition of N-linked carbohydrates, subsequently allowing conformational changes and transition between distinct virus morphologies. (4) Virus particles bud from the Golgi apparatus and upon release from the cell, display further morphological changes resulting in full infectivity.

cells are more dense particles, named type 2 intracellular particles, which possess a more angular and structured envelope. The third form is the infectious extracellular particle, which also has an angular external contour, but also exhibits closely packed surface spike projections that hint at a potentially icosahedral symmetry. Transition between the forms was thought to involve modification to the Gn/Gc heterodimer, and specifically, transition between type 1 and type 2 intracellular forms was thought to involve the acquisition of endo-H resistant carbohydrate in the *trans*-Golgi compartment. This modification was suggested to be required for correct heterodimer folding and also for promoting crucial lateral interactions between adjacent heterodimers within the envelope. The structural change that allows transition of type 2 viruses into the fully-infectious extracellular form is less well characterized, but is though to occur on exit from the cell.

Recent work has provided insight into the detailed three-dimensional structure of infectious bunyaviruses, through direct visualization of purified extracellular viruses using cryo electron tomography. For UUKV,[138] this work revealed that mature virus particles exhibited striking $T = 12$ icosahedral symmetry, due to the arrangement of the external glycoprotein spikes. Interestingly, these spikes exhibited pH-dependant conformations, being pointed at pH 7 and more flattened at pH 6, and these changes were thought to relate to changes in the Gc moiety and cause conformational changes to the heterodimer that would facilitate virus attachment. The internal RNP was found to partially interact with the viral membrane, but to be otherwise relatively unorganized, and suggested that lateral contacts between glycoprotein spikes likely played a major role in determining virion morphology. The lack of RNP organization is in contrast to *Influenza virus*, which exhibits an RNP array in which seven segments consistently surround an eighth.[139,140]

For RVFV, another phlebovirus, the virus particles were also shown to exhibit $T = 12$ symmetry,[141,142] and as for UUKV this was due to the ordered arrangement of the glycoprotein heterodimer. Icosahedral averaging revealed the particle composed of 12 pentamers and 110 hexamers, each forming a hollow cylinder and likely composing of 5 and 6 Gn/Gc heterodimers, respectively. Whether the observed regularity of the phlebovirus virus particle is a conserved feature of other bunyaviruses remains to be determined.

7. Bunyavirus Evasion of the Host Cell Anti-Viral Response

7.1. *Induction of interferon gene expression*

One of the initial lines of host defense against viral infection is the interferon system, which is induced by a variety of non-self components *via* interactions with a range of host cell-encoded pattern recognition receptors that detect pathogen-associated molecular patterns (PAMPs). The main PAMP of negative-stranded RNA viruses such as the bunyaviruses is thought to be either double-stranded RNA or un-capped 5′ termini of single-stranded RNA molecules (Fig. 4). The subsequent activation of transcription factors including NF-κB and IRF-3 leads to cooperative stimulation of IFN-β gene expression, and the subsequent expression of many anti-viral gene products including the Mx GTPases and protein kinase R (PKR). The MxA protein has been shown to inhibit the replication of orthobunyaviruses, hantaviruses, phleboviruses and nairoviruses through directly associating with bunyavirus nucleocapsids, rapidly following infection.[143–145] This association has been proposed to sequester critical components away from their intended location, thereby disrupting viral multiplication. In contrast, PKR activation leads to phosphorylation of translation initiation factor eIF-2, and a general down regulation of both viral and host cell translation.

7.2. *Suppression of IFN expression*

The primary role of the orthobunyavirus and phlebovirus NSs proteins is to act as potent inhibitors of the host cell anti-viral response,[146–148] and recent work suggests that this may also be the case for the NSs protein expressed from new world hantaviruses.[37] In the case of the orthobunyavirus BUNV, NSs-mediated IFN antagonism was shown to act at the stage of IFN transcription, achieved by inhibiting specific phosphorylation of serine-2 in the characteristic heptapeptide repeat (YSPTSPS) of the RNA polymerase-II C-terminal domain (Fig. 4), which non-specifically down-regulates host cell mRNA synthesis.[149] Critical in this inhibition is the interaction of a C-terminal domain of NSs with the MED8 component of the Mediator complex,[150] which plays an important role in the regulation of RNA polymerase II activity. Interestingly, more recent evidence suggests that an additional N-terminal domain within NSs may also be important for overcoming the host cell anti-viral response, possibly using a different mechanism.[151] The NSs protein of the closely related *La Crosse virus* (LACV) is similarly proficient at blocking the IFN response, and investigation of NSs effects in arthropod cells indicates that the expression of NSs is redundant in terms of establishing a persistent infection. The main role of NSs within the dual-host LACV life cycle was concluded to be in blocking the effects of IFN in the context of animal infections.[146]

A similar strategy of global host cell transcription inhibition has also been elucidated for the NSs protein of RVFV; however, the mechanism of action appears quite different. Within infected cells, the RVFV NSs protein is found in the nucleus within large filamentous structures that co-localize with the p44 and XPB subunits of

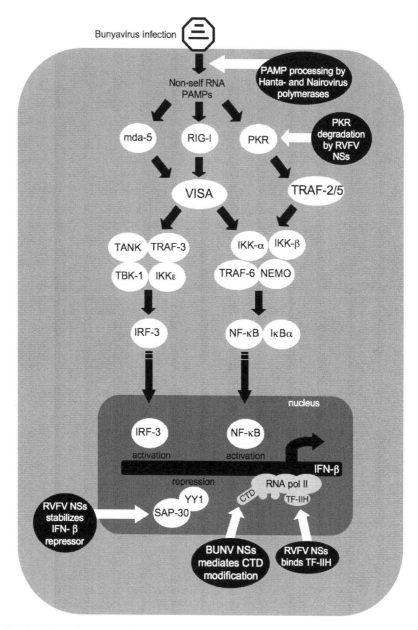

Fig. 4. Schematic summary of the major components involved in activation of the interferon-β promoter during the host cell anti-viral response. Currently identified steps within these pathways that are targeted by bunyavirus gene products are shown in black.

TFIIH, preventing the formation of the complete functional TFIIH complex (Fig. 4). More recently, the RVFV NSs protein has also been shown to specifically repress IFN-β expression through association with SAP-30, YY1 and Sin3A-associated cofactors within the nuclear filaments. Chromatin immunoprecipitation experiments showed that these components are recruited directly to the IFN-β promoter, inhibiting histone modification and thus preventing transcriptional activation.[152] It has been suggested that the NSs-associated filaments may represent nuclear compartments that lead to suppression of the specific subset of genes whose promoters interact with SAP-30 or YY1. In addition to this global transcriptional repression, the RVFV NSs protein also targets the anti-viral response by specifically degrading PKR (Fig. 4), and it has been suggested that this additional function may correspond with the enhanced pathogenicity of the wild-type RVFV.[153] The fact that these two non-structural proteins that share virtually no sequence homology, are disparate in size, are expressed through fundamentally different coding strategies and yet have very similar functions is remarkable.

An additional strategy to evade innate cellular immunity is demonstrated by members of the *Hantavirus* and *Nairovirus* genera, which is to trim the 5′ triphosphate moiety from protruding 5′ termini of genomic RNA strands, thus avoiding detection by the RNA helicase RIG-I.[154] The mechanism responsible for generating these novel RNA termini is likely "prime and re-align",[97,99] mediated by the viral polymerase during RNA replication, which is described above in more detail. It is interesting to note that all of the nairoviruses and many of the hantaviruses do not make an NSs protein. Perhaps genome end modification has arisen as an alternative strategy to evade cellular innate immunity in the absence of this additional coding capacity.

7.3. *RNA interference*

The NSs protein of members of the plant-infecting *Tospovirus* genus is also involved in disrupting the host cell anti-viral response, but instead of targeting IFN that plant cells are unable to express, the tospovirus NSs protein targets the RNA-interference (RNAi) mechanism that represents a powerful hurdle to pathogen infiltration. TSWV infection of plants harboring a silenced GFP transgene was able to effectively reverse RNAi-mediated silencing of GFP expression, and transient over-expression of only the NSs protein was able to reproduce this effect.[155,156] The mechanism by which NSs achieves this effect within the plant cell is currently unknown, as is the mechanism of its observed RNAi suppression within the thrip vector.[157]

While mammalian cells have an active RNAi system, it is widely considered that its effect is masked by the more powerful and wide-ranging IFN system. In contrast, within insect cells the RNAi circuit plays an important role in pathogen immunity.[158] Comparisons of NSs sequences of TSWV and LACV show a considerable degree of amino acid similarity, prompting the possibility that the NSs protein of vertebrate-infecting bunyaviruses may also possess anti-RNAi activity. Studies to investigate this possibility appear conflicting, with transient over-expression of LACV NSs protein appearing to disrupt RNAi signaling in mammalian cells,[159] whereas in the context of infectious viruses generated by reverse genetics, the LACV NSs protein was shown to possess little or no effect on the RNAi circuit in neither mammalian nor insect cells.[146]

7.4. *Apoptosis*

Apoptosis or programmed cell death is an evolutionarily conserved mechanism that rapidly leads to destruction of the cell in a manner that is distinct from necrosis. In the context of a virus infection, induction of apoptosis early in the infectious cycle can limit virus assembly, and thus minimizes virus release and spread throughout neighboring cells. However, for viruses with rapid multiplication cycles, induction of apoptosis may favour the release and spread of progeny particles, while also invoking reduced immunoinflammatory reactions. Thus, it may be that the timing of apoptosis in relation to virus assembly is critical in determining the outcome of infection, and it is not unreasonable to assume that viruses have acquired mechanisms to tip this balance in their favour. There is evidence to suggest that virus infection induces both anti- and pro-apoptotic pathways, and the balance between their progression can determine whether the outcome of infection is either acute or persistent,[160] with important consequences for pathogenesis.

Apoptosis is triggered by a diverse range of stimuli and involves well-characterized extrinsic and intrinsic pathways that lead to activation of a class of cysteine proteases called caspases.[161–163] These pathways are tightly regulated by checkpoint proteins, which possess anti-apoptotic activities that are able to halt further progression of the cascade. If unchecked, these pathways lead to appearance of classical apoptosis characteristics including DNA fragmentation, membrane alterations, cleavage of a range of specific cellular substrates and ultimately cell death.

Several bunyaviruses have been implicated in the induction of apoptosis within a variety of cell types and small animal models. The first bunyavirus reported to modulate apoptosis was the orthobunyavirus member LACV, which was shown to induce apoptosis in some mouse neuronal cell lines and infected animals.[164]

More recent work showed that the LACV NSs protein-induced apoptosis by reversing the activity of the checkpoint protein Hsc70 by binding to an activating protein, Scythe.[165] Interestingly, the LACV NSs protein shows sequence similarity to the baculovirus Reaper protein, which has also been shown to inhibit Hsc70 anti-apoptotic activity using a similar Scythe-mediated mechanism. The orthobunyavirus BUNV also induces apoptosis; however, deletion of the NSs protein from the BUNV genome advances the appearance of the apoptotic response,[166] suggesting that NSs has anti-apoptotic effects early in the infectious cycle. By analysing the effect of BUNV infection on well-characterized effector molecules involved in apoptosis signaling, it was shown that both WT and NSs-deleted viruses activated IRF-3, which induces apoptosis by driving transcription of the IFN α/β genes. This finding established IRF-3 as an important component of BUNV-induced apoptosis, and suggested that the anti-apoptotic effect of NSs was mediated downstream of IRF-3 activation. IRF-3 activation has since been shown to be one of the key effector molecules involved in apoptosis induction for many negative-stranded viruses, and for *Sendai virus* is thought to play a key role in determining the acute or persistent outcome of infection.[160,167] This finding may have important consequences for the bunyaviruses that display alternative lytic/persistent disease outcomes in their vertebrate or insect host cells.

Several members of the *Hantavirus* genus including *Hantaan* and *Prospect Hill viruses*,[168] *Andes* and *Seoul viruses*,[169] *Tula*[170,171] and *Oropouche viruses*[172] have also been shown to induce apoptosis in a variety of cell types including Vero E6 and human embryonic kidney (HEK) cells, and also in *Puumala virus* infected patients.[173] There is some conflicting information surrounding the ability of *Hantaan virus* to induce apoptosis, and this may be due to cell culture-specific effects that can significantly affect the outcome of these experiments.[170]

The apoptotic effect of *Hantaan virus* infection was induced in Vero E6 cells relatively late following infection, required virus gene expression, and was accompanied by a reduction of levels of the pro-survival protein Bcl-2.[168] In common with *Hantaan virus*, *Oropouche virus* (OROV) was shown to induce apoptosis in HeLa cells in a manner that depended on protein expression by live replicating virus. This suggested that the apoptotic effect was performed by a newly synthesized, nonstructural viral component, and the observation that the OROV NSs protein possessed considerable sequence identity with the reaper protein makes it an attractive candidate for this role.[172] Induction of apoptosis by Tula hantavirus also requires virus replication and was accompanied by up-regulation of TNF-α rather than IFN-α pathways and the induction of caspase-8 pathways,[171] a finding in agreement with previous studies of *Hantaan virus* in HEK293 cells.[169] Hantavirus replication was found to induce up-regulated expression of ER-stress markers such as Grp78/BiP, with the subsequent activation of multiple death pathways leading to apoptosis.[174] These authors suggested that mis-folding of the membrane glycoprotein within the ER was responsible for activating this effect. Interestingly, several other reports have implicated the hantavirus nucleocapsid protein as playing a major role in apoptosis signaling. The *Hantaan virus* N protein causes retention of NF-κB in the cytoplasm of transfected cells,[175] possibly through direct interaction, an effect that was reversed with several N protein mutants.[61] As NF-κB trafficking to the nucleus is one way that leads to transcriptional activation of apoptosis, it was proposed that the N protein expressed during hantavirus infection is able to down-regulate the induction of apoptosis and thus represents an important pathogenicity factor. The interaction between the *Puumala virus* N protein and the Daxx apoptosis enhancer indicates that the involvement of the hantavirus N protein in apoptosis signaling may be a widespread phenomenon.

8. Concluding Remarks

The advent of reverse genetics systems capable of generating infectious bunyaviruses has allowed a rapid expansion of understanding all stages of the bunyavirus life cycle. While the availability of such systems has principally benefited the *Orthobunyavirus* and *Phlebovirus* genera, no doubt future successes in developing efficient rescue systems for the remaining three bunyavirus genera will provide the means for similar advances. Combining these reverse genetics techniques with high-resolution structural information, recent advances in electron microscopy and increased understanding of host cell biology will pave the way for an exciting period of research that will provide further detailed insights into the molecular and cellular biology of this important group of pathogens.

References

1. Guelmino, D.J. and M. Jevtic, An epidemiological and hematological study of sandfly fever in Serbia. *Acta Trop*, 1955. **12**: 179–82.
2. Pinheiro, F.P., A.P. Travassos da Rosa, M.L. Gomes, J.W. LeDuc and A.L. Hoch, Transmission of Oropouche virus from man to hamster by the midge *Culicoides paraensis*. *Science*, 1982. **215**: 1251–3.
3. Padula, P.J., *et al.*, Hantavirus pulmonary syndrome outbreak in Argentina: Molecular evidence for person-to-person transmission of Andes virus. *Virology*, 1998. **241**: 323–30.
4. Mardani, M., M. Keshtkar-Jahromi, B. Ataie and P. Adibi, Crimean-Congo hemorrhagic fever virus as a nosocomial pathogen in Iran. *Am J Trop Med Hyg*, 2009. **81**: 675–8.

5. Suleiman, M.N., *et al.*, Congo/Crimean haemorrhagic fever in Dubai. An outbreak at the Rashid Hospital. *Lancet*, 1980. **2**: 939–41.

6. Adam, I. and M.S. Karsany, Case report: Rift Valley Fever with vertical transmission in a pregnant Sudanese woman. *J Med Virol*, 2008. **80**: 929.

7. Tesh, R.B. and G.B. Modi, Maintenance of Toscana virus in *Phlebotomus perniciosus* by vertical transmission. *Am J Trop Med Hyg*, 1987. **36**: 189–93.

8. Watts, D.M., S. Pantuwatana, G.R. DeFoliart, T.M. Yuill and W.H. Thompson, Transovarial transmission of LaCrosse virus (California encephalitis group) in the mosquito, *Aedes triseriatus. Science*, 1973. **182**: 1140–1.

9. Thompson, W.H. and B.J. Beaty, Venereal transmission of La Crosse (California encephalitis) arbovirus in *Aedes triseriatus* mosquitoes. *Science*, 1977. **196**: 530–1.

10. Khan, A.S., *et al.*, Hantavirus pulmonary syndrome: The first 100 US cases. *J Infect Dis*, 1996. **173**: 1297–303.

11. Beaty, B.J., D.R. Sundin, L.J. Chandler and D.H. Bishop, Evolution of bunyaviruses by genome reassortment in dually infected mosquitoes (*Aedes triseriatus*). *Science*, 1985. **230**: 548–50.

12. Gerrard, S.R., L. Li, A.D. Barrett and S.T. Nichol, Ngari virus is a Bunyamwera virus reassortant that can be associated with large outbreaks of hemorrhagic fever in Africa. *J Virol*, 2004. **78**: 8922–6.

13. Chandler, L.J., *et al.*, Reassortment of La Crosse and Tahyna bunyaviruses in *Aedes triseriatus* mosquitoes. *Virus Res*, 1991. **20**: 181–91.

14. Reese, S.M., *et al.*, Potential for La Crosse virus segment reassortment in nature. *Virol J*, 2008. **5**: 164.

15. Schmaljohn, C.S. and S.T. Nichol, Bunyaviridae. D. Knipe Editor. *Virology*, 2006. **2**: 1741–89.

16. Bergeron, E., C.G. Albarino, M.L. Khristova and S.T. Nichol, Crimean-Congo hemorrhagic fever virus-encoded ovarian tumor protease activity is dispensable for virus RNA polymerase function. *J Virol*, 2010. **84**: 216–26.

17. Bridgen, A. and R.M. Elliott, Rescue of a segmented negative-strand RNA virus entirely from cloned complementary DNAs. *Proc Natl Acad Sci U S A*, 1996. **93**: 15400–4.

18. Dunn, E.F., D.C. Pritlove, H. Jin and R.M. Elliott, Transcription of a recombinant bunyavirus RNA template by transiently expressed bunyavirus proteins. *Virology*, 1995. **211**: 133–43.

19. Flick, K., *et al.*, Rescue of Hantaan virus minigenomes. *Virology*, 2003. **306**: 219–24.

20. Flick, K., *et al.*, Functional analysis of the noncoding regions of the Uukuniemi virus (Bunyaviridae) RNA segments. *J Virol*, 2004. **78**: 11726–38.

21. Flick, R., F. Elgh and R.F. Pettersson, Mutational analysis of the Uukuniemi virus (Bunyaviridae family) promoter reveals two elements of functional importance. *J Virol*, 2002. **76**: 10849–60.

22. Flick, R., K. Flick, H. Feldmann and F. Elgh, Reverse genetics for crimean-congo hemorrhagic fever virus. *J Virol*, 2003. **77**: 5997–6006.

23. Kohl, A., A.C. Lowen, V.H. Leonard and R.M. Elliott, Genetic elements regulating packaging of the Bunyamwera orthobunyavirus genome. *J Gen Virol*, 2006. **87**: 177–87.

24. Lopez, N., R. Muller, C. Prehaud and M. Bouloy, The L protein of Rift Valley fever virus can rescue viral ribonucleoproteins and transcribe synthetic genome-like RNA molecules. *J Virol*, 1995. **69**: 3972–9.

25. Prehaud, C., N. Lopez, M.J. Blok, V. Obry and M. Bouloy, Analysis of the 3′ terminal sequence recognized by the Rift Valley fever virus transcription complex in its ambisense S segment. *Virology*, 1997. **227**: 189–97.

26. Raju, R. and D. Kolakofsky, The ends of La Crosse virus genome and antigenome RNAs within nucleocapsids are base paired. *J Virol*, 1989. **63**: 122–8.

27. Obijeski, J.F., D.H. Bishop, E.L. Palmer and F.A. Murphy, Segmented genome and nucleocapsid of La Crosse virus. *J Virol*, 1976. **20**: 664–75.

28. Pettersson, R.F. and C.H. von Bonsdorff, Ribonucleoproteins of Uukuniemi virus are circular. *J Virol*, 1975. **15**: 386–92.

29. Barr, J.N., R.M. Elliott, E.F. Dunn and G.W. Wertz, Segment-specific terminal sequences of Bunyamwera bunyavirus regulate genome replication. *Virology*, 2003. **311**: 326–38.

30. Gauliard, N., A. Billecocq, R. Flick and M. Bouloy, Rift Valley fever virus noncoding regions of L, M and S segments regulate RNA synthesis. *Virology*, 2006. **351**: 170–9.

31. Lowen, A.C. and R.M. Elliott, Mutational analyses of the nonconserved sequences in the Bunyamwera Orthobunyavirus S segment untranslated regions. *J Virol*, 2005. **79**: 12861–70.

32. Lowen, A.C., A. Boyd, J.K. Fazakerley and R.M. Elliott, Attenuation of bunyavirus replication by rearrangement of viral coding and noncoding sequences. *J Virol*, 2005. **79**: 6940–6.

33. Fujii, K., *et al.*, Importance of both the coding and the segment-specific noncoding regions of the influenza A virus NS segment for its efficient incorporation into virions. *J Virol*, 2005. **79**: 3766–74.

34. Fujii, Y., H. Goto, T. Watanabe, T. Yoshida and Y. Kawaoka, Selective incorporation of influenza virus RNA segments into virions. *Proc Natl Acad Sci U S A*, 2003. **100**: 2002–7.

35. Mohamed, M., A. McLees and R.M. Elliott, Viruses in the Anopheles A, Anopheles B, and Tete serogroups in the Orthobunyavirus genus (family Bunyaviridae) do not encode an NSs protein. *J Virol*, 2009. **83**: 7612–8.

36. Plyusnin, A. Genetics of hantaviruses: Implications to taxonomy. *Arch Virol*, 2002. **147**: 665–82.

37. Jaaskelainen, K.M., *et al.*, Tula and Puumala hantavirus NSs ORFs are functional and the products inhibit activation of the interferon-beta promoter. *J Med Virol*, 2007. **79**: 1527–36.

38. Mohl, B.P. and J.N. Barr, Investigating the specificity and stoichiometry of RNA binding by the nucleocapsid protein of Bunyamwera virus. *RNA*, 2009. **15**: 391–9.

39. Leonard, V.H., A. Kohl, J.C. Osborne, A. McLees and R.M. Elliott, Homotypic interaction of Bunyamwera virus nucleocapsid protein. *J Virol*, 2005. **79**: 13166–72.

40. Eifan, S.A. and R.M. Elliott, Mutational analysis of the Bunyamwera orthobunyavirus nucleocapsid protein gene. *J Virol*, 2009. **83**: 11307–17.

41. Overby, A.K., R.F. Pettersson and E.P. Neve, The glycoprotein cytoplasmic tail of Uukuniemi virus (Bunyaviridae) interacts with ribonucleoproteins and is critical for genome packaging. *J Virol*, 2007. **81**: 3198–205.

42. Overby, A.K., V.L. Popov, R.F. Pettersson and E.P. Neve, The cytoplasmic tails of Uukuniemi Virus (Bunyaviridae) G(N) and G(C) glycoproteins are important for intracellular targeting and the budding of virus-like particles. *J Virol*, 2007. **81**: 11381–91.

43. Ribeiro, D., J.W. Borst, R. Goldbach and R. Kormelink, Tomato spotted wilt virus nucleocapsid protein interacts with both viral glycoproteins Gn and Gc in planta. *Virology*, 2009. **383**: 121–30.

44. Alfadhli, A., *et al.*, Hantavirus nucleocapsid protein oligomerization. *J Virol*, 2001. **75**: 2019–23.

45. Kaukinen, P., V. Koistinen, O. Vapalahti, A. Vaheri and A. Plyusnin, Interaction between molecules of hantavirus nucleocapsid protein. *J Gen Virol*, 2001. **82**: 1845–53.

46. Kaukinen, P., *et al.*, Oligomerization of Hantavirus N protein: C-terminal alpha-helices interact to form a shared hydrophobic space. *J Virol*, 2004. **78**: 13669–77.

47. Kaukinen, P., A. Vaheri and A. Plyusnin, Mapping of the regions involved in homotypic interactions of Tula hantavirus N protein. *J Virol*, 2003. **77**: 10910–6.

48. Mir, M.A. and A.T. Panganiban, Trimeric hantavirus nucleocapsid protein binds specifically to the viral RNA panhandle. *J Virol*, 2004. **78**: 8281–8.

49. Boudko, S.P., R.J. Kuhn and M.G. Rossmann, The coiled-coil domain structure of the Sin Nombre virus nucleocapsid protein. *J Mol Biol*, 2007. **366**: 1538–44.

50. Wang, Y., *et al.*, NMR structure of the N-terminal coiled coil domain of the Andes hantavirus nucleocapsid protein. *J Biol Chem*, 2008. **283**: 28297–304.

51. Le May, N., N. Gauliard, A. Billecocq and M. Bouloy, The N terminus of Rift Valley fever virus nucleoprotein is essential for dimerization. *J Virol*, 2005. **79**: 11974–80.

52. Kainz, M., P. Hilson, L. Sweeney, E. Derose and T.L. German, Interaction between Tomato spotted wilt virus N protein monomers involves nonelectrostatic forces governed by multiple distinct regions in the primary structure. *Phytopathology*, 2004. **94**: 759–65.

53. Uhrig, J.F., *et al.*, Homotypic interaction and multimerization of nucleocapsid protein of tomato spotted wilt tospovirus: Identification and characterization of two interacting domains. *Proc Natl Acad Sci U S A*, 1999. **96**: 55–60.

54. Raymond, D.D., M.E. Piper, S.R. Gerrard and J.L. Smith, Structure of the Rift Valley fever virus nucleocapsid protein reveals another architecture for RNA encapsidation. *Proc Natl Acad Sci U S A*, 2010. **107**(26): 11769–74.

55. Overby, A.K., V. Popov, E.P. Neve and R.F. Pettersson, Generation and analysis of infectious virus-like particles of uukuniemi virus (bunyaviridae): A useful system for studying bunyaviral packaging and budding. *J Virol*, 2006. **80**: 10428–35.

56. Mir, M.A. and A.T. Panganiban, The hantavirus nucleocapsid protein recognizes specific features of the viral RNA panhandle and is altered in conformation upon RNA binding. *J Virol*, 2005. **79**: 1824–35.

57. Mir, M.A. and A.T. Panganiban, Characterization of the RNA chaperone activity of hantavirus nucleocapsid protein. *J Virol*, 2006. **80**: 6276–85.

58. Ogg, M.M. and J.L. Patterson, RNA binding domain of Jamestown Canyon virus S segment RNAs. *J Virol*, 2007. **81**: 13754–60.

59. Osborne, J.C. and R.M. Elliott, RNA binding properties of bunyamwera virus nucleocapsid protein and selective binding to an element in the 5′ terminus of the negative-sense S segment. *J Virol*, 2000. **74**: 9946–52.

60. Mir, M.A., W.A. Duran, B.L. Hjelle, C. Ye and A.T. Panganiban, Storage of cellular 5′ mRNA caps in P bodies for viral cap-snatching. *Proc Natl Acad Sci U S A*, 2008. **105**: 19294–9.

61. Ontiveros, S.J., Q. Li and C.B. Jonsson, Modulation of apoptosis and immune signaling pathways by the Hantaan virus nucleocapsid protein. *Virology*, 2010. **401**: 165–78.

62. Shi, X. and R.M. Elliott, Analysis of N-linked glycosylation of hantaan virus glycoproteins and the role of oligosaccharide side chains in protein folding and intracellular trafficking. *J Virol*, 2004. **78**: 5414–22.

63. Andersson, A.M. and R.F. Pettersson, Targeting of a short peptide derived from the cytoplasmic tail of the G1 membrane glycoprotein of Uukuniemi virus (Bunyaviridae) to the Golgi complex. *J Virol*, 1998. **72**: 9585–96.

64. Gerrard, S.R. and S.T. Nichol, Characterization of the Golgi retention motif of Rift Valley fever virus G(N) glycoprotein. *J Virol*, 2002. **76**: 12200–10.

65. Matsuoka, Y., S.Y. Chen and R.W. Compans, A signal for Golgi retention in the bunyavirus G1 glycoprotein. *J Biol Chem*, 1994. **269**: 22565–73.

66. Snippe, M., L. Smeenk, R. Goldbach and R. Kormelink, The cytoplasmic domain of tomato spotted wilt virus Gn glycoprotein is required for Golgi localisation and interaction with Gc. *Virology*, 2007. **363**: 272–9.

67. Bertolotti-Ciarlet, A., *et al.*, Cellular localization and antigenic characterization of crimean-congo hemorrhagic fever virus glycoproteins. *J Virol*, 2005. **79**: 6152–61.

68. Haferkamp, S., L. Fernando, T.F. Schwarz, H. Feldmann and R. Flick, Intracellular localization of Crimean-Congo Hemorrhagic Fever (CCHF) virus glycoproteins. *Virol J*, 2005. **2**: 42.

69. Shi, X. and R.M. Elliott, Golgi localization of Hantaan virus glycoproteins requires coexpression of G1 and G2. *Virology*, 2002. **300**: 31–8.

70. Deyde, V.M., A.A. Rizvanov, J. Chase, E.W. Otteson and S.C. St Jeor, Interactions and trafficking of Andes and Sin Nombre Hantavirus glycoproteins G1 and G2. *Virology*, 2005. **331**: 307–15.

71. Altamura, L.A., *et al.*, Identification of a novel C-terminal cleavage of Crimean-Congo hemorrhagic fever virus PreGN that leads to generation of an NSM protein. *J Virol*, 2007. **81**: 6632–42.

72. Bergeron, E., M.J. Vincent and S.T. Nichol, Crimean-Congo hemorrhagic fever virus glycoprotein processing by the endoprotease SKI-1/S1P is critical for virus infectivity. *J Virol*, 2007. **81**: 13271–6.

73. Shi, X., *et al.*, Requirement of the N-terminal region of orthobunyavirus nonstructural protein NSm for virus assembly and morphogenesis. *J Virol*, 2006. **80**: 8089–99.

74. Fontana, J., N. Lopez-Montero, R.M. Elliott, J.J. Fernandez and C. Risco, The unique architecture of Bunyamwera virus factories around the Golgi complex. *Cell Microbiol*, 2008. **10**: 2012–28.

75. Kormelink, R., M. Storms, J. Van Lent, D. Peters and R. Goldbach, Expression and subcellular location of the NSM protein of tomato spotted wilt virus (TSWV), a putative viral movement protein. *Virology*, 1994. **200**: 56–65.

76. Gerrard, S.R., B.H. Bird, C.G. Albarino and S.T. Nichol, The NSm proteins of Rift Valley fever virus are dispensable for maturation, replication and infection. *Virology*, 2007. **359**: 459–65.

77. Won, S., T. Ikegami, C.J. Peters and S. Makino, NSm protein of Rift Valley fever virus suppresses virus-induced apoptosis. *J Virol*, 2007. **81**: 13335–45.

78. Muller, R., O. Poch, M. Delarue, D.H. Bishop and M. Bouloy, Rift Valley fever virus L segment: Correction of the sequence and possible functional role of newly identified

regions conserved in RNA-dependent polymerases. *J Gen Virol*, 1994. **75**(Pt 6): 1345–52.

79. Poch, O., B.M. Blumberg, L. Bougueleret and N. Tordo, Sequence comparison of five polymerases (L proteins) of unsegmented negative-strand RNA viruses: Theoretical assignment of functional domains. *J Gen Virol*, 1990. **71**(Pt 5): 1153–62.

80. Aquino, V.H., M.L. Moreli and L.T. Moraes Figueiredo, Analysis of oropouche virus L protein amino acid sequence showed the presence of an additional conserved region that could harbour an important role for the polymerase activity. *Arch Virol*, 2003. **148**: 19–28.

81. Dias, A., *et al.*, The cap-snatching endonuclease of influenza virus polymerase resides in the PA subunit. *Nature*, 2009. **458**: 914–8.

82. Fechter, P., *et al.*, Two aromatic residues in the PB2 subunit of influenza A RNA polymerase are crucial for cap binding. *J Biol Chem*, 2003. **278**: 20381–8.

83. Guilligay, D., *et al.*, The structural basis for cap binding by influenza virus polymerase subunit PB2. *Nat Struct Mol Biol*, 2008. **15**: 500–6.

84. Li, M.L., P. Rao and R.M. Krug, The active sites of the influenza cap-dependent endonuclease are on different polymerase subunits. *EMBO J*, 2001. **20**: 2078–86.

85. Brown, D.D., *et al.*, Rational attenuation of a morbillivirus by modulating the activity of the RNA-dependent RNA polymerase. *J Virol*, 2005. **79**: 14330–8.

86. Duprex, W.P., F.M. Collins and B.K. Rima, Modulating the function of the measles virus RNA-dependent RNA polymerase by insertion of green fluorescent protein into the open reading frame. *J Virol*, 2002. **76**: 7322–8.

87. Ruedas, J.B. and J. Perrault, Insertion of enhanced green fluorescent protein in a hinge region of vesicular stomatitis virus L polymerase protein creates a temperature-sensitive virus that displays no virion-associated polymerase activity *in vitro*. *J Virol*, 2009. **83**: 12241–52.

88. Shi, X. and R.M. Elliott, Generation and analysis of recombinant Bunyamwera orthobunyaviruses expressing V5 epitope-tagged L proteins. *J Gen Virol*, 2009. **90**: 297–306.

89. de Haan, P., *et al.*, Tomato spotted wilt virus L RNA encodes a putative RNA polymerase. *J Gen Virol*, 1991. **72**(Pt 9): 2207–16.

90. Honig, J.E., J.C. Osborne and S.T. Nichol, Crimean-Congo hemorrhagic fever virus genome L RNA segment and encoded protein. *Virology*, 2004. **321**: 29–35.

91. Marriott, A.C. and P.A. Nuttall, Large RNA segment of Dugbe nairovirus encodes the putative RNA polymerase. *J Gen Virol*, 1996. **77**(Pt 8): 1775–80.

92. van Poelwijk, F., M. Prins and R. Goldbach, Completion of the impatiens necrotic spot virus genome sequence and genetic comparison of the L proteins within the family Bunyaviridae. *J Gen Virol*, 1997. **78**(Pt 3): 543–6.

93. Kinsella, E., *et al.*, Sequence determination of the Crimean-Congo hemorrhagic fever virus L segment. *Virology*, 2004. **321**: 23–8.

94. de Medeiros, R.B., J. Figueiredo, O. Resende Rde and A.C. De Avila, Expression of a viral polymerase-bound host factor turns human cell lines permissive to a plant- and insect-infecting virus. *Proc Natl Acad Sci U S A*, 2005. **102**: 1175–80.

95. Zamoto-Niikura, A., K. Terasaki, T. Ikegami, C.J. Peters and S. Makino, Rift valley fever virus L protein forms a biologically active oligomer. *J Virol*, 2009. **83**: 12779–89.

96. Duijsings, D., R. Kormelink and R. Goldbach, *In vivo* analysis of the TSWV cap-snatching mechanism: Single base complementarity and primer length requirements. *Embo J*, 2001. **20**: 2545–52.

97. Jin, H. and R.M. Elliott, Non-viral sequences at the 5′ ends of Dugbe nairovirus S mRNAs. *J Gen Virol*, 1993. **74**: 2293–7.

98. van Knippenberg, I., M. Lamine, R. Goldbach and R. Kormelink, Tomato spotted wilt virus transcriptase *in vitro* displays a preference for cap donors with multiple base complementarity to the viral template. *Virology*, 2005. **335**: 122–30.

99. Garcin, D., *et al.*, The 5′ ends of Hantaan virus (Bunyaviridae) RNAs suggest a prime-and-realign mechanism for the initiation of RNA synthesis. *J Virol*, 1995. **69**: 5754–62.

100. Ikegami, T., S. Won, C.J. Peters and S. Makino, Rift Valley fever virus NSs mRNA is transcribed from an incoming anti-viral-sense S RNA segment. *J Virol*, 2005. **79**: 12106–11.

101. Barr, J.N., J.W. Rodgers and G.W. Wertz, The Bunyamwera virus mRNA transcription signal resides within both the 3′ and the 5′ terminal regions and allows ambisense transcription from a model RNA segment. *J Virol*, 2005. **79**: 12602–7.

102. Barr, J.N. and G.W. Wertz, Bunyamwera bunyavirus RNA synthesis requires cooperation of 3′- and 5′-terminal sequences. *J Virol*, 2004. **78**: 1129–38.

103. Barr, J.N. and G.W. Wertz, Role of the conserved nucleotide mismatch within 3′- and 5′-terminal regions of Bunyamwera virus in signaling transcription. *J Virol*, 2005. **79**: 3586–94.

104. Kohl, A., E.F. Dunn, A.C. Lowen and R.M. Elliott, Complementarity, sequence and structural elements within the 3′ and 5′ non-coding regions of the Bunyamwera orthobunyavirus S segment determine promoter strength. *J Gen Virol*, 2004. **85**: 3269–78.

105. Jin, H. and R.M. Elliott, Characterization of Bunyamwera virus S RNA that is transcribed and replicated by the L protein expressed from recombinant vaccinia virus. *J Virol*, 1993. **67**: 1396–404.

106. Patterson, J.L. and D. Kolakofsky, Characterization of La Crosse virus small-genome transcripts. *J Virol*, 1984. **49**: 680–5.

107. Bouloy, M., N. Pardigon, P. Vialat, S. Gerbaud and M. Girard, Characterization of the 5′ and 3′ ends of viral messenger RNAs isolated from BHK21 cells infected with Germiston virus (Bunyavirus). *Virology*, 1990. **175**: 50–8.

108. Cunningham, C. and J.F. Szilagyi, Viral RNAs synthesized in cells infected with Germiston Bunyavirus. *Virology*, 1987. **157**: 431–9.

109. Eshita, Y., B. Ericson, V. Romanowski and D.H. Bishop, Analyses of the mRNA transcription processes of snowshoe hare bunyavirus S and M RNA species. *J Virol*, 1985. **55**: 681–9.

110. Barr, J.N., J.W. Rodgers and G.W. Wertz, Identification of the Bunyamwera bunyavirus transcription termination signal. *J Gen Virol*, 2006. **87**: 189–98.

111. Ikegami, T., S. Won, C.J. Peters and S. Makino, Characterization of Rift Valley fever virus transcriptional terminations. *J Virol*, 2007. **81**: 8421–38.

112. Barr, J.N., S.P. Whelan and G.W. Wertz, Transcriptional control of the RNA-dependent RNA polymerase of vesicular stomatitis virus. *Biochim Biophys Acta*, 2002. **1577**: 337–53.

113. Hutchinson, K.L., C.J. Peters and S.T. Nichol, Sin Nombre virus mRNA synthesis. *Virology*, 1996. **224**: 139–49.

114. Barr, J.N., Bunyavirus mRNA synthesis is coupled to translation to prevent premature transcription termination. *RNA*, 2007. **13**: 731–6.

115. Bellocq, C. and D. Kolakofsky, Translational requirement for La Crosse virus S-mRNA synthesis: A possible mechanism. *J Virol*, 1987. **61**: 3960–7.

116. Lopez, N. and M.T. Franze-Fernandez, A single stem-loop structure in Tacaribe arenavirus intergenic region is essential for transcription termination but is not required for a correct initiation of transcription and replication. *Virus Res*, 2007. **124**: 237–44.

117. Pinschewer, D.D., M. Perez and J.C. de la Torre, Dual role of the lymphocytic choriomeningitis virus intergenic region in transcription termination and virus propagation. *J Virol*, 2005. **79**: 4519–26.

118. Blakqori, G., I. van Knippenberg and R.M. Elliott, Bunyamwera orthobunyavirus S-segment untranslated regions mediate poly(A) tail-independent translation. *J Virol*, 2009. **83**: 3637–46.

119. van Knippenberg, I., R. Goldbach and R. Kormelink, Tomato spotted wilt virus S-segment mRNAs have overlapping 3'-ends containing a predicted stem-loop structure and conserved sequence motif. *Virus Res*, 2005. **110**: 125–31.

120. Marzluff, W.F., Histone 3' ends: Essential and regulatory functions. *Gene Expr*, 1992. **2**: 93–7.

121. Raju, R., L. Raju and D. Kolakofsky, The translational requirement for complete La Crosse virus mRNA synthesis is cell-type dependent. *J Virol*, 1989. **63**: 5159–65.

122. van Knippenberg, I., R. Goldbach and R. Kormelink, *In vitro* transcription of Tomato spotted wilt virus is independent of translation. *J Gen Virol*, 2004. **85**: 1335–8.

123. Kikkert, M., *et al.*, Tomato spotted wilt virus particle morphogenesis in plant cells. *J Virol*, 1999. **73**: 2288–97.

124. Kuismanen, E., K. Hedman, J. Saraste and R.F. Pettersson, Uukuniemi virus maturation: Accumulation of virus particles and viral antigens in the Golgi complex. *Mol Cell Biol*, 1982. **2**: 1444–58.

125. Matsuoka, Y., S.Y. Chen and R.W. Compans, Bunyavirus protein transport and assembly. *Curr Top Microbiol Immunol*, 1991. **169**: 161–79.

126. Murphy, F.A., A.K. Harrison and S.G. Whitfield, Bunyaviridae: Morphologic and morphogenetic similarities of Bunyamwera serologic supergroup viruses and several other arthropod-borne viruses. *Intervirology*, 1973. **1**: 297–316.

127. Novoa, R.R., G. Calderita, P. Cabezas, R.M. Elliott and C. Risco, Key Golgi factors for structural and functional maturation of bunyamwera virus. *J Virol*, 2005. **79**: 10852–63.

128. Smith, J.F. and D.Y. Pifat, Morphogenesis of sandfly viruses (Bunyaviridae family). *Virology*, 1982. **121**: 61–81.

129. Shi, X., A. Kohl, P. Li and R.M. Elliott, Role of the cytoplasmic tail domains of Bunyamwera orthobunyavirus glycoproteins Gn and Gc in virus assembly and morphogenesis. *J Virol*, 2007. **81**: 10151–60.

130. Snippe, M., J. Willem Borst, R. Goldbach and R. Kormelink, Tomato spotted wilt virus Gc and N proteins interact *in vivo*. *Virology*, 2007. **357**: 115–23.

131. Salanueva, I.J., *et al.*, Polymorphism and structural maturation of bunyamwera virus in Golgi and post-Golgi compartments. *J Virol*, 2003. **77**: 1368–81.

132. Andersson, I., *et al.*, Role of actin filaments in targeting of Crimean Congo hemorrhagic fever virus nucleocapsid protein to perinuclear regions of mammalian cells. *J Med Virol*, 2004. **72**: 83–93.

133. Ramanathan, H.N., *et al.*, Dynein-dependent transport of the hantaan virus nucleocapsid protein to the endoplasmic reticulum-Golgi intermediate compartment. *J Virol*, 2007. **81**: 8634–47.

134. Ravkov, E.V., S.T. Nichol, C.J. Peters and R.W. Compans, Role of actin microfilaments in Black Creek Canal virus morphogenesis. *J Virol*, 1998. **72**: 2865–70.

135. Simon, M., C. Johansson, A. Lundkvist and A. Mirazimi, Microtubule-dependent and microtubule-independent steps in Crimean-Congo hemorrhagic fever virus replication cycle. *Virology*, 2009. **385**: 313–22.

136. Ravkov, E.V., S.T. Nichol and R.W. Compans, Polarized entry and release in epithelial cells of Black Creek Canal virus, a New World hantavirus. *J Virol*, 1997. **71**: 1147–54.

137. Goldsmith, C.S., L.H. Elliott, C.J. Peters and S.R. Zaki, Ultrastructural characteristics of Sin Nombre virus, causative agent of hantavirus pulmonary syndrome. *Arch Virol*, 1995. **140**: 2107–22.

138. Overby, A.K., R.F. Pettersson, K. Grunewald and J.T. Huiskonen, Insights into bunyavirus architecture from electron cryotomography of Uukuniemi virus. *Proc Natl Acad Sci U S A*, 2008. **105**: 2375–9.

139. Harris, A., *et al.*, Influenza virus pleiomorphy characterized by cryoelectron tomography. *Proc Natl Acad Sci U S A*, 2006. **103**: 19123–7.

140. Noda, T., *et al.*, Architecture of ribonucleoprotein complexes in influenza A virus particles. *Nature*, 2006. **439**: 490–2.

141. Freiberg, A.N., M.B. Sherman, M.C. Morais, M.R. Holbrook and S.J. Watowich, Three-dimensional organization of Rift Valley fever virus revealed by cryoelectron tomography. *J Virol*, 2008. **82**: 10341–8.

142. Sherman, M.B., A.N. Freiberg, M.R. Holbrook and S.J. Watowich, Single-particle cryo-electron microscopy of Rift Valley fever virus. *Virology*, 2009. **387**: 11–5.

143. Bridgen, A., D.A. Dalrymple, F. Weber and R.M. Elliott, Inhibition of Dugbe nairovirus replication by human MxA protein. *Virus Res*, 2004. **99**: 47–50.

144. Kochs, G., C. Janzen, H. Hohenberg and O. Haller, Antivirally active MxA protein sequesters La Crosse virus nucleocapsid protein into perinuclear complexes. *Proc Natl Acad Sci U S A*, 2002. **99**: 3153–8.

145. Frese, M., G. Kochs, H. Feldmann, C. Hertkorn and O. Haller, Inhibition of bunyaviruses, phleboviruses, and hantaviruses by human MxA protein. *J Virol*, 1996. **70**: 915–23.

146. Blakqori, G., *et al.*, La Crosse bunyavirus nonstructural protein NSs serves to suppress the type I interferon system of mammalian hosts. *J Virol*, 2007. **81**: 4991–9.

147. Bridgen, A., F. Weber, J.K. Fazakerley and R.M. Elliott, Bunyamwera bunyavirus nonstructural protein NSs is a nonessential gene product that contributes to viral pathogenesis. *Proc Natl Acad Sci U S A*, 2001. **98**: 664–9.

148. Muller, R., *et al.*, Characterization of clone 13, a naturally attenuated avirulent isolate of Rift Valley fever virus, which is altered in the small segment. *Am J Trop Med Hyg*, 1995. **53**: 405–11.

149. Thomas, D., *et al.*, Inhibition of RNA polymerase II phosphorylation by a viral interferon antagonist. *J Biol Chem*, 2004. **279**: 31471–7.

150. Leonard, V.H., A. Kohl, T.J. Hart and R.M. Elliott, Interaction of Bunyamwera Orthobunyavirus NSs protein with mediator protein MED8: A mechanism for inhibiting the interferon response. *J Virol*, 2006. **80**: 9667–75.

151. van Knippenberg, I., C. Carlton-Smith and R.M. Elliott, The N-terminus of Bunyamwera virus NSs protein is essential for interferon antagonism. *J Gen Virol*, 2010. **91**: 2002–6.

152. Le May, N., *et al.*, A SAP30 complex inhibits IFN-beta expression in Rift Valley fever virus infected cells. *PLoS Pathog*, 2008. **4**: e13.

153. Habjan, M., *et al.*, NSs protein of rift valley fever virus induces the specific degradation of the double-stranded RNA-dependent protein kinase. *J Virol*, 2009. **83**: 4365–75.

154. Habjan, M., *et al.*, Processing of genome 5′ termini as a strategy of negative-strand RNA viruses to avoid RIG-I-dependent interferon induction. *PLoS One*, 2008. **3**: e2032.

155. Takeda, A., *et al.*, Identification of a novel RNA silencing suppressor, NSs protein of Tomato spotted wilt virus. *FEBS Lett*, 2002. **532**: 75–9.

156. Bucher, E., T. Sijen, P. De Haan, R. Goldbach and M. Prins, Negative-strand tospoviruses and tenuiviruses carry a gene for a suppressor of gene silencing at analogous genomic positions. *J Virol*, 2003. **77**: 1329–36.

157. Garcia, S., *et al.*, Viral suppressors of RNA interference impair RNA silencing induced by a Semliki Forest virus replicon in tick cells. *J Gen Virol*, 2006. **87**: 1985–9.

158. Wang, X.H., *et al.*, RNA interference directs innate immunity against viruses in adult Drosophila. *Science*, 2006. **312**: 452–4.

159. Soldan, S.S., M.L. Plassmeyer, M.K. Matukonis and F. Gonzalez-Scarano, La Crosse virus nonstructural protein NSs counteracts the effects of short interfering RNA. *J Virol*, 2005. **79**: 234–44.

160. Peters, K., S. Chattopadhyay and G.C. Sen, IRF-3 activation by Sendai virus infection is required for cellular apoptosis and avoidance of persistence. *J Virol*, 2008. **82**: 3500–8.

161. Boatright, K.M., *et al.*, A unified model for apical caspase activation. *Mol Cell*, 2003. **11**: 529–41.

162. Boatright, K.M. and G.S. Salvesen, Mechanisms of caspase activation. *Curr Opin Cell Biol*, 2003. **15**: 725–31.

163. Riedl, S.J. and Y. Shi, Molecular mechanisms of caspase regulation during apoptosis. *Nat Rev Mol Cell Biol*, 2004. **5**: 897–907.

164. Pekosz, A., J. Phillips, D. Pleasure, D. Merry and F. Gonzalez-Scarano, Induction of apoptosis by La Crosse virus infection and role of neuronal differentiation and human bcl-2 expression in its prevention. *J Virol*, 1996. **70**: 5329–35.

165. Colon-Ramos, D.A., *et al.*, Inhibition of translation and induction of apoptosis by Bunyaviral nonstructural proteins bearing sequence similarity to reaper. *Mol Biol Cell*, 2003. **14**: 4162–72.

166. Kohl, A., *et al.*, Bunyamwera virus nonstructural protein NSs counteracts interferon regulatory factor 3-mediated induction of early cell death. *J Virol*, 2003. **77**: 7999–8008.

167. Chattopadhyay, S., *et al.*, Viral apoptosis is induced by IRF-3-mediated activation of Bax. *EMBO J*, 2010. **29**: 1762–73.

168. Kang, J.I., S.H. Park, P.W. Lee and B.Y. Ahn, Apoptosis is induced by hantaviruses in cultured cells. *Virology*, 1999. **264**: 99–105.

169. Markotic, A. Immunopathogenesis of hemorrhagic fever with renal syndrome and hantavirus pulmonary syndrome. *Acta Med Croatica*, 2003. **57**: 407–14.
170. Hardestam, J., J. Klingstrom, K. Mattsson and A. Lundkvist, HFRS causing hantaviruses do not induce apoptosis in confluent Vero E6 and A-549 cells. *J Med Virol*, 2005. **76**: 234–40.
171. Li, X.D., *et al.*, Tula hantavirus infection of Vero E6 cells induces apoptosis involving caspase 8 activation. *J Gen Virol*, 2004. **85**: 3261–8.
172. Acrani, G.O., *et al.*, Apoptosis induced by Oropouche virus infection in HeLa cells is dependent on virus protein expression. *Virus Res*, 2010. **149**: 56–63.
173. Klingstrom, J., *et al.*, Loss of cell membrane integrity in puumala hantavirus-infected patients correlates with levels of epithelial cell apoptosis and perforin. *J Virol*, 2006. **80**: 8279–82.
174. Li, X.D., H. Lankinen, N. Putkuri, O. Vapalahti and A. Vaheri, Tula hantavirus triggers pro-apoptotic signals of ER stress in Vero E6 cells. *Virology*, 2005. **333**: 180–9.
175. Taylor, S.L., N. Frias-Staheli, A. Garcia-Sastre and C.S. Schmaljohn, Hantaan virus nucleocapsid protein binds to importin alpha proteins and inhibits tumor necrosis factor alpha-induced activation of nuclear factor kappa B. *J Virol*, 2009. **83**: 1271–9.

Chapter 14

Ebolaviruses: What We Know and Where We Are on Potential Therapeutics

Peter Halfmann*, Gabriele Neumann*
and Yoshihiro Kawaoka*,†,‡,§

1. An Introduction to Ebolavirus

The first recorded outbreak of Ebolavirus occurred in 1976, when a mysterious disease killed 280 of 318 infected individuals in Zaire (now the Republic of Congo). This outbreak resulted in a case fatality rate of 88%, a figure that alarmed health care workers and scientists. The causative agent was eventually identified as Ebolavirus, named after the Ebola River that runs through Zaire. As then, there have been additional sixteen outbreaks of Ebolaviruses, resulting in 2299 confirmed cases with 1556 deaths.

Ebolavirus belongs to the family *Filoviridae* (order *Mononegavirales*) along with the closely related Marburgvirus.[1] Within the genus *Ebolavirus*, five species are currently recognized: *Zaire ebolavirus*, *Sudan ebolavirus*, *Ivory Coast ebolavirus*, *Bundibugyo ebolavirus*, and *Reston ebolavirus*. The recently discovered *Bundibugyo ebolavirus* (named after the Bundibugyo district of Uganda where the outbreak occurred in 2007) shows a 32% nucleotide difference from its closest phylogenetic relative, *Ivory Coast ebolavirus*, which is similar in nucleotide divergence to the other Ebolavirus species (which ranges from 35–45%).[2]

*Department of Pathobiological Sciences, School of Veterinary Medicine, University of Wisconsin-Madison, 2015 Linden Drive, Madison, WI 53706, USA.

†Division of Virology, Department of Microbiology and Immunology, Institute of Medical Science, University of Tokyo, Tokyo 108-8639, Japan.

‡International Research Center for Infectious Diseases, Institute of Medical Science, University of Tokyo, Tokyo 108-8639, Japan.

§ERATO Infection-Induced Host Responses Project, Saitama 332-0012, Japan.

Table 1.   Known Ebolavirus cases and outbreaks.

Year	Country	Species	Cases	Deaths	Case fatality (%)
1967	DCR[1]	Zaire	318	280	88
1976	Sudan	Sudan	284	151	53
1977	DCR	Zaire	1	1	100
1979	Sudan	Sudan	34	22	65
1989–1990	USA	Reston	0	0	0
1989–1990	Philippines	Reston	0	0	0
1992	Italy	Reston	0	0	0
1994	Gabon	Zaire	52	31	60
1994	Ivory Coast	Ivory Coast	1	0	0
1995	DCR	Zaire	315	250	81
1996	Gabon	Zaire	37	21	57
1996	Gabon	Zaire	60	45	74
1996	South Africa	Zaire	2	1	50
1996	USA	Reston	0	0	100
1996	Philippines	Reston	0	0	100
2000	Uganda	Sudan	425	224	53
2001	Gabon	Zaire	65	83	82
2001	RC[2]	Zaire	59	44	75
2002	RC	Zaire	143	128	89
2003	RC	Zaire	35	29	83
2004	Sudan	Sudan	17	7	41
2007	RC	Zaire	264	187	71
2007	Uganda	Bundibugyo	149	37	25
2008	Philippines	Reston	6	0	0
2000–2009	RC	Zaire	32	15	47

[1]Democratic Republic of Congo.
[2]Republic of Congo.

The mortality rate in humans depends on the causative species of Ebolavirus (see Table 1). For *Zaire ebolavirus* and *Sudan ebolavirus*, case fatality rates of 55–100% have been reported,[1] whereas an outbreak caused by *Ivory Coast ebolavirus* was associated with severe disease but not fatalities.[3] To date, *Reston ebolavirus* has not caused morbidity or mortality in humans.[4] The outbreak caused by *Bundibugyo ebolavirus* in 2007 resulted in a case fatality rate of 25%.[2]

Symptoms of Ebolavirus hemorrhagic fever appear abruptly and include fever, myalgias, headache, and nausea.[5] As these symptoms resemble other diseases commonly found in Africa, such as malaria, Ebolavirus infections can be misdiagnosed in their early stages. Incubation times range from 4 to 10 days, and death typically occurs 6–16 days after the onset of symptoms.[1] Fatal cases are typically characterized by an increase in viremia over the course of infection, leading to higher viral loads in the blood than in non-fatal cases.[6,7] At the late stages of infection, organ failure, shock and disseminated intravascular coagulation are common.[1,5]

Most outbreaks of Ebolavirus have occurred in Central Africa[8]; however, several US outbreaks were reported between 1989 and 1996 at primate facilities in Reston, Virginia, and Alice, Texas.[4,9] Cynomolgus macaques imported to these facilities from the Philippines were infected with what was a new species of virus, *Reston ebolavirus*. Recently, *Reston ebolavirus* was discovered in domestic swine in the Philippines.[10] An increase in pig mortality in 2008 prompted an investigation into the causative agent. Pigs tested positive for porcine reproductive and respiratory syndrome virus; however, six of 28 swine samples also tested positive for *Reston ebolavirus,* as detected by use of a panviral microarray.[10,11] Moreover, 6 of 141 tested individuals were positive for antibodies to *Reston ebolavirus*.[10] The Ebolaviruses isolated from pigs were at least 95.5% identical to the *Reston ebolaviruses* isolated from non-human primates (NHPs) in Reston, Virginia. The finding of *Reston ebolavirus* in swine raised concerns over the potential emergence of this virus in the human food chain, and its potential transmission from pigs to humans.[10]

Ebolavirus infections are not only a public health concern but also a threat to the wild-life population of Africa. Fatal Ebolavirus infections have taken a significant toll on the gorilla and chimpanzee populations in Central Africa, as nearly 5000 gorillas may have succumbed to Ebolavirus infections in 2002–2003 alone,[12–14] These infections, together with factors such as poaching, have resulted in gorillas being listed as threatened species in Africa.

2.   The Reservoir of Ebolaviruses

The animal reservoir of Ebolaviruses has been elusive since the first outbreak in 1976. Several extensive surveillance studies in non-human vertebrates and arthropod vectors were conducted after major outbreaks; however, these studies did not identify a definitive reservoir.[15–17] In another study, the organs of 242 small mammals were tested for Ebolaviruses.[18] Neither live virus nor viral antigens were found. However, the sequences of the Ebolavirus glycoprotein (GP) or polymerase gene were detected in RNA samples of seven animals, representing two genera of rodents.[18]

A survey of 1030 small vertebrates collected between 2001 and 2003 in Gabon and the Republic of Congo suggested fruit bats as a reservoir for Ebolaviruses.[19] For three different species of fruit bats, on average, 5% of tested animals were PCR-positive and 8% were antibody-positive, although these animals showed no signs of Ebolavirus infection.[19] Interestingly, no sample was both PCR- and antibody-positive, likely due to the kinetics of Ebolavirus replication and immune responses in animals. A larger surveillance study in the epidemic regions of Gabon and the Congo that included over 2100 bats detected antibodies to Ebolavirus in 4% of the tested animals, which represented nine different species.[20] Evidence now suggests that consumption of infected fruits bats caused the index case of the Kikwit outbreak of 1995.[21] Although live Ebolavirus has not been isolated from bats, live Marburgvirus was found in bats tested in a cave in Uganda.[22] Collectively, these data point to fruit bats as the potential reservoir of filoviruses, although additional ecological studies are needed to assess Ebolavirus transmission from bats to swine, rodents, NHPs and potentially humans.

3.   The Gene Products of Ebolaviruses

Filoviruses are so-named because of their filamentous morphology (Fig. 1). They are enveloped and contain a non-segmented, negative-sense RNA genome with seven genes that encode eight (Ebolaviruses) or seven (Marburgviruses) proteins.[1] The nucleocapsid protein (NP), viral protein (VP)35, VP30, and the viral poly-merase protein, L, are necessary for the replication and transcription of the viral genome, which occurs in the cytoplasm of infected cells.[23] The primary matrix protein, VP40, plays a critical role in virion formation and budding.[24–26] The roles of the remaining VPs and that of VP35 will be discussed below.

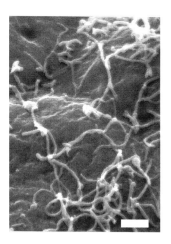

Fig. 1.

3.1. *The viral GP*

The fourth open-reading frame in the viral genome encodes two viral GP: the secreted GP (sGP) and the full-length GP. The primary GP mRNA (which accounts for ~80% of transcripts) encodes the non-structural sGP. The full-length, structural GP is produced from the edited version of the GP mRNA, which possesses an additional adenosine at a stretch of seven adenosine residues.[27,28] This coding strategy allows sGP (364 amino acids) and GP (676 amino acids) to share their N-terminal 295 amino acids, but differs in their C-terminal amino acids.[29] The role of sGP in the viral life cycle is not known, but it may play a role in viral pathogenesis.[30] Detectable amounts of sGP in the serum of infected patients[31] suggest it may function as a decoy protein that absorbs neutralizing antibodies to GP.

Viral entry into host cells is mediated by the only viral surface protein, GP. Ebolaviruses can infect many different types of cells, with the exception of T and B lymphocytes.[33–35] Several cell surface proteins have been suggested as potential receptors or co-receptors.[36–41] These cellular proteins include folate receptor-α[37] and C-type lectins like dendritic cell-specific ICAM-3 grabbing non-integrin (DC-SIGN) on macrophages and dendritic cells, as well as L-SIGN on endothelial cells.[40,41] Recently, members of the tyrosine kinase receptor family (such as Axl, Dtk, Mer) were reported to mediate Ebolaviruses entry.[36] Collectively, these findings suggest that Ebolaviruses may not be limited to a single receptor or receptor-type.

Viral infectivity can be affected by the post-translational cleavage of the GP, resulting in the exposure of a so-called fusion peptide.[42] One such example is influenza virus: a multi-basic motif at the hemagglutinin cleavage site is cleaved by ubiquitous proteases, resulting in systemic infections. By contrast, a single basic amino acid at the cleavage site results in infections restricted to the respiratory or intestinal tract.[42] Ebolavirus GP also possesses a multi-basic stretch of amino acids, $_{497}$R-R-T-R-R$_{501}$, at its GP cleavage site, which is recognized by the ubiquitous cellular protease furin.[43] However, an engineered virus with a single basic amino acid at this site grew efficiently in cell culture and was pathogenic in NHPs, suggesting that cleavage of GP by furin may not be required for Ebolavirus infection.[44,45]

Viral pathogenicity may also be affected by the glycosylation pattern of the GP.[46–48] The GP of Ebolaviruses is heavily glycosylated by both O-linked and N-linked sugars.[49] The recently resolved crystal structure of GP[50] suggests that the so-called mucin domain, a highly O- and N-glycosylated domain, may act as a "glycan shield" that prevents the binding of neutralizing antibodies. In addition, the mucin domain is associated with the induction of proinflammatory cytokines and vascular endothelial injury.[51,52] Additional studies on the role of the mucin domain have been hampered by an inability to generate Ebolavirus that lacks this region, although pseudo-typed viruses lacking the Ebolavirus GP mucin domain can be generated.[53]

3.2. *VP35 and VP24, Ebolavirus interferon antagonists*

Typically, double-stranded RNA (dsRNA) produced during viral infection is detected in the cytoplasm by cellular sensors, such as retinoic acid-inducible gene I (RIG-I)[54] and melanoma differentiation-associated gene 5.[55] These sensors activate transcription factors, such as interferon regulatory factor (IRF)-3, IRF-7, and NFκB, leading to the transcriptional up-regulation of type I interferons (IFNα/β).[56,57] Upon binding to their receptors, type I interferons activate several signal transduction pathways, including the JAK-STAT pathway. After sequential phosphorylation of downstream effectors, phosphorylated transcription factors translocate to the nucleus, where they bind to interferon-stimulated response elements that are located in the promoter regions of interferon-stimulated genes (ISGs).[56,58] ISGs encode anti-VPs, such as $2'5'$-oligoadenylate synthetase, dsRNA-dependent protein kinase R, and Mx proteins, whose expression leads to an antiviral state in the cell.[58] Ebolaviruses encode two known interferon antagonists, VP35 and VP24, which inhibit the induction of RIG-I signaling, the synthesis of type I interferon, and the cellular responses to type I interferon. In this manner, they allow the virus to escape from the host antiviral responses.[59–67]

VP35 binds dsRNA through sequences in its C-terminus,[68,69] a region that is also critical to inhibit interferon responses.[64,66,70] In particular, Arg312 of VP35 is essential for dsRNA binding and inhibition of IFN responses, suggesting dsRNA-dependent IFN inhibition.[71] The recently resolved crystal structure of VP35 bound to dsRNA suggested that Arg312 forms direct contacts with the phosphodiester backbone of the dsRNA, and that the interaction of VP35 with the dsRNA shields the RNA from recognition by RIG-I.[64,68,69] Ebolavirus with mutations in the dsRNA-binding domain of VP35 is not significantly attenuated in interferon-defective cells (such as Vero cells), but is avirulent in the guinea pig model.[72]

VP35 was originally identified as an interferon antagonist that suppresses the transcriptional up-regulation of IFNβ mRNA.[65] VP35 executes this function by blocking the phosphorylation of IRF-3 through the inhibition of two upstream kinases, IKKepsilon and TBK-1.[73] The inhibition of IRF-3 phosphorylation prevents its dimerization and nuclear translocation,[66] and hence the up-regulation of IFNβ mRNA.

In addition to inhibiting type I interferon synthesis, Ebolaviruses also inhibit the cellular response to type I interferons. Ebolavirus-infected endothelial cells have an impaired interferon response and show little to no induction of ISGs when stimulated with type I interferons, unlike uninfected cells.[60] The inhibition of IFN signaling is mediated by the Ebolavirus VP24 protein,[62] which interacts with several cellular nuclear import factors (karyopherins α1, α5, and α6) to block the nuclear localization of the transcription factor STAT-1 and thus the up-regulation of ISGs.[62,63,74]

4. Prevention and Treatment of Ebolavirus Infections

Currently, there are no licensed compounds against Ebolaviruses, prohibiting the treatment of infected individuals. Protection from and treatment of Ebolavirus infections are desirable not only for populations in affected areas of Africa, but also for health care and laboratory workers, and for military personnel.

4.1. *Antiviral compounds*

Ebolavirus infections are characterized by impaired coagulation, which can be attributed to an overexpression of the procoagulant tissue factor.[75] Nematode anticoagulant protein c2 (NAPc2) inhibits the tissue factor pathway and is thus an interesting candidate for the treatment of Ebolavirus infections. In NHPs, the gold-standard of filovirus infection studies, NAPc2 provided partial protection against Ebolavirus.[76]

Recently, several systems have been developed to screen compound libraries for small molecules that interfere with the Ebolavirus life cycle; among these are retro-, lenti-, or vesicular stomatitis viruses (VSVs) pseudo-typed with Ebolavirus GP,[34,77] Ebolavirus reporter assays,[78,79] and a biologically contained Ebolavirus[80] that can be handled under non-BSL-4 conditions. These screens have identified small molecules such as FGI-103[81] and FGI-106,[82] which have antiviral activity against Ebolavirus in cell culture and/or mice. One such molecule, LJ001, was shown to be effective against numerous enveloped viruses including arenaviruses, bunyaviruses, flaviviruses, paramyxoviruses, poxviruses, and filoviruses.[83] This compound inhibits virus–cell fusion but not cell–cell fusion, making it an attractive candidate to block enveloped viruses.[83] Further studies are clearly needed to test the mode of action and efficacy of these molecules in NHPs.

4.2. *Vaccines*

The first efforts to develop an Ebolavirus vaccine were based on the classical approach of inactivated virus; however, sucrose-purified, γ-irradiated virus protected only 25% of NHPs from lethal challenge.[84] Immunized animals had poor antibody responses to GP and low to non-existent neutralizing antibody titers,[84] probably because virus inactivation disrupted the protein structure.[85] As then, other vaccine approaches have shown various degrees of success, three of these candidates are discussed below.

4.2.1. *Virus-like particles*

Virus-like particles (VLPs) are an attractive platform for Ebolavirus vaccine development because of their overall safety and their efficacy in disease models.[86,87] The expression in mammalian cells of three Ebolavirus proteins, NP, VP40, and GP, resulted in VLPs that were indistinguishable in size and shape from authentic Ebolaviruses.[26,88,89] Ebolavirus VLPs produced in insect cells have been shown to protect mice and guinea pigs.[90] More importantly, three immunizations with a mammalian cell produced VLP together with an adjuvant elicited antibody and cellular immune responses in NHPs and protected animals against lethal challenge with *Ebolavirus*.[91] Hence, VLPs appear to have a protective efficacy. A multiple dose regimen may, however, be required, which would not be ideal.

4.2.2. *A replication-defective adenovirus-based vaccine*

Adenoviruses have been studied intensely as vaccine and gene therapy platforms (review Refs. 92–94), and their potential as Ebolavirus vaccines has been demonstrated in numerous studies,[95–98] including a study that tested post-exposure treatment.[99] However, some of these studies were carried out in mice and the efficacy of the test vaccine needs to be confirmed in NHPs. To date, most studies have been carried out with adenovirus serotype 5, against which widespread immunity exists in human populations.[87] This shortcoming may be overcome by the use of other serotypes, such as serotype 35.[100]

4.2.3. *A replication-competent VSV-based vaccine*

A replication-competent VSV expressing Ebolavirus GP instead of its own GP (VSV-EbGP) completely protected NHPs against lethal challenge.[101–103] Depending on when it was administered, the vaccine also conferred full to partial protection as a post-exposure treatment.[104,105] This vaccine was administered to a researcher who accidentally pricked herself with an Ebolavirus-contaminated needle (http://www.promedmail.org/pls/otn/f?p=2400:1001:8376707730348). She did not develop symptoms of Ebolavirus infection; however, it was not clear if infection through the needle stick had, in fact, occurred. Nevertheless, this experimental vaccine was thus tested in a real-world scenario, and no severe adverse effects to it were observed (http://www.promedmail.org/pls/otn/f?p=2400:1001:8376707730348). Although these findings are promising, safety concerns have been raised regarding VSV as it can cause latent and persistent disease in livestock.[106]

In conclusion, multiple concerns including safety, potency, and overall efficacy exist with the current approaches, underscoring the need for better and more reliable therapeutic options.

5. Concluding Remarks

Ebolavirus and its close relative Marburgvirus continue to cause outbreaks in humans, NHPs, and also, as recently discovered, in pigs. With their high case fatality rates and their potential use as bioterrorist agents, filoviruses pose a clear threat. More research is needed on their natural reservoir, their mode of transmission to mammalian species, key factors that contribute to sporadic outbreaks, and the determinants and mechanisms of pathogenicity.

Acknowledgments

The authors thank Susan Watson for editing the manuscript. The original work from our laboratory was supported by National Institute of Allergy and Infectious Diseases Public Health Service research grants R01 AI055519-05 and RO1AI077593 and by grants-in-aid from the Ministries of Education, Culture, Sports, Science, Japan. The author acknowledge membership within and support from the Region V "Great Lakes" Regional Center for Excellence, U54AI57153.

References

1. Sanchez A, Geisbert TW, Feldmann H. (2007) In, Knipe DM, Howley PM, Griffin DE, Martin MA, Lamb RA, Roizman B, Straus SE (eds). *Fields Virology*, pp. 1409–1448. Lippincott, Williams & Wilkins, Philadelphia.
2. Towner JS, Sealy TK, Khristova ML, Albarino CG, Conlan S, Reeder SA, Quan PL, Lipkin WI, Downing R, Tappero JW, *et al.* (2008) *PLoS. Pathog* **4**: e1000212.
3. Le GB, Formenty P, Wyers M, Gounon P, Walker F, Boesch C. (1995) *Lancet* **345**: 1271–1274.
4. Rollin PE, Williams RJ, Bressler DS, Pearson S, Cottingham M, Pucak G, Sanchez A, Trappier SG, Peters RL, Greer PW, *et al.* (1999) *J Infect Dis* **179** (Suppl 1): S108–S114.
5. Bwaka MA, Bonnet MJ, Calain P, Colebunders R, De RA, Guimard Y, Katwiki KR, Kibadi K, Kipasa MA, Kuvula KJ, *et al.* (1999) *J Infect Dis* **179** (Suppl 1): S1–S7.

6. Sanchez A, Lukwiya M, Bausch D, Mahanty S, Sanchez AJ, Wagoner KD, Rollin PE. (2004) *J Virol* **78**: 10370–10377.

7. Towner JS, Rollin PE, Bausch DG, Sanchez A, Crary SM, Vincent M, Lee WF, Spiropoulou CF, Ksiazek TG, Lukwiya M, *et al.* (2004) *J Virol* **78**: 4330–4341.

8. Peterson AT, Bauer JT, Mills JN. (2004) *Emerg Infect Dis* **10**: 40–47.

9. Miranda ME, Ksiazek TG, Retuya TJ, Khan AS, Sanchez A, Fulhorst CF, Rollin PE, Calaor AB, Manalo DL, Roces MC, *et al.* (1999) *J Infect Dis* **179** (Suppl 1): S115–S119.

10. Barrette RW, Metwally SA, Rowland JM, Xu L, Zaki SR, Nichol ST, Rollin PE, Towner JS, Shieh WJ, Batten B, *et al.* (2009) *Science* **325**: 204–206.

11. Cyranoski D. (2009) *Nature* **457**: 364–365.

12. Bermejo M, Rodriguez-Teijeiro JD, Illera G, Barroso A, Vila C, Walsh PD. (2006) *Science* **314**: 1564.

13. Le Gouar PJ, Vallet D, David L, Bermejo M, Gatti S, Levrero F, Petit EJ, Menard N. (2009) *PLoS One* **4**: e8375.

14. Vogel G. (2007) *Science* **317**: 1484.

15. Breman JG, Johnson KM, van der GG, Robbins CB, Szczeniowski MV, Ruti K, Webb PA, Meier F, Heymann DL, (1999) *J Infect Dis* **179**(Suppl 1): S139–S147.

16. Leirs H, Mills JN, Krebs JW, Childs JE, Akaibe D, Woollen N, Ludwig G, Peters CJ, Ksiazek TG. (1999) *J Infect Dis* **179** (Suppl 1): S155–S163.

17. Reiter P, Turell M, Coleman R, Miller B, Maupin G, Liz J, Kuehne A, Barth J, Geisbert J, Dohm D, *et al.* (1999) *J Infect Dis* **179** (Suppl 1): S148–S154.

18. Morvan JM, Deubel V, Gounon P, Nakoune E, Barriere P, Murri S, Perpete O, Selekon B, Coudrier D, Gautier-Hion A, *et al.* (1999) *Microbes Infect* **1**: 1193–1201.

19. Leroy EM, Kumulungui B, Pourrut X, Rouquet P, Hassanin A, Yaba P, Delicat A, Paweska JT, Gonzalez JP, Swanepoel R. (2005) *Nature* **438**: 575–576.

20. Pourrut X, Souris M, Towner JS, Rollin PE, Nichol ST, Gonzalez JP, Leroy E. (2009) *BMC Infect Dis* **9**: 159.

21. Leroy EM, Epelboin A, Mondonge V, Pourrut X, Gonzalez JP, Muyembe-Tamfum JJ, Formenty P. (2009) *Vector Borne Zoonotic Dis* **9**: 723–728.

22. Towner JS, Amman BR, Sealy TK, Carroll SA, Comer JA, Kemp A, Swanepoel R, Paddock CD, Balinandi S, Khristova ML, *et al.* (2009) *PLoS Pathog* **5**: e1000536.

23. Muhlberger E, Weik M, Volchkov VE, Klenk HD, Becker S. (1999) *J Virol* **73**: 2333–2342.

24. Yamayoshi S, Noda T, Ebihara H, Goto H, Morikawa Y, Lukashevich IS, Neumann G, Feldmann H, Kawaoka Y. (2008) *Cell Host Microbe* **3**: 168–177.

25. Yasuda J, Nakao M, Kawaoka Y, Shida H. (2003) *J Virol* **77**: 9987–9992.

26. Noda T, Sagara H, Suzuki E, Takada A, Kida H, Kawaoka Y. (2002) *J Virol* **76**: 4855–4865.

27. Sanchez A, Trappier SG, Mahy BW, Peters CJ, Nichol ST. (1996) *Proc Natl Acad Sci USA* **93**: 3602–3607.

28. Volchkov VE, Becker S, Volchkova VA, Ternovoj VA, Kotov AN, Netesov SV, Klenk HD. (1995) *Virology* **214**: 421–430.

29. Sanchez A, Yang ZY, Xu L, Nabel GJ, Crews T, Peters CJ. (1998) *J Virol* **72**: 6442–6447.

30. Feldmann H, Jones S, Klenk HD, Schnittler HJ. (2003) *Nat Rev Immunol* **3**: 677–685.

31. Sanchez A, Ksiazek TG, Rollin PE, Miranda ME, Trappier SG, Khan AS, Peters CJ, Nichol ST. (1999) *J Infect Dis* **179** (Suppl 1): S164–S169.

32. Ito H, Watanabe S, Takada A, Kawaoka Y. (2001) *J Virol* **75**: 1576–1580.

33. Chan SY, Speck RF, Ma MC, Goldsmith MA. (2000) *J Virol* **74**: 4933–4937.

34. Takada A, Robison C, Goto H, Sanchez A, Murti KG, Whitt MA, Kawaoka Y. (1997) *Proc Natl Acad Sci U S A* **94**: 14764–14769.

35. Wool-Lewis RJ, Bates P. (1998) *J Virol* **72**: 3155–3160.

36. Shimojima M, Takada A, Ebihara H, Neumann G, Fujioka K, Irimura T, Jones S, Feldmann H, Kawaoka Y. (2006) *J Virol* **80**: 10109–10116.

37. Chan SY, Empig CJ, Welte FJ, Speck RF, Schmaljohn A, Kreisberg JF, Goldsmith MA. (2001) *Cell* **106**: 117–126.

38. Takada A, Fujioka K, Tsuiji M, Morikawa A, Higashi N, Ebihara H, Kobasa D, Feldmann H, Irimura T, Kawaoka Y. (2004) *J Virol* **78**: 2943–2947.

39. Lin G, Simmons G, Pohlmann S, Baribaud F, Ni H, Leslie GJ, Haggarty BS, Bates P, Weissman D, Hoxie JA, *et al.* (2003) *J Virol* **77**: 1337–1346.

40. Marzi A, Moller P, Hanna SL, Harrer T, Eisemann J, Steinkasserer A, Becker S, Baribaud F, Pohlmann S. (2007) *J Infect Dis* **196** (Suppl 2): S237–S246.

41. Alvarez CP, Lasala F, Carrillo J, Muniz O, Corbi AL, Delgado R. (2002) *J Virol* **76**: 6841–6844.

42. Klenk HD, Garten W. (1994) *Trends Microbiol* **2**: 39–43.

43. Volchkov VE, Feldmann H, Volchkova VA, Klenk HD. (1998) *Proc Natl Acad Sci U S A* **95**: 5762–5767.

44. Neumann G, Feldmann H, Watanabe S, Lukashevich I, Kawaoka Y. (2002) *J Virol* **76**: 406–410.

45. Wool-Lewis RJ, Bates P. (1999) *J Virol* **73**: 1419–1426.

46. Beasley DW, Whiteman MC, Zhang S, Huang CY, Schneider BS, Smith DR, Gromowski GD, Higgs S, Kinney RM, Barrett AD. (2005) *J Virol* **79**: 8339–8347.

47. Reading PC, Pickett DL, Tate MD, Whitney PG, Job ER, Brooks AG. (2009) *Respir Res* **10**: 117.

48. Montefiori DC, Robinson Jr., WE, Mitchell WM. (1988) *Proc Natl Acad Sci U S A* **85**: 9248–9252.

49. Jeffers SA, Sanders DA, Sanchez A. (2002) *J Virol* **76**: 12463–12472.

50. Lee JE, Fusco ML, Hessell AJ, Oswald WB, Burton DR, Saphire EO. (2008) *Nature* **454**: 177–182.

51. Yang ZY, Duckers HJ, Sullivan NJ, Sanchez A, Nabel EG, Nabel GJ. (2000) *Nat Med* **6**: 886–889.

52. Okumura A, Pitha PM, Yoshimura A, Harty RN. (2010) *J Virol* **84**: 27–33.

53. Medina MF, Kobinger GP, Rux J, Gasmi M, Looney DJ, Bates P, Wilson JM. (2003) *Mol Ther* **8**: 777–789.

54. Yoneyama M, Kikuchi M, Natsukawa T, Shinobu N, Imaizumi T, Miyagishi M, Taira K, Akira S, Fujita, T. (2004) *Nat Immunol* **5**: 730–737.

55. Andrejeva J, Childs KS, Young DF, Carlos TS, Stock N, Goodbourn S, Randall RE. (2004) *Proc Natl Acad Sci U S A* **101**: 17264–17269.

56. Goodbourn S, Didcock L, Randall RE. (2000) *J Gen Virol* **81**: 2341–2364.

57. Weber F, Kochs G, Haller O. (2004) *Viral Immunol* **17**: 498–515.

58. Stark GR, Kerr IM, Williams BR, Silverman RH, Schreiber RD. (1998) *Annu Rev Biochem* **67**: 227–264.

59. Harcourt BH, Sanchez A, Offermann MK. (1998) *Virology* **252**: 179–188.

60. Harcourt BH, Sanchez A, Offermann MK. (1999) *J Virol* **73**: 3491–3496.

61. Kash JC, Muhlberger E, Carter V, Grosch M, Perwitasari O, Proll SC, Thomas MJ, Weber F, Klenk HD, Katze MG. (2006) *J Virol* **80**: 3009–3020.

62. Reid SP, Leung LW, Hartman AL, Martinez O, Shaw ML, Carbonnelle C, Volchkov VE, Nichol ST, Basler CF. (2006) *J Virol* **80**: 5156–5167.

63. Reid SP, Valmas C, Martinez O, Sanchez FM, Basler CF. (2007) *J Virol* **81**: 13469–13477.

64. Cardenas WB, Loo YM, Gale M, Jr., Hartman AL, Kimberlin CR, Martinez-Sobrido L, Saphire EO, Basler CF. (2006) *J Virol* **80**: 5168–5178.

65. Basler CF, Wang X, Muhlberger E, Volchkov V, Paragas J, Klenk HD, Garcia-Sastre A, Palese P. (2000) *Proc Natl Acad Sci U S A* **97**: 12289–12294.

66. Basler CF, Mikulasova A, Martinez-Sobrido L, Paragas J, Muhlberger E, Bray M, Klenk HD, Palese P, Garcia-Sastre A. (2003) *J Virol* **77**: 7945–7956.

67. Schumann M, Gantke T, Muhlberger, E. (2009) *J Virol* **83**: 8993–8997.

68. Leung DW, Prins KC, Borek DM, Farahbakhsh M, Tufariello JM, Ramanan P, Nix JC, Helgeson LA, Otwinowski Z, Honzatko RB, *et al.* (2010) *Nat Struct Mol Biol* **17**: 165–172.

69. Kimberlin CR, Bornholdt ZA, Li S, Woods Jr., VL, MacRae IJ, Saphire EO. (2010) *Proc Natl Acad Sci U S A* **107**: 314–319.

70. Hartman AL, Towner JS, Nichol ST. (2004) *Virology* **328**: 177–184.

71. Hartman AL, Ling L, Nichol ST, Hibberd ML. (2008) *J Virol* **82**: 5348–5358.

72. Prins KC, Delpeut S, Leung DW, Reynard O, Volchkova VA, Reid SP, Ramanan P, Cardenas WB, Amarasinghe GK, Volchkov VE, *et al.* (2010) *J Virol* **84**: 3004–3015.

73. Prins KC, Cardenas WB, Basler CF. (2009) *J Virol* **83**: 3069–3077.

74. Mateo M, Reid SP, Leung LW, Basler CF, Volchkov VE. (2010) *J Virol* **84**: 1169–1175.

75. Geisbert TW, Young HA, Jahrling PB, Davis KJ, Kagan E, Hensley LE. (2003) *J Infect Dis* **188**: 1618–1629.

76. Geisbert TW, Hensley LE, Jahrling PB, Larsen T, Geisbert JB, Paragas J, Young HA, Fredeking TM, Rote WE, Vlasuk GP. (2003) *Lancet* **362**: 1953–1958.

77. Quinn K, Brindley MA, Weller ML, Kaludov N, Kondratowicz A, Hunt CL, Sinn PL, McCray Jr., PB, Stein CS, Davidson BL, *et al.* (2009) *J Virol* **83**: 10176–10186.

78. Jasenosky LD, Neumann G, Kawaoka Y. (2010) *Antimicrob Agents Chemother*. doi: 10.1128/AAC.00138-10.

79. McCarthy SE, Licata JM, Harty RN. (2006) *J Virol Methods* **137**: 115–119.

80. Halfmann P, Kim JH, Ebihara H, Noda T, Neumann G, Feldmann H, Kawaoka Y. (2008) *Proc Natl Acad Sci U S A* **105**: 1129–1133.

81. Warren TK, Warfield KL, Wells J, Enterlein S, Smith M, Ruthel G, Yunus AS, Kinch MS, Goldblatt M, Aman MJ, *et al.* (2010) *Antimicrob Agents Chemother* **54**: 2152–2159.

82. Aman MJ, Kinch MS, Warfield K, Warren T, Yunus A, Enterlein S, Stavale E, Wang P, Chang S, Tang Q, *et al.* (2009) *Antiviral Res* **83**: 245–251.

83. Wolf MC, Freiberg AN, Zhang T, kyol-Ataman Z, Grock A, Hong PW, Li J, Watson NF, Fang AQ, Aguilar HC, *et al.* (2010) *Proc Natl Acad Sci U S A* **107**: 3157–3162.

84. Geisbert TW, Pushko P, Anderson K, Smith J, Davis KJ, Jahrling PB. (2002) *Emerg Infect Dis* **8**: 503–507.
85. Warfield KL, Swenson DL, Olinger GG, Kalina WV, Viard M, Aitichou M, Chi X, Ibrahim S, Blumenthal R, Raviv Y, *et al.* (2007) *J Infect Dis* **196** (Suppl 2): S276–S283.
86. Harper DM. (2008) *Gynecol Oncol* **110**: S11–S17.
87. Garland SM, Hernandez-Avila M, Wheeler CM, Perez G, Harper DM, Leodolter S, Tang GW, Ferris DG, Steben M, Bryan J, *et al.* (2007) *N Engl J Med* **356**: 1928–1943.
88. Licata JM, Johnson RF, Han Z, Harty RN. (2004) *J Virol* **78**: 7344–7351.
89. Timmins J, Scianimanico S, Schoehn G, Weissenhorn W. (2001) *Virology* **283**: 1–6.
90. Warfield KL, Posten NA, Swenson DL, Olinger GG, Esposito D, Gillette WK, Hopkins RF, Costantino J, Panchal RG, Hartley JL, *et al.* (2007) *J Infect Dis* **196** (Suppl 2): S421–S429.
91. Warfield KL, Swenson DL, Olinger GG, Kalina WV, Aman MJ, Bavari S. (2007) *J Infect Dis* **196** (Suppl 2): S430–S437.
92. Thacker EE, Timares L, Matthews QL. (2009) *Expert Rev Vaccines* **8**: 761–777.
93. Gallo P, Dharmapuri S, Cipriani B, Monaci P. (2005) *Gene Ther* **12** (Suppl 1): S84–S91.
94. Ghosh SS, Gopinath P, Ramesh A. (2006) *Appl Biochem Biotechnol* **133**: 9–29.
95. Sullivan NJ, Geisbert TW, Geisbert JB, Xu L, Yang ZY, Roederer M, Koup RA, Jahrling PB, Nabel GJ. (2003) *Nature* **424**: 681–684.
96. Swenson DL, Wang D, Luo M, Warfield KL, Woraratanadharm J, Holman DH, Dong JY, Pratt WD. (2008) *Clin Vaccine Immunol* **15**: 460–467.
97. Hensley LE, Mulangu S, Asiedu C, Johnson J, Honko AN, Stanley D, Fabozzi G, Nichol ST, Ksiazek TG, Rollin PE, *et al.* (2010) *PLoS Pathog* **6**: e1000904.
98. Sullivan NJ, Sanchez A, Rollin PE, Yang ZY, Nabel GJ. (2000) *Nature* **408**: 605–609.
99. Richardson JS, Yao MK, Tran KN, Croyle MA, Strong JE, Feldmann H, Kobinger GP. (2009) *PLoS One* **4**: e5308.
100. Sheets RL, Stein J, Bailer RT, Koup RA, Andrews C, Nason M, He B, Koo E, Trotter H, Duffy C, *et al.* (2008) *J Immunotoxicol* **5**: 315–335.
101. Jones SM, Feldmann H, Stroher U, Geisbert JB, Fernando L, Grolla A, Klenk HD, Sullivan NJ, Volchkov VE, Fritz EA, *et al.* (2005) *Nat Med* **11**: 786–790.
102. Geisbert TW, ddario-DiCaprio KM, Lewis MG, Geisbert JB, Grolla A, Leung A, Paragas J, Matthias L, Smith MA, Jones SM, *et al.* (2008) *PLoS Pathog* **4**: e1000225.
103. Geisbert TW, ddario-DiCaprio KM, Geisbert JB, Reed DS, Feldmann F, Grolla A, Stroher U, Fritz EA, Hensley LE, Jones SM, *et al.* (2008) *Vaccine* **26**: 6894–6900.
104. Feldmann H, Jones SM, ddario-DiCaprio KM, Geisbert JB, Stroher U, Grolla A, Bray M, Fritz EA, Fernando L, Feldmann F, *et al.* (2007) *PLoS Pathog* **3**: e2.
105. Geisbert TW, ddario-DiCaprio KM, Williams KJ, Geisbert JB, Leung A, Feldmann F, Hensley LE, Feldmann H, Jones SM, (2008) *J Virol* **82**: 5664–5668.
106. Letchworth GJ, Barrera JC, Fishel JR, Rodriguez L. (1996) *Virology* **219**: 480–484.

Index